American Science in an Age of Anxiety

Jessica Wang

American Science in an Age of Anxiety

Scientists, Anticommunism, and the Cold War

• • •

The University of North Carolina Press

Chapel Hill and London

Set in Baskerville and Meta types
by Keystone Typesetting, Inc.
Manufactured in the United States of America

The paper in this book meets the guidelines for permanence and durability of the Committee on Production Guidelines for Book Longevity of the Council on Library Resources.

Library of Congress Cataloging-in-Publication Data
Wang, Jessica.
American science in an age of anxiety: scientists, anticommunism, and the cold war / Jessica Wang.
p. cm.
Includes bibliographical references and index.
ISBN 0-8078-2447-X (alk. paper). — ISBN 0-8078-4749-6 (pbk.: alk. paper)
1. Science — United States — History — 20th century. 2. Science and state — United States — History — 20th century. I. Title.
Q127.U6W36 1999
509.73 — dc21 98-14441
CIP

03 02 01 00 99 5 4 3 2 1

Chapter 1 contains materials previously published in "Liberals, the Progressive Left, and the Political Economy of Postwar American Science: The National Science Foundation Debate Revisited," *Historical Studies in the Physical and Biological Sciences* 26, no. 1 (Fall 1995): 139–66.

Chapter 3 contains materials previously published in "Science, Security, and the Cold War: The Case of E. U. Condon," *Isis* 83 (June 1992): 238–69.

To my parents

CONTENTS

ILLUSTRATIONS

ACKNOWLEDGMENTS

It is sobering to realize that a decade has passed since I began to contemplate writing an undergraduate thesis on "something to do with scientists and postwar politics." What started out as a senior thesis about Edward U. Condon eventually became this book. During that time, as I made my own journey from undergraduate to assistant professor, I received aid and comfort from more people than I can possibly thank in these pages, but I will do my best. To those whose names are unjustly absent, I apologize and hope they know they are ever in my thoughts.

I was fortunate to receive financial and institutional support for research and writing from the following sources: the Cornell Undergraduate Research Fund, the MacArthur Foundation and the MIT Center for International Studies, MIT's Program in Science, Technology, and Society, the Dibner Institute for the History of Science and Technology, the National Air and Space Museum, the National Science Foundation and the Program in the History of Science and Technology at the University of Minnesota, and the University of California at Los Angeles.

The historical profession could not function without talented and dedicated archivists. Sadly, space does not permit me to list individual names, but I am deeply grateful for the assistance of archivists at the following places: the American Association for the Advancement of Science, the American Institute of Physics, the American Philosophical Society, the Department of Energy, the Federal Bureau of Investigation, Harvard University, the Library of Congress, the Johns Hopkins University, the Massachusetts Institute of Technology, the National Academy of Sciences, the National Archives, the National Security Archive, Princeton University, the State University of New York at Albany, the Harry S. Truman Library, the University of California at Berkeley, the University of California at San Diego, the University of Chicago, the University of Minnesota, and the University of Wisconsin, Madison. Everywhere I have conducted research, I have benefited from the talents of able and enthusiastic individuals who were glad to meet my ill-formulated requests for material and happy to share their expert knowledge of their respective collections.

I am grateful to Dr. Robert H. Vought for his gracious consent to be interviewed and to share his records with me, and I thank him and his wife, Mrs. Eleanor Vought, for their warm hospitality during my visit. I also enjoyed the privilege of an interview with Dr. John P. Blewett and access to his personal files, and Dr. M. Hildred Blewett kindly responded to my inquiries with a thoughtful letter.

I thank Richard Polenberg and Walter LaFeber for leading me into the historical profession, and I am grateful to Walt for directing my undergraduate thesis work. Russell Olwell, David Guston, and David Hart provided much fruitful discussion and inspiration about the doing of history and the nature of twentieth-century political economy during my years in graduate school and afterward. I must thank Russ in particular for having the patience and tenacity to request the FBI's files on the Federation of American Scientists and the grace to share them with me. Chapter 2 could not have been written without his generosity. Betty Anderson, Barton Bernstein, Ronald Edsforth, Loren Graham, Kenneth Keniston, Peter Kuznick, Stuart W. Leslie, and Charles Weiner responded to various drafts of the manuscript with insight and care, and Ron, Russ, Sigmund Diamond, and Athan Theoharis kindly volunteered reactions to an early version of chapter 2. I am especially indebted to Ron, who gave so selflessly of his time and energy to train me in American history during a difficult interval in his own career. To a great extent, I owe my professional life and livelihood to him. I deeply admire his enormous talents as a teacher and a scholar, and I am extremely fortunate to know him as both a mentor and a friend.

I thank my editor, the indefatigable Lewis Bateman, for the wisdom of his counsel and the good cheer of his heart. He has made being a first-time author a relatively painless and even pleasurable experience. I know his thoughtful and diligent nature is appreciated not just by me but by much of the historical profession.

My parents, James and Sophia Wang, were initially skeptical about my chosen career path. But they possessed the grace to accept my profession with understanding and pride. For my own part, after two and a half years on the job market, I learned to appreciate their reservations and concerns. This book is for them.

Los Angeles, California
November 1997

ABBREVIATIONS

AAAS: American Association for the Advancement of Science
AAPG: American Association of Petroleum Geologists
AASW: American Association of Scientific Workers
ACLU: American Civil Liberties Union
ADA: Americans for Democratic Action
AEC: Atomic Energy Commission
AFL: American Federation of Labor
ANYS: Association of New York Scientists
AORES: Association of Oak Ridge Engineers and Scientists
ASC: Atomic Scientists of Chicago
BAS: *Bulletin of the Atomic Scientists*
CCNY: City College of New York
CIO: Congress of Industrial Organizations
FAS: Federation of American Scientists
FBI: Federal Bureau of Investigation
GAC: General Advisory Committee
GE: General Electric Company
HUAC: House Committee on Un-American Activities
JCAE: Joint [Congressional] Committee on Atomic Energy
NAS: National Academy of Sciences
NCAI: National Committee on Atomic Information
NCAS: Northern California Association of Scientists
NDRC: National Defense Research Committee
NRC: National Research Council
NSF: National Science Foundation
ONR: Office of Naval Research
OSRD: Office of Scientific Research and Development
PCA: Progressive Citizens of America
SAC: Special Agent in Charge
SCAE: [Senate] Special Committee on Atomic Energy
SCLP: Scientists' Committee on Loyalty Problems
SCLS: Scientists' Committee on Loyalty and Security
UE: United Electrical, Radio, and Machine Workers of America
UPAC: United People's Action Committee

American Science in an Age of Anxiety

INTRODUCTION

More than any other event in twentieth-century history, the atomic bombing of Hiroshima has come to symbolize the contradictions and dilemmas of science in the modern age. To war-weary Americans, Hiroshima's agony brought World War II to a long-awaited end. But sentiments of triumph melded disquietingly with feelings of profound ambivalence. The scenes of devastation and human misery at both Hiroshima and Nagasaki represented a harsh and terrifying coda that mocked victory celebrations and diminished hopes for a calm and comforting future. In the weeks following Japan's surrender, Americans contemplated uneasily the possible meanings of the atomic bomb, the war's end, and what they implied about the role of American power in the postwar world. The war left the United States the foremost military and economic power in the world, while other industrialized nations lay in ruins. Yet American might seemed no guarantee for stability and prosperity. Instead, the destructive force of the atom suggested that any future international conflict could prove so cataclysmic as to be too horrible to comprehend.[1]

The anxieties of the postwar world raised solemn questions for American scientists. In the military, science had found a new and powerful patron during the war. Would that patronage continue afterward? Should it continue? The atomic bomb suggested that science and the state were a potent but all too deadly combination. But if not the military, what was the alternative? Could the relationship between science and the state be restructured to ensure both high levels of support for research and the responsible application of scientific knowledge for peaceful, socially beneficial purposes? During the war, scientists had become expert at applying their craft toward building ever more creative and potent means of carnage. With the war over, many scientists, especially those who had worked on the atomic bomb, wondered whether they now had the obligation to shift their talents toward the maintenance of international stability in an age in which a future war might mean nuclear holocaust. Following a path with few antecedents for American science, ordinary scientists turned to the political realm, where they explored a new synthesis of science, mass-based politics, and the legislative process in order to shape the contours of postwar nuclear policy and the institutional structure of science. As they did so, scientists began to recast their own political identity, pursue fundamental changes in the science-government partnership, and rethink the basic nature of the relationship between science and society.

Scientists faced a rapidly shifting political environment as they wrestled with reshaping their social role and re-creating the political economy of science in the early postwar era. As U.S.-Soviet relations deteriorated, the ideology of anticommunism began to dominate American politics. Communism became a bogeyman and a scapegoat for a host of deeper conflicts. Anticommunism and apprehension over the atom proved mutually reinforcing, as real and imagined anxieties about growing U.S.-Soviet tensions, Soviet-sponsored conspiracies, atom spies, and the possibility of a future nuclear holocaust pervaded the nation's consciousness. Together, the Cold War, domestic anticommunism, and nuclear fear merged in a way that placed scientists among the first victims of the post–World War II red scare. Wartime security clearances, originally conceived as a temporary expedient, became a permanent feature of Cold War science. Within the newly formed Atomic Energy Commission, a central locus of the former Manhattan Project scientists' attempts to reconstruct the postwar relationship between science and the state, security investigations led to critical political conflicts between scientists, the commission, and Congress. Scientists not engaged in military research had to contend with other intrusive loyalty inquiries launched by the federal loyalty program, the House Committee on Un-American Activities, and the Federal Bureau of Investigation. Through this array of loyalty and security investigations, anticommunism came to pose a major obstacle to the political goals of scientists dedicated to nonmilitary sponsorship of basic research, the pursuit of science within a cooperative international world order, and other positions that contradicted the underlying assumptions of the Cold War state. Anticommunism also threatened to undermine reputations, disrupt personal lives, and destroy careers.

This book provides an account of scientists' encounters with domestic anticommunism in the postwar decade. It examines individual cases of scientists denied security clearances or subjected to attacks on their loyalty, the response of scientists' organizations to the politics of anticommunism, and the relationship between Cold War ideology and postwar science policy. My aim is not only to illuminate the effects of anticommunism on scientists but to understand better the integration of scientists into the Cold War political consensus. Domestic anticommunism did more than interfere with individual scientists' lives; it affected the entire scientific enterprise. The Cold War transformed the politics of the scientific profession, the relationship of scientists to the state, and the bureaucratic order devoted to scientific research.

The Cold War has scarcely lacked scholarly attention over the last half century. Questions about the wisdom of Cold War foreign policy, the precise nature of the Soviet Union's postwar intentions, the myths and realities behind the perceived communist threat, and the bearers of responsibility for the postwar red scare continue to be the subjects of intense, and sometimes bitter, debate. Yet, despite the importance of the atomic bomb to postwar foreign policy, the Cold War emphasis on the need for technological advances to maintain American military superiority, and the postwar relationship between nuclear fear and domestic anticommunism, until recently few U.S. historians have paid close attention to American scientists. When Cold War scholars have addressed scientists at all, they have portrayed scientists as exceptional in their vigorous resistance to the rise of the Cold War, the evolution of American strategic dependence on weapons of mass destruction, and the spread of domestic anticommunism. In such historical writing, J. Robert Oppenheimer is cast as the central tragic figure, the foremost symbol of the victimization of scientists by Cold War hysteria and the general anti-intellectual atmosphere of the 1950s.[2]

In the past decade, however, historians of science interested in the political development of American science have formulated a very different picture. A small but growing group of historians of American science has come to emphasize the role of military patronage in redirecting the research priorities of the physical sciences during the Cold War. Although physicists and scientists in related disciplines claimed and believed they were engaged in value-neutral basic research, the content and direction of their research agendas were swayed heavily by the technological needs of the Cold War, and their own political prominence and influence depended on their building close ties to the Cold War national security state.[3]

Unfortunately, most Cold War historians have yet to recognize this provocative body of work produced by specialists in the history of American science. At the same time, the political history of American science, despite the recent emphasis on the relationship between the national security state and the content of postwar research, remains insufficiently grounded in the context of Cold War politics. By focusing only on elites within the scientific profession, recent research has unfortunately created the impression of a monolithic scientific community, an autonomous actor that wholeheartedly bought into the requirements of the national security state. What needs to be recognized is that American scientists, like other groups, were deeply divided by political differences during the Cold War. They were, simultaneously, partially responsible for but also subject to the ideological constraints of the postwar era. Conservative, liberal, and leftist scientists articulated competing visions for the role of science in the postwar political order, and their ranks included both winners and losers.[4]

Anticommunism was hardly new to the United States in the 1940s. It and other forms of nativism traced a periodic cycle of ebb and flow throughout the late nineteenth and twentieth centuries as American society responded to wrenching social and economic dislocations and perceptions of upheaval overseas.[5] The railroad strike of 1877, the 1886 Haymarket Riot, the antiradical clauses of the Naturalization Act of 1903 and Naturalization Act of 1906, the suppression of the Industrial Workers of the World throughout the 1910s, the passage of the Espionage Act of 1917 and the Sedition Act of 1918, the post–World War I red scare, and the passage of restrictive immigration laws in the 1920s constituted key moments in the long-term development of a worldview that targeted the importation of foreign ideas as the primary cause of domestic unrest and disorder. In the 1930s, nativists turned to what they saw as the threat from within, as the Dies committee (later named the House Committee on Un-American Activities), state investigatory committees, and other reactionary elements attacked the New Deal and the revived labor movement as hotbeds of dangerous radicalism. Long before the onset of the Cold War, antiradical nativism was already a familiar part of American politics.

The anticommunism of the post–World War II decade, however, outpaced earlier periods of antiradical nativism in length and intensity. The startling new geopolitical conditions of the postwar world elevated the stakes of U.S.-Soviet conflict. The balance of world power had shifted away from Europe, and for the first time, the United States stood as a global superpower. According to the dictates of anticommunist ideology, the price of insufficient vigilance against radicalism was no longer mere domestic instability but the spread of totalitarianism worldwide.

American science also faced a mixture of old and new in the postwar years. As much as they might have liked to think otherwise, scientists were not political neophytes in the early Cold War era. The linkage between science and politics is as old as science itself; indeed, it has become somewhat of a cliché for historians of science to note the political entanglements of Newton and Galileo. In late-nineteenth- and early-twentieth-century America, the growth of a federal science establishment, the rise of corporate research laboratories, and the creation of powerful philanthropic foundations that sponsored research provided scientists with increased social capital and political influence.[6] Historians have assumed that for the most part, prior to the 1930s, scientists allied themselves with entrenched forces of wealth and power and acted to reinforce the prevailing social and economic order.[7] But throughout the twentieth century, scientists have uneasily balanced accommodation and rebellion, boosterism and criticism, and the differing political implications of these varied attitudes.

In the early twentieth century, some scientific elites shared the reform impulses of their day. During the Progressive Era, engineers began to consider the need for greater social responsibility on the part of their profession, and early social scientists hoped scientific rigor might provide solutions to the pressing urban problems of a rapidly industrializing nation.[8] In the 1930s, spurred by the privations of the Great Depression and a growing unease that science could no longer be equated inevitably with progress, scientists began to experiment with political activism in larger numbers. Through the science and society movement of the American Association for the Advancement of Science (AAAS), for example, scientists discussed how their work could be tailored to better serve society's needs and help alleviate the persistent economic crisis. Some scientists adopted a more radical critique of American society. Through recently founded, depression-era organizations such as the American Association of Scientific Workers, scientists on the progressive left (the left-most end of American liberalism)[9] directly questioned the wisdom of industrial capitalism and its capacity to produce social justice. With war looming abroad, they also joined the broader antifascist movement to counter the threat Nazi Germany posed to democratic governments worldwide.[10]

The discussions of the 1930s provided important precedents for scientists' political engagements after World War II. As we shall see, scientists' battles for civilian control of atomic energy and a National Science Foundation to support basic research reflected progressive left assumptions about political economy and the optimal relationship between science and society. In the postwar debate over science policy, liberal-left scientists articulated what I call a progressive left politics of science—

commitment to an internationalist world order, international scientific cooperation on research in atomic energy, greater responsiveness to public needs and the application of social good as a criterion for federal funding of science, and an insistence that the American public should benefit freely from publicly funded research. Furthermore, many of the older scientists who entered the postwar public policy debates, such as Harlow Shapley and Harold C. Urey, had themselves played central roles in the dialogues of the 1930s.

The activism of scientists in the postwar period, however, also constituted a profound departure from the past. The atomic scientists' movement that flourished immediately after the war was led primarily by scientists whose formative life experience and primary source of political motivation was Hiroshima and Nagasaki, not the turmoil of the 1930s. Strategies and methods also differed. The science and society movement of the earlier period, in particular, was more an intellectual than a political movement, a mood rather than a concrete program of action. Progressive left scientists proved more energetic than the AAAS during the depression years and generated a lively series of fervent discussions, studies, fund-raising efforts, public statements and petitions, and large public gatherings. But they appealed for the most part within the scientific, academic, and intellectual communities, and they developed relatively little influence over the nation's political life. By contrast, scientists engaged in political activism immediately after World War II plunged directly into the domain of national-level legislative politics. In a manner that was simultaneously savvy, improvised, and chaotic, they aggressively cultivated relations with members of Congress and the media, and they appealed to the general public for mass-based support.

By the end of World War II, scientists had also developed another, more direct route to political power. Although scientists had mobilized during the First World War, the short duration of U.S. involvement and the rapid demobilization of American resources in that conflict meant that World War I yielded few long-term changes in the politics of science. During World War II, however, prominent scientists joined the top ranks of the policymaking hierarchy and fostered a new relationship between science and the state. As a result, American science came of age during the war. Henceforth, elite scientist-administrators regularly entered the inner sanctum of government, where they discussed the foreign policy implications of new military technologies in addition to dispensing advice about technical matters. They also rubbed shoulders regularly with military leaders, high-level public officials, and, in the case of Vannevar Bush, the president himself.

As the respective wartime leaders of the Office of Scientific Research and Development, the National Defense Research Committee, and Los Alamos, heart of the Manhattan Project, scientist-administrators such as Bush, James B. Conant, and J. Robert Oppenheimer combined a number of talents. Through managerial skill, agility at mediating relationships between scientific personnel, the military, and civilian officials, and the ability to oversee impressive advances in weapons and medical technology, they demonstrated the efficacy of science-military cooperation and the usefulness of scientists to the policy process. After World War II, Bush, Conant, and Oppenheimer became celebrated public figures, and they continued to serve as top advisers to civilian leaders, the military, and the Atomic Energy Commission. They forged their political power through an alliance with the emerging national security state, and, despite moments of profound ambivalence, they perpetuated the continued mobilization of the physical sciences for military purposes during the early Cold War years. Conant, in particular, became an important architect of Cold War policy despite his deep despair over nuclear weapons and the U.S.-Soviet arms race.[11]

Thus American scientists both challenged and perpetuated the development of the Cold War political order. Liberal and progressive left scientists proposed imaginative alternatives to nuclear proliferation and military dominance of scientific research, and for a short time it seemed as if their ideas might triumph. But the political repression that accompanied the postwar revival of domestic anticommunism soon placed the progressive left politics of science on the defensive.[12] As scientists tested the nature of their influence after the war, they eventually found little room to maneuver outside the Cold War consensus. Within only a few years, they had to retreat from pursuing a complex legislative agenda for science and international relations and instead concentrate on the more basic issues of preserving their livelihoods and their right to political dissent.

The scientific community's response to domestic anticommunism was mixed. Scientists' public resistance to anticommunism was tempered by both fear of increased repression and the politically safer route of quietly building alliances within the Atomic Energy Commission and other receptive government agencies in order to mitigate the adverse effects of the loyalty-security system. Scientists achieved some short-term successes in limiting the impact of security investigations on scientists, but ultimately the safer strategy was self-defeating. By earning concessions through unobtrusive, backroom negotiations with government officials and deliberately avoiding public scrutiny, scientists failed to develop public political means of countering the growing ideological demands of the Cold War era. Their alliance with federal agencies bought time, but eventually time ran out. In 1949, scientists and the AEC lost a key political battle with Congress over the AEC fellowship program. Thereafter, the scope of anticommunist attacks on scientists expanded steadily for the next several years. During the late 1940s, few scientists other than those who were federal employees or engaged in classified research had much reason to fear harm from the postwar red scare. By the 1950s, however, all politically active scientists, no matter what their positions, were vulnerable.

As political elites whose expertise was essential to the military basis for Cold War foreign policy, scientists had, in theory, more freedom of action than other groups to defy the politics of anticommunism. But the very structure of modern bureaucratic government, combined with the specific political conditions of postwar America, restrained scientists' inclination to challenge the dominant order. Cold War ideology transformed the very nature of American politics. The political culture of Cold War liberalism celebrated the potential of the bureaucratic state, privileging procedural reform over principle, and process over fundamental questions of value. In the late 1940s, many scientists began to confine their actions within the rhetorical and political boundaries of Cold War liberalism. Increasingly, scientists turned away from a progressive left rhetorical style that emphasized fundamental civil libertarian principles. Instead, they relied on more limited procedural reforms, most notably the protection of due process rights, to counter the threat to political freedom posed by the loyalty-security system.

Some readers will undoubtedly question my decision not to discuss events of the 1950s, especially the 1954 Oppenheimer case, in as much detail as the immediate postwar years. I have chosen to focus primarily on the period from 1945 to 1950 out of the belief that it constituted a critical phase in twentieth-century U.S. history. It was during these pivotal five years that the Cold War political consensus was built, and a domestic and international order based on U.S.-Soviet conflict became the defining political reality for the next forty years. The McCarthyism of the 1950s was a consequence of the political consensus built in the late 1940s, not a causal phenomenon in its own right. For American scientists, the late 1940s were the crucial period in which they simultaneously established and discovered the extent and limit of their own political power.

In terms of vastly increased funding, science benefited from the Cold

War. Integration into the national security state, however, had a high price tag. Cold War anticommunism narrowed scientists' immediate political strategies and ultimately their visions for expanding their own social role and capacity to challenge the status quo. Scientists abandoned a nascent public political style that attempted to rework the relationship between science and society and instead grew increasingly reliant on internal negotiations within government agencies to achieve more limited policy goals.

World War II ended with the tragedy of atomic devastation at Hiroshima and Nagasaki. In the months after the war, Manhattan Project scientists sought to turn that tragedy into a creative reordering of international relations, as well as a new political function for American scientists. For a brief, exciting, tumultuous period, the atomic scientists moved to the center of American politics as the leaders of a fledgling political movement, launched at a turbulent time as the prewar past of New Deal reform collided with the looming perils of the postwar world.

CHAPTER 1

Competing Political Visions for Postwar Science

Scientists and Science Legislation, 1945–1947

• • •

From the perspective of the present, the Cold War seems to have had a certain inevitability, but from the vantage point of the end of World War II, the future shape of American foreign relations and domestic politics was far from clear. By September 1945, U.S.-Soviet relations were already uneasy, as they had been throughout the war, but they had not yet broken down irreparably. Winston Churchill would not make his "Iron Curtain" speech for another six months, and President Truman's announcement of the Truman Doctrine lay a year and a half away. The general public would not know Alger Hiss's name for nearly three more years; Joseph McCarthy's debut on the national scene would wait for almost another five years. At war's end geopolitics appeared to dictate that the United States and the Soviet Union would form the new loci of power in global affairs. Beyond that, nothing seemed clear except the impermanence and ambiguity that defined international relations in the postwar world.

Social and political tensions dominated American life in the first months after the war. As doubts grew about Soviet intentions abroad, demobilization created domestic economic hardship and labor unrest. Massive shortages of housing and basic consumer goods, inflation, and lost jobs marked the transition from wartime to peacetime. A record strike wave began in the fall of 1945 and continued throughout 1946, as working-class Americans tried to hold on to and extend economic gains they had made during the war. On top of these immediate economic crises, the newly perceived peril of the atomic bomb cast an even gloomier pall over the postwar peace.

This combination of international tension, domestic economic instability, and nuclear fear provided the backdrop for the post–World War II resurgence of antiradical nativism. But several years would pass before Cold War anticommunism became an all-encompassing force in American politics. In the meantime, the transitive flux of international and domestic uncertainty provided a climate for more than the mere expression of apprehension and reflexive nativism. Americans did not simply react to impersonal social forces in a deterministic manner; they participated in the creation of events and responses to them. The unsettled nature of the postwar order and the need to come to terms with the nuclear age supplied an open window of opportunity for intense public discussion and political debate about the future of international relations and domestic political economy in the aftermath of World War II.

Many Manhattan Project scientists believed that the atomic bomb had to become the public's paramount concern. After the destruction of Hiroshima and Nagasaki, a dark sense of foreboding led them to weigh the meaning of the nuclear age for the postwar world. Soon they matched thoughtful discussion with action. By September 1945, scientists who had worked on the bomb began to organize what later became known as the atomic scientists' movement. They worked to influence the nuclear policies of the United States and, by extension, the shape of postwar international relations. Through the establishment of a civilian Atomic Energy Commission (AEC) to oversee atomic energy research and development, and the formation of a system of international control of atomic energy, they hoped to ensure the worldwide use of the atom for peaceful purposes rather than nuclear destruction. Meanwhile, other scientists continued a debate that had begun during the war over the general future of postwar science and the proper relationship between science and government. In their proposals for a National Science Foundation, liberal and progressive left scientists sought to create a political structure for postwar science that would tie basic research more closely to the general public welfare, reconcile the tensions between expert rule and democratic control, and avoid dominance of science by military imperatives.

Together, these scientists defined a progressive left politics of science. Through their legislative proposals, they espoused a vision of internationalism and public access to science within the framework of a New Deal political economy. Much of their political vision was vague and inchoate, but they made a creative and genuine attempt to wrestle with difficult questions about the social role of science and the place of science within the evolving political economy of postwar America. Internationalism, opposition to military funding of science, and New Deal notions about social and economic equity, however, soon came into conflict with countervailing political tendencies. Ultimately, the progressive left conception of science policy would not withstand the exigencies of the Cold War.

Containing the Atom: The Atomic Scientists' Movement

The atomic scientists' movement was a watershed in the political history of American science.[1] It constituted scientists' first and most successful effort to influence national-level politics from a mass political base. For a brief period in 1945 and 1946, it seemed as if the movement had the potential to become a permanent force in American politics. During that time, the former Manhattan Project scientists' carefully formulated case for the integral connections between civilian control of atomic energy, freedom in science, and world peace found a receptive audience. But by advocating internationalism and insisting that there were no atomic secrets, the atomic scientists promoted ideas that, over time, grew increasingly at odds with the development of Cold War ideology and an evolving perception of atomic weapons as the keystone of American national security. The atomic scientists' movement created a new political role for American scientists, but with the rise of the Cold War, it also provided a major avenue for anticommunist attacks against them.

With the exceptions of Niels Bohr and a group of scientists at the University of Chicago's Metallurgical Laboratory, few scientists connected to the Manhattan Project gave serious thought to the social, political, and moral implications of the atomic bomb during the war. Without reservation, the scientific leaders of the Manhattan Project recommended the use of nuclear weapons on Japan, while scientific personnel at Los Alamos and Oak Ridge generally remained silent. Seduced by the technical problems posed by their task and swept up by the war-induced momentum to bring their undertaking to technological fruition, most Manhattan Project scientists had little inclination to consider the consequences of the atomic bomb's use or the long-range impact of weapons of mass destruction.[2]

Immediately after the bombings of Hiroshima and Nagasaki, however, the atomic scientists could talk about little else. Knowing what their handiwork had wrought, they began to envision a terrifying future conflict that would end with the indiscriminate use of atomic weapons and the collapse of civilization. It was difficult for them to walk city streets and pass scenes of everyday life without imagining them transformed into nightmarish landscapes of death and rubble at ground zero. That science could someday prove responsible for slaughter and ruin so complete as to be beyond all description was unbearable. Humanity, it seemed, now possessed the technological means for easy self-annihilation. Without drastic, intelligent, informed action, the former Manhattan Project scientists feared, global disaster would result.

Pushed out of their complacency by images of nuclear devastation and a tormented sense that scientific development had outpaced humanity's ability to cope with it, scientists at Chicago, Los Alamos, and Oak Ridge began to discuss the long-term implications of atomic weapons and possible ways to remove the threat of nuclear warfare. They created formal organizations: the Atomic Scientists of Chicago, the Association of Oak Ridge Scientists (later the Association of Oak Ridge Engineers and Scientists), and the Association of Los Alamos Scientists. The scientists reached a general consensus that long-term security could not be based on atomic weapons. Since discoveries made by scientists in one country could be made independently in another, eventual loss of the U.S. atomic monopoly was inevitable. In the absence of international agreements, a nuclear arms race between the United States and the Soviet Union would promote international instability and increase the risk of worldwide nuclear holocaust. Given the destructive capability of the atomic bomb, the atomic scientists concluded that only some form of international control of atomic energy and a postwar international order based on cooperation between nations could protect the world from the danger of nuclear obliteration.[3]

This emphasis on internationalism tied the atomic scientists' agenda to the progressive left. Popular front liberals, as Mary S. McAuliffe has observed, embraced "an identification with common people and antipathy toward big business, a faith in popular government, and a belief in progress and man's capacity for improvement, if not perfectibility."[4] For the postwar period, the progressive left hoped for an expansion of New Deal social and economic reform and an international order characterized by a commitment to peaceful coexistence between the United States and the Soviet Union. The atomic scientists remained vague concerning economic structures and the role of the state in areas outside science. Given this lack of interest in general matters of political economy, the atomic scientists should not be thought of as part of the progressive left. But their commitment to international cooperation in science and a postwar system of international relations based on a cooperative relationship between the Soviet Union and the United States, their optimistic faith in the capacity of people to act rationally for the greater good, and their belief that science could serve as a model for reasoned decision making and greater trust between nations together defined a more moderate liberalism that shared some of the central tenets of the progressive left.

Meanwhile, other developments soon threatened to preempt the atomic scientists' attempt to rethink international relations. The May-Johnson bill, announced on October 3, proposed a very different conception of atomic energy's political future. Drafted hastily by the War Department in the weeks following Japan's surrender, the May-Johnson bill left all aspects of atomic energy open to military control, required strict secrecy and security regulations, and mandated heavy penalties for any security violations. Three prominent scientist-administrators—Vannevar Bush, James B. Conant, and J. Robert Oppenheimer—endorsed the bill. Scientists back at the Manhattan Project sites were outraged by the scientific elite's apparent capitulation to military authority. Security policy had been a major source of contention between scientists and the army during the war, and for many scientists, scientific freedom and the need to minimize external controls on research more than justified opposition to the bill. The atomic scientists objected to May-Johnson for reasons beyond pure self-interest and a desire for scientific autonomy, however. They opposed the bill's exclusive emphasis on military applications of atomic energy, and they felt the secrecy regulations left little room for information exchange between nations, international cooperation on peaceful uses of atomic energy, and the development of international control schemes. Heavy secrecy restrictions were not simply a drag on scientific research; they constituted a misguided and futile attempt to preserve the U.S. atomic monopoly. Furthermore, if the United States pursued an atomic energy policy based purely on the military uses of the atom, other nations would distrust American motives, with disastrous consequences for international relations. Atomic energy, the atomic scientists insisted, had to be placed under the authority of a civilian agency in order to convince the world that the United States was serious about promoting the peaceful development of atomic energy and averting its military use.[5]

The support of Bush, Conant, and Oppenheimer for the May-Johnson bill, as opposed to the hostility of the working scientists at the Manhattan Project sites, is indicative of the political divisions within the American scientific community during the early postwar years. The elite scientist-administrators tended to be elder statesmen of science and team players when it came to matters of policy, and only Oppenheimer might have been described as something other than conservative. Bush's conservatism was particularly strong and well formulated. Although personally fond of Franklin D. Roosevelt, he had no use for the New Deal. Distrustful of state expansion, he preferred Herbert Hoover's vision of voluntary associationalism. Conant, a lifelong Republican, sometimes demonstrated a tepid liberalism, but when faced with potential controversy, he usually allowed political pressure and pragmatic considerations to hold sway over his personal beliefs. As for Oppenheimer, although he had been involved in radical politics in the 1930s, his own political identity was rather hazy. By the late 1940s, he labeled his earlier interest in the Communist Party an example of youthful experimentation long outgrown. After the war he became reluctant, for the most part, to challenge publicly prevailing government policy, and when he did dissent, he was guided more by a generalized ethos about scientific freedom than a commitment to any particular ideology, whether conservative, liberal, or radical.[6]

It would be simplistic and wrong to suggest that Bush, Conant, and Oppenheimer therefore had nothing in common with the atomic scientists' movement. To the contrary, Bush and Conant had devoted extensive thought to the need for international control of atomic energy since the war, and in 1946, Oppenheimer served as one of the primary architects of the Acheson-Lilienthal report, the first American working proposal for international control. At the same time, however, these three scientific leaders were slow to question the conventional wisdom of high-level officials or to endorse measures that required going outside the decision-making hierarchy. As part of the rarefied social circle of the policy elite, they also had few qualms about forging close postwar ties between science and the military. By contrast, the atomic scientists viewed strict limits on military authority as critical to maintaining both the autonomy of science and the stability of the postwar world order. They also searched for political alternatives beyond the upper ranks of government and attempted to circumvent the policy elite by bringing mass-based politics to science policy.

Members of the scientists' associations at the Manhattan Project sites quickly launched a lively and energetic campaign against the May-Johnson bill. Most were younger working scientists, such as William A. Higinbotham and Katharine Way, who came from the lower ranks of the Manhattan Project. Only in his mid-thirties at the end of the war, Higinbotham had been a group leader in electronics at Los Alamos despite never having earned a doctorate. After the war he became a leading figure in the atomic scientists' movement, and throughout his postwar career as a physicist at Brookhaven National Laboratory, nuclear arms control would remain a central part of his life. Katharine Way had been a physicist at the University of Chicago's Metallurgical Laboratory during the war, where she contributed to reactor design and theoretical work on the decay of fission products. She also compiled data on nuclear cross sections and created what became known as the "Kay Way tables." As part of the atomic scientists' movement, Way directed publicity for the Atomic Scientists of Chicago during the fight over the May-Johnson bill, and she coedited the book *One World or None* (1946), a best-selling collection of essays about the dangers of the nuclear age. She would later protest restrictive security clearance measures while working at Oak Ridge. Although most of the participants in the scientists' movement had worked on the atomic bomb during the war, a few had no connection to the Manhattan Project. Richard L. Meier, for example, earned his Ph.D. from the University of California at Los Angeles in 1944 and was working as a research chemist at the California Research Corporation in Richmond, California, when he first became involved in the scientists' movement. Some of the atomic scientists, such as Philip Morrison and Robert R. Wilson, would later become well-known and highly respected scientists. Most, however, would lead successful but quiet careers, known and recognized by their colleagues but with names largely unfamiliar to historians.[7]

Despite their individual anonymity, the atomic scientists soon established a powerful presence in American political life. They appealed directly to the public through a long series of media interviews, articles, radio addresses, and public speaking engagements in which they discussed both the specific legislation at hand and the general political and social implications of atomic energy. In mid-October, members of the Atomic Scientists of Chicago, the Association of Oak Ridge Scientists, the Association of Los Alamos Scientists, and the Association of Manhattan Project Scientists (centered at Columbia University) began to travel to Washington, D.C., on a rotating basis. Between October and December 1945, some thirty-odd scientists went to Washington, where, in a whirlwind of social and political activity, they built influence in excess of their numbers. As Washington's newest lobby, they arranged for federal lawmakers to be flooded with telegrams from scientists and other citizens, spoke to and dined with members of Congress, and testified about atomic energy in hearings on the proposed National Science Foundation. With the formation of the Senate Special Committee on Atomic Energy (SCAE) in late October and Congress's entry into the public debate over atomic policy, they found especially strong allies in Chair-

John A. Simpson of the Atomic Scientists of Chicago, testifying before the Senate Committee on Atomic Energy, December 13, 1945. (New York World Telegram and Sun Collection, Library of Congress; reproduced by permission of Corbis-Bettmann)

man Brien McMahon (D.-Conn.) and committee counsel James R. Newman, head of the science section at the Office of War Mobilization and Reconversion. Through the committee, the atomic scientists gained the opportunity to participate directly in the formulation of atomic energy policy. At the end of October, they attained an official presence within the Senate committee when Edward U. Condon, a well-known theoretical physicist who had worked on both the wartime radar and atomic bomb projects and who was about to become director of the National Bureau of Standards, became the SCAE's scientific adviser.[8]

In their public appeals, the atomic scientists continually stressed that there existed no fundamental secret of the atomic bomb and no defense against it. It was a jarring message for those who believed atomic might guaranteed U.S. preeminence in the postwar world. According to the scientists' analysis, although aspects of atomic bomb design rightly remained strictly classified, a reliance on secrecy would not preserve the U.S. atomic monopoly. Other nations would eventually develop their own atomic bombs, and American dependence on nuclear weapons would inevitably lead to a ruinous arms race. Therefore, international control of atomic energy was an utter necessity to forestall a disastrous future of nuclear proliferation. Over and over, the atomic scientists stressed a three-part analysis, quickly summarized by journalist Louis Falstein as follows: "1. The A-Bomb is no secret. 2. We cannot long have a monopoly of its manufacture. 3. International control is the only solution." Without international control, the scientists argued, the world risked nuclear catastrophe.[9]

As a result of the atomic scientists' public campaign, by mid-December Congress recognized that the May-Johnson bill faced too much opposition and it needed new legislation. On December 20, Senator McMahon introduced S.R. 1717 in the Senate. Senate Resolution 1717, also known as the McMahon bill, was drafted by SCAE counsel Newman's assistant, Byron Miller. In working on the bill, Miller consulted Condon; Chicago Metallurgical Laboratory physicist Leo Szilard, the Manhattan Project gadfly who became a tireless crusader for arms control in the postwar period; members of the Atomic Scientists of Chicago; and Edward Levi, a professor at the University of Chicago Law School who had worked with the Atomic Scientists of Chicago to draft an atomic energy bill in November. The proposed legislation contained everything the scientists' associations desired. It placed atomic energy under the control of a civilian Atomic Energy Commission (AEC), emphasized the peaceful development of atomic energy rather than its military uses, explicitly stated that atomic energy would be administered in a way consistent with future international agreements, and specified that the AEC would undertake no research in military applications of atomic energy if such research violated future international agreements. A provision for study of the social, political, and economic implications of atomic energy satisfied scientists' concerns about the need to match scientific advances with social development. The McMahon bill also contained guarantees for maximizing the free exchange of scientific information and carried far less stringent penalties than the May-Johnson bill for violations of security regulations.[10] For Newman and Miller, both New Dealers committed to the ideal of economic planning, provisions for government ownership of all patents dealing with atomic energy, government control of fissionable materials, and heavy regulation of a projected nuclear power industry met their vision of rational planning by an expanded state. This combination of internationalism, concern for the social consequences of science, and New Deal political economy defined the central features of the bill and its essential progressive left character.

By this time, the scientists' groups had organized more formally. Early in November, the Atomic Scientists of Chicago, the Association of Oak Ridge Scientists, the Association of Los Alamos Scientists, and the Association of Manhattan Project Scientists combined to form the Federation of Atomic Scientists. In the middle of the month, scientists' groups from around the country met in Washington, D.C., to expand the new

(*Left to right*) H. H. Goldsmith of the Atomic Scientists of Chicago, Irving Kaplan of the Association of Manhattan Project Scientists, Lyle B. Borst of the Association of Oak Ridge Scientists, and W. M. Woodward of the Association of Los Alamos Scientists announce the formation of the Federation of Atomic Scientists, November 11, 1945. (National Archives; reproduced by permission of Corbis-Bettmann)

federation into a national organization with the task of "furthering world peace and the general welfare of mankind." In January 1946, the organization was renamed the Federation of American Scientists (FAS) to reflect the growth of its membership beyond Manhattan Project scientists. Meanwhile, scientists in Chicago had started their own journal, the *Bulletin of the Atomic Scientists of Chicago* (later renamed the *Bulletin of the Atomic Scientists*). Although the *Bulletin* was never officially published by the FAS and always insisted it was independent, for all practical purposes it represented the views of the scientists' movement.[11]

At the time Senator McMahon announced S.R. 1717, the scientists' public relations efforts had also been set up to run more smoothly and reach a large audience. On November 16, 1945, the Federation of Atomic Scientists held a joint conference with representatives from fifty national-level professional, educational, labor, and religious organizations to discuss ways to inform the public about atomic energy. Participating groups included the American Bar Association, the American Council on Education, the American Federation of Labor, and the Congress of Industrial Organizations. On December 18, the groups met again and formed the National Committee on Atomic Information (NCAI) to provide a link between the scientists and the public. The committee would perform this vital function by receiving information from the atomic scientists and publicizing it "to promote the widest possible understanding of the facts and implications of developments in the field of atomic energy."[12] The NCAI, it appeared, could reach an audience of millions—its member organizations had an estimated combined membership of ten million people. With this potential mass political base ready to be tapped, the FAS began to lobby for the McMahon bill.

At first, the bill's prospects appeared healthy. It seemed to have public support as well as the FAS's endorsement. The War and Navy Departments objected to the bill's exclusion of the military, but the Truman administration backed the concept of civilian control.[13] Growing U.S.-Soviet tension intervened, however. On February 16, news broke of the arrests of twenty-two people in Ottawa for passing secret information, including information about the atomic bomb, to the Soviet Union. On February 20, in what was perhaps a clumsy attempt to defuse the issue, the Soviet government revealed that it had received what it called "insignificant secret data" about atomic energy from Canadian citizens.[14]

In historical retrospect, it appears the Soviet Union only obtained samples of slightly enriched uranium. At the time, however, vague and contradictory reports of a Canadian spy ring unleashed widespread public fears that the secrets of the atomic bomb had been lost to espionage. Despite the scientists' insistence that there were no real atomic secrets to be lost, the news from Canada revived barely submerged beliefs that espionage posed the most serious threat to the U.S. atomic monopoly. Although politicians and the public easily understood and accepted the scientists' reasoning on an intellectual level, they also quickly discarded scientists' arguments amid reports of atom spies. Fear of atomic espionage would become an oft repeated and all too familiar theme over the next few years.[15]

Had the Canadian spy scare been an isolated incident, the furor it caused might have gradually died out. But it transpired as relations between the United States and the Soviet Union worsened dramatically. Late in World War II and immediately after, President Truman and his advisers believed and hoped that despite disagreements over the postwar settlement, dependence on American funds for postwar reconstruction combined with the American atomic monopoly would provide leverage over the Soviet Union and guarantee the U.S. conception of a postwar world based on self-determination for individual nations and open markets favorable to American economic expansion. Instead, the U.S.-Soviet relationship deteriorated rapidly early in 1946. On February 9, Stalin publicly announced a new five-year plan that praised Soviet industrial capacity and expressed the faith that domestic production would be sufficient to expand the Soviet Union's postwar economy. Although the speech itself was relatively innocuous, it ended American hopes for controlling future Soviet actions through economic aid. Many American observers interpreted Stalin's speech as a call to arms against capitalist nations. The reports of atomic espionage in Canada came only a week later. Then, at the beginning of March, Stalin refused to withdraw troops from Iran on the date previously agreed on by the wartime allies. Days later, on March 5, Winston Churchill made his well-known "Iron Curtain" speech in which he pointed to growing Soviet dominance in Eastern Europe and warned of the likely need for the United States to commit itself unhesitatingly to the resistance of Soviet expansionism. The "Iron Curtain" speech provided a stark statement on the widening rift between the United States and the Soviet Union and constituted an early step in preparing public opinion for a postwar bipolar international order built around U.S.-Soviet confrontation.[16]

Caught in the storm of U.S.-Soviet tensions and growing fears of atomic espionage, the McMahon bill rapidly lost support. The steady flow of mail to the Senate Committee on Atomic Energy in favor of the bill dropped precipitously, and backing for a military presence in the proposed Atomic Energy Commission emerged within the committee. The SCAE soon adopted an amendment submitted by Senator Arthur Vandenberg (R.-Mich.) that vastly expanded the military's prerogatives.[17]

The atomic scientists objected strenuously to the Vandenberg amendment and quickly mobilized against it. The Federation of American Scientists emphasized that the reassertion of military control implied by the amendment would have deleterious consequences for international relations, and the FAS continued to criticize the assumption that America's postwar security could be maintained by possession of the atomic bomb. In the March 15 issue of the *Bulletin of the Atomic Scientists*, Eugene Rabinowitch and Edward U. Condon carefully outlined the interdependence of civilian control, scientific freedom, and international control. Rabinowitch sharply criticized arguments that military control of atomic energy guaranteed American security. Military control, he insisted, could not keep the peace. Instead, given the tense state of relations between the United States and the Soviet Union, military control would wreak severe damage on international relations. Rabinowitch declared, "Permanent military control of atomic energy in America will signify to the world that America is basing its long-range policies on the assumption that a new war is inevitable, and this will help to make it inevitable."[18] In Rabinowitch's view, assigning the military a role in the control of atomic energy would solidify perceptions overseas that the United States was bent on a course of confrontation with the Soviet Union.

Condon felt the same way. A firm believer in internationalism, he opposed military control of science as detrimental to both science and international relations. An emphasis on the military applications of science, Condon argued, meant restrictions on information exchanges between scientists of different nations, leading to an atmosphere of distrust. He wrote presciently: "It is sinister indeed how one evil step leads to another. Having created an air of suspicion and mistrust, there will be persons among us who think other nations can know nothing except what is learned by espionage. So, when other countries make atom bombs, these persons will cry 'treason' at our scientists, for they will find it inconceivable that another country could make a bomb in any other way except by aid from Americans." Rather than continue on such a destructive course, Condon asked his fellow Americans to "chase this isolationist, chauvinist poison from our minds," and he offered international control of atomic energy and international cooperation on peaceful applications of the atom as ways to promote world peace.[19]

To scientists such as Rabinowitch and Condon, the larger implications of internationalism in science led to the conclusion that scientific freedom and international relations were intimately connected. Secrecy and military dominance of atomic energy held dangers far greater than damage to scientific research. Secrecy constituted a mind-set that created a false sense of security that defense of the United States could be maintained by the preservation of the atomic monopoly. Reliance on the atomic monopoly in turn provoked a dangerous confrontational foreign policy on the part of the United States. The best way to maintain peace was not through wielding the atomic bomb but through civilian control of atomic energy and a policy of international cooperation. Although Rabinowitch and Condon did not see themselves as part of the progressive left, by linking domestic atomic policy and scientific freedom to international relations, they shared common ground with the progressive left's insistence that domestic prosperity depended on a stable world order.

With these considerations in mind, the FAS launched a vigorous campaign against the Vandenberg amendment. Through its public organizational network, the National Committee on Atomic Information and the newly created Emergency Committee for Civilian Control of Atomic Energy, the FAS successfully arranged for the Senate Committee on Atomic Energy to be inundated with mail from the general public. The federation also leveled pressure on Vandenberg directly by organizing rallies in his home state. Constituents' objections to the amendment soon reached the senator. Faced with public criticism, Vandenberg became amenable to changes, and the FAS dispatched Chicago Metallurgical Laboratory chemist Thorfin R. Hogness to negotiate a compromise.[20]

Hogness and Vandenberg agreed to limitations on the military's power and jurisdiction within the AEC, and the SCAE accepted the new Vandenberg amendment. Members of the FAS, however, were unhappy with the compromise, and the FAS council ultimately declined to endorse it.[21] But the Vandenberg amendment provided only the first of many concessions required to pass the McMahon bill. During late March, McMahon's staff made additional changes to satisfy the objections of some SCAE members. Among other alterations, the section entitled "Dissemination of Information" was changed to "Control of Information," the original insistent language on freedom of information exchange was dropped, and the penalties for secrecy and security violations were strengthened. The SCAE also deleted the proposed commission's mandate to study the social, political, and economic implications of atomic energy. After making other, less significant revisions, the committee reported the bill to the Senate on April 19.[22]

The FAS council disliked the revisions, especially the new secrecy provisions and the Vandenberg amendment, but council members reluctantly resigned themselves to the bill as the best possible given the political circumstances.[23] Their perception was probably correct. McMahon anticipated strong conservative opposition to his bill in the House, especially from members of the Military Affairs Committee. He therefore tailored the bill to win quick acceptance from the Senate, with the hope that it could survive a fight in the House. Eugene Rabinowitch warned that if S.R. 1717 died in the House, it would be another year before scientists could expect new legislation, and in the intervening time it was possible that the military would gain permanent control of atomic energy.[24] Any bill at all that preserved the concept of civilian control would be preferable to having a more acceptable version of S.R. 1717 die in Congress.

The McMahon bill passed the Senate easily on June 1, but on the other side of Capitol Hill, it immediately met stringent opposition from House conservatives. Most of their objections centered around a distaste for what they perceived as an extension of New Deal state-sponsored bureaucracy and, more important, fear of losing the so-called secret of the atom. The bill's assailants expressed both criticisms through anti-communist rhetoric that foreshadowed the internal divisions and external political repression that would plague scientists with the rise of the postwar red scare.

Conservative Republicans and southern Democrats objected to the McMahon bill's stipulations relating to ownership of fissionable materials and industrial development as an unacceptable expansion of state power that constituted "state socialism." To these members of Congress, the features of S.R. 1717 that mandated government authority over the development of industrial aspects of atomic energy conjured up images of Soviet-style government. At the same time, however, their opposition to excessive state regulation was tempered by their feeling that the destructive potential of nuclear weapons necessitated extra precautions and an unusual level of government control.[25]

The military significance of atomic energy led congressional conservatives to downplay their antipathy toward the economic-planning aspects of the McMahon bill, but military concerns also elevated their zeal for security. Conservatives perceived the McMahon bill's provisions for dissemination of basic scientific information as tantamount to giving away the "secret of the atom," despite the fact that S.R. 1717 expressly prohibited the exchange of information related to the manufacture of atomic weapons and production of fissionable material. Representative Charles H. Elston (R.-Ohio) objected to transferring "the secrets of the atomic bomb and of atomic energy" from the military to a civilian agency, a move that he labeled "a dangerous thing."[26] Representative Frederick C. Smith (R.-Ohio) similarly viewed a civilian agency as more unreliable than the military and commented, "I have never been able to understand why anybody should want to give this secret away. It is safe where it is. Why turn it over to this Commission that will be set up under this bill and run the risk of losing it?"[27] Representative John E. Rankin (D.-Miss.), a member of the House Committee on Un-American Activities (HUAC) who was notorious for his vituperative statements about the dangers of communism (as well as his ugly racial bigotry and anti-

Semitism), referred to the Canadian situation and attempted, without evidence, to link it to possible espionage within the United States. He declared with portentous hyperbole, "You know that little Canada, glorious Canada, arrested those spies that had been down here, down to Oak Ridge, stealing the secrets of the atomic bomb, in order to use it against you and me and the rest of the American people."[28] In these and other statements on the House floor, conservative members of Congress repeatedly linked American security to the protection of secrets, despite the scientists' arguments that attempts to maintain the atomic monopoly would fail. Even though the McMahon bill did not propose to release specific technical information related to atomic weapons, opponents of S.R. 1717 equated the bill's provisions for exchanges of basic scientific information with a lax attitude toward security.

The McMahon bill made it through the House, and President Truman signed the Atomic Energy Act of 1946 into law on August 1. The principle of civilian control survived, but much of the original intent behind the act was lost. The early emphasis on the free dissemination of information and its implicit connection to international agreements gave way to an insistence on restrictions more conducive to preserving "the secret." References to prohibitions abounded in the act's section on control of information, and the maximum punishment for improper use of restricted data, the death penalty, went beyond even the May-Johnson bill.[29] Furthermore, in the face of worsening U.S.-Soviet relations and the rise of the Cold War, civilian control did not guarantee a commitment to civilian goals. The language of the McMahon Act still reflected progressive left intentions, so much so that James R. Newman and Byron S. Miller later proudly declared that the act had done "nothing less than establish in the midst of our privately controlled economy a socialist island with undefined and possibly expanding frontiers."[30] But the Cold War soon took priority over the act's civilian goals. Not long after the AEC's creation, weapons research, development, and production came to dominate the commission's agenda. With the rapid political change of the early postwar years, the atomic scientists' triumphant foray into mass-based politics ultimately came to a disappointing end.

Adventures in Political Economy: The NSF Debate

As the atomic scientists directed their primary attention to the vital problems posed by the atomic bomb, other scientists continued an older discussion about the future of postwar science. While former Manhattan

Project scientists geared up to lobby against the May-Johnson bill and military control of atomic energy, a subcommittee of the Senate Military Affairs Committee initiated extensive hearings on a more general topic of interest to scientists: the possible establishment of a National Science Foundation to sponsor basic research. Unlike the May-Johnson bill, proposals for science legislation were not drafted hastily after the end of World War II. The Atomic Energy Commission emerged from a sense of immediate urgency and crisis following Hiroshima and Nagasaki. To the atomic scientists, the finer points of political theory were not completely unimportant, but understandably they ran a distant second to the pressing need to prevent a postwar arms race and nuclear Armageddon. The National Science Foundation debate, in contrast, lacked the element of life-or-death urgency and instead provided a more protracted opportunity to deal with deep questions about the relationships between science, political economy, and democratic society.

Plans for the postwar organizational structure of scientific research originated from wartime discussions set in a New Deal framework. In 1942, freshman senator Harley M. Kilgore (D.-W.Va.) began to apply his populist, antimonopolist sentiments to the problem of efficient wartime and postwar mobilization of scientific research. His first effort at drafting science legislation, a 1942 bill to establish an Office of Technological Mobilization, proposed a heavy centralization of research efforts in both the public and private sectors that would be enforced by government coercion. The bill satisfied almost no one.[31] Six months later, Kilgore tried again and introduced S.R. 702, the Science Mobilization bill.

The new bill set forth the issues that would define later clashes over the political economy of postwar science. It called for the creation of a permanent Office of Scientific and Technological Mobilization that would study and coordinate federal research activities. Reflecting Kilgore's populist suspicion of large-business interests, the bill proposed federal sponsorship of fellowships for science education in order to forestall universities' dependence on industry, and it mandated public ownership of patents derived from publicly funded research in order to maximize freedom in research and protect the public from corporate monopolies maintained through patent pools. The bill also called for administration of the new research agency by a board representing not only scientists but also a wide variety of other interest groups, including industry, labor, small business, and agriculture.[32] With this administrative design, Kilgore's proposal began to crystallize into a structure for science driven by more than just the antimonopoly and planning impulses of New Deal political thought. His vision also began to address the tensions between expertise and popular rule, and the place of science within the political economy of a democratic nation. Kilgore sought to free science and its potential social and economic benefits from what he perceived as domination by big business, bring science within the domain of the larger political realm, and recognize the stake nonscientists had in the sponsorship and direction of scientific research.

The Science Mobilization bill received support from prominent New Dealers, including Thurman Arnold and Vice President Henry A. Wallace, who both reserved special praise for the bill's antimonopoly features. Among scientists, the progressive left American Association of Scientific Workers (AASW) applauded Kilgore's efforts. (By 1948, the AASW would be little more than "a paper organization" and largely inactive, but it would remain a proponent of progressive left politics throughout the Cold War years.)[33] Deep-seated opposition rose from other quarters, however. In addition to representatives from industry and the military, elite leaders of science, including Vannevar Bush, James B. Conant, National Academy of Sciences president Frank B. Jewett, and MIT president Karl T. Compton, also objected strongly. These scientists argued that S.R. 702 meant bureaucratic interference that would shackle science to the state and stifle freedom in scientific research. Although it would later support a more moderate version of Kilgore's ideas, the American Association for the Advancement of Science (AAAS) also opposed the bill for allowing too broad an imposition of state power on science.[34]

Bush relayed his objections to Kilgore at length and began to formulate his own plan for postwar research. In July 1945, he completed *Science, the Endless Frontier*, his well-known manifesto for a program of basic research firmly under the authority of scientists and insulated from lay control.[35] As Bush was finishing his tract, Oscar Ruebhausen, counsel at the Office of Scientific Research and Development (OSRD), and Carroll L. Wilson, Bush's top assistant at the OSRD, drafted a bill based on Bush's principles, and Wilson arranged to have it introduced in Congress. On July 19, the day *Science, the Endless Frontier* was released, Senator Warren G. Magnuson (D.-Wash.) introduced S.R. 1285, Bush's bill for a National Science Foundation. Kilgore countered Bush and Magnuson with another bill of his own and introduced S.R. 1297 four days later.

As contemporary observers and historians have both pointed out, the Kilgore and Magnuson bills contained three basic differences. These differences arose not simply from matters of legislative detail but signified widely divergent views about the proper relationship between science and government and the optimal political structure for postwar science. First, S.R. 1297, the Kilgore bill, called for a presidentially appointed director, whereas the Magnuson bill granted the power to appoint a director to a part-time board consisting of individuals, presumably scientists, chosen "solely on the basis of their demonstrated interest in, and capacity to promote, the purposes of the Foundation." Kilgore's board lacked the specific interest group configuration of the earlier Science Mobilization bill, but by placing representatives from eight federal agencies, most of them cabinet level, in addition to eight public members appointed by the president and subject to confirmation by the Senate, Kilgore maintained the concept of a diverse board that would assure public access to science policy. Second, although the Kilgore bill did not call for the NSF to coordinate federal research directly, it preserved a commitment to the New Deal planning ideal by specifying at great length means by which the NSF could "assist government agencies in achieving maximum effectiveness in their research and development activities, and in efficiently programing and coordinating such activities." The Magnuson bill called on the foundation "to develop and promote a national policy for scientific research and scientific education" but did not elaborate further. Third, S.R. 1297 included Kilgore's insistence on government ownership and licensing of patents derived from federally funded research in order to ensure that the benefits of such research would be dedicated freely to the public, while the Magnuson bill avoided any mention of patent reform.

After extensive hearings on S.R. 1285 and S.R. 1297, Kilgore revised his legislation in December. The new bill, S.R. 1720, delineated two additional differences between the Kilgore and Bush approaches to science legislation that further attested to Kilgore's devotion to creating a political structure for science that would tie research to social needs. Kilgore's S.R. 1720 included a division for the social sciences in the NSF, in order to assist in linking basic research to social and economic needs; Magnuson's S.R. 1285 ignored such concerns and left the social sciences out. In order to avoid concentration of funds in just a few elite institutions, S.R. 1720 also mandated that 10 percent of research and development funds be apportioned to the states in equal shares and 15 percent divided among the states on the basis of population. Senate Resolution 1285 eschewed such formulas on the grounds that funds should be distributed purely according to the merits of individual proposals.[36] Together, all five of these issues spelled out the fundamental differences between Kilgore's inclination to create a governmental system for science amenable to democratic access and Bush's desire to insulate science from political influence.

The revised Kilgore bill showed striking similarities to the original legislation for civilian control of atomic energy, so much so that the FAS viewed the fight for the McMahon bill and the NSF debate as part of the same struggle over postwar science policy.[37] Both bills articulated a particular conception of the interrelationship between science and society and the need to integrate science into the political life of the nation. The original McMahon bill called for civilian control of atomic energy, a government monopoly on the development of atomic energy, and government control of patents. Civilian control was the analogue to Kilgore's insistence on proper democratic control and oversight of science. In the debate over atomic energy, it embodied the notion that in a democracy a civilian agency provided the proper locus for authority over science. Centralized government control over the development of atomic energy, like the coordination role projected for the NSF, reflected a commitment to the planning ideal derived from the New Deal. Compulsory federal licensing of patents related to atomic energy and government control over fissionable materials were required not simply to protect the national security; they also evinced the New Deal philosophy that power constituted both a public resource and a natural monopoly that required heavy state regulation. Although the patent provisions in the McMahon Act were less controversial than in the NSF legislation because of Congress's perception that atomic energy required exceptional government oversight, both pieces of legislation linked patents to larger economic reform.[38]

The addition of the social sciences to S.R. 1720 and the McMahon bill's original mandate to study the social, political, and economic implications of atomic energy (ultimately deleted from the final act) also represented shared aspirations. The provision in the McMahon bill reflected the feeling, especially among the atomic scientists, that scientific knowledge had to be matched by an understanding of human society in order to head off the destructive potential of science. Henry A. Wallace presented a similar justification in hearings over the NSF legislation and argued that inclusion of the social sciences was necessary because progress in the physical sciences, especially in the aftermath of Hiroshima and Nagasaki, had dangerously outpaced "the science of humanity."[39]

The scientific community was strongly divided over whether to support Kilgore or Magnuson. Kilgore's proposals for the postwar organization of science alienated conservative scientists but attracted those on the progressive left. Scientists who subscribed to the antimonopoly and planning tenets of New Deal liberalism favored Kilgore's provisions for patent reform and coordination of scientific research. Moderately liberal scientists also tended to favor Kilgore's ideas as an attempt to bring science into the public sphere. But the senator's ideas were anathema to conservatives.

National Academy of Sciences president Frank B. Jewett, who objected to both the Bush and Kilgore plans, was the most steadfast conservative among American scientists. He firmly opposed the prospect of expanded state power and political oversight in American science. Federal funding, he felt, inevitably meant some form of political intrusion, no matter what sort of administrative or legislative devices were created to maximize scientists' autonomy. In hearings over the 1945 Kilgore and Magnuson bills, S.R. 1297 and S.R. 1285, Jewett was the only witness to reject the basic idea of a National Science Foundation. Instead, he advocated a return to the prewar days, with funding solicited from the private sector and any necessary coordination done under the auspices of the National Research Council (NRC), a branch of the National Academy of Sciences. Jewett essentially proposed a form of Hoover-era voluntary associationalism, in which scientific research would remain under the control of the private sector, with limited public access via the quasi-governmental NRC.[40]

Vannevar Bush was also politically conservative, but unlike Jewett, he believed that the massive wartime growth of funding for science could continue only through federal support. The primary challenge, in Bush's mind, involved creating a system for government patronage of pure science that would avoid the pitfalls of New Deal centralization and place control in the hands of the appropriate experts, namely, scientists themselves. Bush opposed the continuation of the OSRD after the war on the grounds that it was an overly regimented organization suited for the wartime emergency but not free scientific research. He also felt the military was not a suitable sponsor for basic research in peacetime. At the same time, however, he viewed the OSRD as an example of an efficiently run organization directed by the proper experts and dedicated to the support of the best research.

Bush was determined to prevent what he considered as inappropriate external political influence over science, and he designed his proposed administrative structure to insulate the NSF as much as possible from political authority. He firmly opposed schemes for geographical dis-

Vannevar Bush, 1943.
(National Academy of Sciences)

tribution of funds in order to preserve scientists' decision-making autonomy, as well as to ensure that the NSF distributed awards purely on the basis of scientific merit. In Bush's view, formulas for geographical distribution constituted pork-barrel politics, rather than means for building up less privileged institutions. Instead, Bush believed the best research would come by supporting universities that were already at the top, a strategy that the Rockefeller Foundation had followed by "making the peaks higher" in the prewar era. Above all else, Bush insisted that matters extraneous to science had to be kept out of the NSF legislation.

This insistence on keeping external issues at bay overrode any reform impulses Bush might have had. Bush did not reject changes in the patent system per se. He even shared some of Kilgore's concerns, but he thought that Kilgore's widespread dedication of patent rights to the public went too far and would stifle incentives for innovation. More significant, he insisted that patent reform should be the province of a separate congressional committee and did not belong in the NSF legislation. Likewise, he felt the social sciences properly belonged in their own agency, not in a foundation dedicated to the natural sciences. Although Nathan Reingold has characterized Bush's plan as "a new political economy for research and development," Bush was really, in a sense, proposing an agency outside the political economy altogether. His philosophy for the NSF contemplated a vision of science firmly under expert control, separate from the polity, and separate from society. His conception of postwar science could not have been more different from Kilgore's evolving notion of a democratic political structure for science that assumed the intertwined nature of science and society.[41]

Passions ran high in discussions of the NSF proposals. Conservative scientists softened their public statements, but in private they lashed out against Kilgore and his scientific supporters in anti–New Deal and antiradical terms.[42] Scientists who favored Kilgore's ideas similarly targeted the conservatism of the Bush camp; Harlow Shapley once described Bush's supporters as "kept men of the industrial friends of science."[43] It was not surprising that their differences proved impossible to bridge. Scientists on the progressive left lauded Kilgore's efforts for the very reasons that conservatives like Bush and Jewett despised them. Lacking enthusiasm for the private sector and the merits of voluntary associationalism, progressive left scientists placed their faith in the economic-planning and antimonopoly wings of New Deal thought as means to guarantee a stable and just political and economic order.[44]

Members of the American Association of Scientific Workers backed Kilgore's original wartime proposals, especially the provisions for coordination and patent reform. Writing in *Science,* chemist K. A. C. Elliot, chairman of the Philadelphia branch of the AASW, and physiologist Harry Grundfest, national secretary of the AASW, praised S.R. 702 as "an admirable illustration of the manner in which science and its fullest utilization both for war and for peace can be 'planned' or coordinated for the benefit of the community without adversely affecting the research freedom of science."[45] A 1944 pamphlet by Grundfest, geologist and AASW president Kirtley F. Mather, and physicist and AASW treasurer Melba Phillips also stressed the need for coordination.[46] In addition, Grundfest applauded Kilgore's principle that patents derived from publicly funded research should be dedicated to the public welfare. At a 1945 conference on the relationship between science and democracy, Grundfest argued that private industry tended to use patents to stifle scientific research. Drawing on the work of Thurman Arnold, Grundfest contended, "The problem is not in the individual patent, but in the corporative control of patents. . . . Patents are used for monopoly control, for limitation of production, for limitation of invention, for adulteration of results."[47] Geneticist L. C. Dunn, a member of the AASW and, in 1944, a consultant to Kilgore's committee, also supported the Science Mobilization bill and suggested that industrial scientists' opposition sprang from fears that the patent provisions of S.R. 702 "strike at the basis of private monopoly control based on exclusive private patent rights."[48]

Scientists who favored Kilgore's ideas predicated their support on more than just commitments to economic planning and patent reform. Progressive left scientists embedded their vision for the NSF in a larger worldview that emphasized scientific cooperation with other nations and a domestic economy based on full employment. Mather, Grundfest, and Phillips observed, "An expanding economy based on international collaboration can bring to the American people, and to the people of the entire world, higher standards of well-being, health, and culture. To achieve these goals, which must be based on full employment and a high level of productivity, it is necessary to harness scientific and technological research for the aims of peace to the same extent that produced the miracles of scientific and technological development in the war effort."[49] The Boston-Cambridge chapter of the AASW also called for science legislation that would "be in harmony with a progressive policy in the fields of national and international relations."[50] The leaders of the AASW firmly believed postwar science policy could not be formed in a vacuum. Rather, science policy had to be tied directly to broader objectives, especially the internationalist and domestic economic goals of the progressive left.

More generally, the progressive left conception of postwar science policy implied the need to open science to public access and input. L. C. Dunn gave a terse expression of this sentiment when he jotted down on a piece of paper, "Scientists not to be exempt from democratic control. Freedom."[51] In another set of handwritten notes, he elaborated what he meant. Unlike Bush, who feared public interference, Dunn insisted that the public have a voice in the promotion of scientific research. He noted, "Essential issue is public or private control—actually not whether public shall take complete control but whether it shall expand its area and share more equally with the Foundations & Industry the responsibility for finance & direction of scientific research." Dunn indicated that the reality of foundation and industry support for research belied the myth that prewar science was protected from outside interests. Given that context, public participation in science policy did not constitute undue external influence but allowed the public to take its rightful place alongside private interests in the governance of research. Dunn then added that because science had to be "conducted for general good & must be responsible to society as a whole," he was "not afraid of public control." What he did fear was "centralized *private* control for profit." Politics, on the other hand, was a natural part of a responsible relationship between science and society ("Political—of course!"), not a form of outside interference that threatened the integrity of research.[52]

Engineer Morris L. Cooke spoke eloquently of the relationship between science and society and the need to link research to larger social goals. As Edwin T. Layton has put it, Cooke believed that the engineer was supposed to "make public interest the master test of his work." Cooke's dedication to serving the public interest dated from the Progressive Era. During the heyday of the Progressive years he served as director of public works in Philadelphia, and in the 1930s he headed the Rural Electrification Administration. Like Bush, Cooke believed in the power of expert authority. Unlike Bush, however, Cooke felt that the scientific expert's authority had to be derived from the consent of the majority.[53] Democratic control of science was not a form of political interference but a necessary part of the social roles of science and technology. Testifying on behalf of the Independent Citizens' Committee of the Arts, Sciences, and Professions in the 1945 hearings on the Kilgore and Magnuson bills, Cooke urged Congress to adopt the administrative structure of the Kilgore plan, which he felt appropriately placed government authority and public accountability over science. He observed that scientists and engineers all too often felt themselves apart from politics and society. Their detachment, however, was an illusion. He stated, "While scientists dread what they call 'political pressures' without knowing too much about them first-hand, most scientists actually do live fairly happily and work quite effectively in an environment where the quite comparable academic pressures operate forcefully and continuously. . . . [Industry] too has its pressures which, when encountered, have to be analyzed and frequently resisted."[54]

Like Dunn, Cooke pointed to the existing social pressures that negated claims that scientific research proceeded only according to its own intellectual dynamic. To that extent, the federal government was no different in principle from the forces already operating on science and technology. Furthermore, Cooke contended, recognition of the public sphere's claim to science would produce a healthier relationship between science and society. He declared, "It will be hard going for a while for those brought up in cloistered atmospheres, but the sooner science learns to take it like all the rest of us, the better it will be. After many years of the closest possible association with engineers and scientists I am strongly of the opinion that both groups will profit by a more intimate association with government and other community institutions." Prevailing corporate interests, Cooke implied, tied science to the predominant goal of profit. The NSF legislation, on the other hand, would harness science to higher aims: "This proposed legislation if passed and implemented with adequate appropriations, will make it possible for the first time in history to develop a balanced system for the sciences based on total human needs."[55]

The FAS and other more moderately liberal scientists' groups did not, by and large, share the AASW's commitment to New Deal political economy, but they did favor Kilgore's approach to administrative structure, patent policy, and the social sciences. The federation was preoccupied with atomic energy legislation in 1946, but the Washington Association of Scientists devoted what resources it could toward aiding Harlow Shapley's efforts on behalf of the progressive left to support Kilgore. Reflecting its political moderation, the FAS reserved internationalist arguments for its support of civilian control of atomic energy and did not view the National Science Foundation as having significant foreign policy ramifications. The federation also avoided the antimonopoly reasoning employed by the AASW. Instead, the federation emphasized the proper relationship between science and government. A 1947 analysis by the Washington Association of Scientists study group on the NSF legislation viewed the differences between the Magnuson and Kilgore approaches as "a sharp cleavage between two opposed philosophies of the relation of science to government and society."[56] The importance of the ethos behind the Kilgore bill lay in its potential to place basic research on a democratic foundation:

> The original Kilgore Bill, concurred in by the President and his advisers as well as by many scientists, is based on the premise that science is a national resource, that its raw material is the Nation's scientific manpower, and that, as a vital national resource, its furtherance should be entrusted to an authority directly responsible to the elected representatives of the people—the Congress and the President. The proponents of this philosophy place primary emphasis upon long-range planning for the whole field of science to ensure the development of scientific potential on the widest possible basis throughout the country. They seek guarantees which will deny to special interests a disproportionate influence in formulation of Foundation policy, or disproportionate benefits from its activities. They insist upon a patent policy which will permit free public access to discoveries made with public funds.[57]

To the FAS, the administrative structure of the NSF, provisions for geographical distribution of funds, and patent policy were part of a larger vision of a new political economy for postwar science. The study group continued, "The basic issue is none other than the proper role of the Federal Government in regulating those areas of our national life which are intimately related to the public welfare and security, in this instance the shape and scope of science."[58] According to the FAS, federal sponsorship of basic research had to function through democratic mechanisms of representation and operate in such a way as to spread the benefits of research as widely as possible. As will be discussed shortly, the federation also viewed the establishment of a National Science Foundation as crucial to preventing the military domination of basic research.

The post–World War II political divide between conservative and liberal-left scientists reflected the much older problem of the role of expertise in American society. The notion that the scientific values of rationality and efficiency could be used to order political affairs had attracted American intellectuals since the late nineteenth century. Technocratic progressives, including social scientists and engineers, promoted the concentration of power in the hands of experts as a solution to the inefficiency and corruption that they felt characterized political life. Fears about the undemocratic nature of expert authority did not concern these reformers, who assumed that the obvious advantages to be gained through the objective application of scientific approaches to policymaking would easily earn vigorous popular assent. That their own interests and value-laden assumptions might sway their decisions, and that their proposed solutions to social and economic problems might prove as controversial as those arrived at through normal political processes, rarely occurred to them.[59] Bush's approach to the NSF, with its emphasis on expert authority, was firmly grounded in this technocratic worldview.

The elitist tendencies of the efficiency ideal were not unique to conservative thinkers, however. Most late-nineteenth- and early-twentieth-century technocratic progressives were espousing a new form of liberalism, one that advocated the use of expertise to mitigate the deleterious effects of the unrestrained free market of classical liberalism (now known popularly as conservatism). As a political program, the gospel of efficiency, to invoke Samuel Hays's elegant turn of phrase,[60] was a malleable concept that crossed conservative, liberal, and radical lines. For example, with respect to the NSF legislation, left-wing geneticist H. J. Muller, his faith in rational planning by the state tempered by his 1930s encounters with the Soviet government, proved just as suspicious of the threat of political interference in science as Vannevar Bush.[61] The economic-planning ideal supported by progressive left scientists as part of a more democratic political structure for basic research was itself more typically associated with a preference for expert-directed policymaking backed by the authority of the state.

The ultimate significance of liberal and progressive left scientists' vision for postwar science, then, lies less in their specific position on the political spectrum than in their willingness to effect a reconciliation between expert authority and popular will and to create a political structure for the former that would give precedence to the latter. Theirs was an unusual populist position, an unabashed synthesis of faith in expertise and confidence in the wisdom of the public. By recognizing the political within science, liberal and progressive left scientists hoped to create a healthier relationship between science and society, from which both scientists and the public would prosper.

In the end, neither Kilgore's nor Bush's supporters realized their vision for the NSF. Bush and other elite scientist-administrators, as well as leading members of the National Academy of Sciences, favored the Magnuson bill, whereas members of the American Association for the Advancement of Science, the AASW, and the FAS preferred Kilgore's legislation.[62] In a compromise brokered between the two camps, Kilgore and Magnuson jointly introduced S.R. 1850 on February 21, 1946. The new bill was virtually identical to Kilgore's S.R. 1720 and left Bush and his supporters to hope for concessions as it wended its way through the Senate and House. Had scientists maintained a public semblance of unity, S.R. 1850 might well have passed. In May, however, when House Republicans introduced a slightly revised version of the original Magnuson bill, Bush broke ranks and threw his support behind the House proposal. The compromise fell apart. Senate Resolution 1850 easily cleared the Senate on July 3 but died in the House Committee on Interstate and Foreign Commerce sixteen days later. Scientists and other contemporary observers blamed its death on their own lack of unity. As the major constituency that would be affected by the establishment of a National Science Foundation, they could not afford fractious infighting at a time when a relatively weak president could not easily persuade Congress to create a new federal agency.[63] Four more years would pass before the creation of the National Science Foundation, and the final product would prove a far smaller and less significant institution than either Bush's or Kilgore's partisans had originally hoped.

Containing the Military: The FAS and Opposition to Military Patronage of Science

Despite the failure to establish a National Science Foundation immediately after the war, basic research suffered no dearth of funds in the postwar period. The Office of Naval Research (ONR), established in August 1946, along with the army and the Atomic Energy Commission, filled the gap. During the postwar decade, well over 90 percent of funding for research in the physical sciences came from agencies devoted to military needs.[64]

Few of the elite leaders of American science feared that defense dollars would affect the content and character of basic research. During the war, men such as Bush, Conant, Oppenheimer, and others had forged a comfortable working relationship with the military. University administrators and department heads also based their plans for postwar expansion on the assumption of continued science-military cooperation.[65] As long as scientists and not military officers directed the distribution of funds, many of the elite scientist-administrators had no qualms that military objectives would dictate the agenda of basic research. Bush went so far as to decry the 1945 controversy over the May-Johnson bill and military control of atomic energy as "utterly unnecessary, and also for that matter entirely unfounded."[66] In a letter to Karl Compton, Bush also told of how he had urged the army and navy to support the NSF legislation in order to head off such irrelevant concerns. He wrote, "If they were to do so they would quiet at once all of this discussion of domination of the universities by the military, which otherwise might lead to a civilian-military fracas such as surrounded the atomic energy legislation, on an entirely false basis."[67] To Bush, wartime cooperation between scientists and military leaders exemplified a system in which decision-making authority lay with the appropriate scientific experts. Although Bush did not favor the peacetime continuation of the OSRD, in the absence of a National Science Foundation he felt the armed forces had "been filling a gap, and . . . on the whole very well."[68] Other scientists agreed that military agencies, especially the ONR, administered funds in a manner that accorded with scientists' chief interest in basic research.

The atomic scientists viewed the military differently and feared that military patronage would adversely affect the content and character of physics research. The FAS had already made clear its opposition to military control of atomic energy during the postwar debate over the McMahon bill. In the NSF legislation, the federation found a positive program for establishing federal patronage of all scientific disciplines on a nonmilitary basis. For the FAS, a National Science Foundation offered a crucial institutional alternative to military sponsorship of basic research.

In the summer of 1946, while lobbying for final passage of the Kilgore-Magnuson compromise bill, the federation specifically highlighted the need to create nonmilitary sources of funding for research. In a letter to Senator Thomas C. Hart (R.-Conn.), Hugh C. Wolfe, secretary of the Association of New York Scientists, and Joseph H. Rush, FAS treasurer, stressed the urgency of passing S.R. 1850 in order to avoid continued army and navy funding of basic research. They declared, "It is essential that the trend toward military domination of our universities be reversed as speedily as possible." In a similar vein, Wolfe warned Representative Virgil Chapman (D.-Ky.) that failure to pass the NSF legislation "would mean a continuation of the process of extending military control over our universities."[69]

After the congressional death of S.R. 1850 and the signing of the McMahon Act, the federation continued to consider the problem of military patronage. In a November 1, 1946, memo to FAS chapters, William Higinbotham discussed the Washington Association of Scientists' plans to study the intricacies of science legislation. He located the problem squarely in the context of military funding, noting high levels of military support for university research. That same day, Eugene Rabinowitch's lead editorial in the *Bulletin of the Atomic Scientists*, titled "Science, a Branch of the Military?," also considered the dangers of military patronage. Six months earlier, during the height of the debate over the atomic energy legislation, Rabinowitch had warned that failure to pass the McMahon bill might allow the military to gain permanent control over atomic energy. Now he feared that in the absence of a National Science Foundation, the military was gaining primary responsibility for the administration of government-sponsored research. He conceded that the army and navy operated "in a reasonable fashion in granting liberal contracts for fundamental research," at least for the moment. Nevertheless, Rabinowitch insisted that "organizational subordination of science to the Armed Forces is an evil thing." In his view, even if military funds were administered in the most exemplary manner imaginable, they violated a proper division between civilian and military roles in American society. Scientists, he concluded, needed to "fight for the liberation of science at large from military sponsorship" and work for the establishment of a National Science Foundation.[70]

In that same issue of the *Bulletin*, Cornell physicist Philip Morrison also discussed the postwar relationship between science and the military.

Morrison was active in the FAS and progressive left politics, and his reasoning reflected the commitments of both. Taking up where Rabinowitch left off, Morrison warned that even if military contracts now seemed benevolent, the situation would quickly change. Morrison recognized the attraction of ONR-funded accelerators that allowed physicists to do "rather fancy" research unrelated to war needs. But, he predicted, "[Soon the] now-amicable contracts will tighten up and the fine print will start to contain talk about results and specific weapons. And science itself will have been bought by war, on the installment plan." In a nod to the progressive left, he contended that the desire to patent results would spread from industrial laboratories to the universities and "destroy the traditional free cooperation of science." From the secrecy produced by patents, Morrison made a transition to the threat of military secrecy and cautioned that the classification of information, security clearance investigations, and other security requirements would make research "narrow, national, and secret." He then called attention to the evolving Cold War context and what the world would make of military sponsorship of American science: "Above all, and in spite of every protestation, American science will appear to the world as the armorer of a new and more frightful war. We are not far from giving that appearance today."[71] Not only would military patronage damage the character of science, but it would confirm other nations' fears that the United States was bent upon pursuing an aggressive foreign policy backed by heavy military buildup.

Morrison then turned to the role of the scientist. He contended the physicist recognized that the situation posed by a military-based political economy for science was "a wrong and dangerous one" but felt "impelled to go along, because he really needs the money." At the same time, Morrison warned, scientists ought not be fooled by the apparent generosity of the ONR: "The ONR contracts are in most cases models of restraint, and assiduously avoid the most common errors of unjustified secrecy or patent restrictions. But either the work is military, and the physicist is working for war purposes, or it is not; and the Navy and the Army ought not to support and, in the long run, control it."[72] No matter how far removed defense-funded research might seem from military application, Morrison argued, it would ultimately serve military purposes.

Morrison proposed the NSF as a necessary alternative to the military. He stated, "Only a National Science Foundation can hope to bring to peacetime fruition the promise science held out in war." The NSF, he emphasized, was not a minor issue in comparison with atomic energy. Every personal and political action had larger repercussions, and all individuals had a duty to take what action they could. Morrison declared, "In every sphere of life we must prepare for peace if we are to have it. . . . Each of us must demobilize in his way if he is to share in the peace. We in science have to demobilize the laboratories, and earnestly to prepare for the peace we won with victory."[73] To Morrison, the establishment of the NSF served the larger purposes of demobilizing science, staving off cold war, and reorienting science toward peaceful purposes.

Other FAS members shared similar concerns. Philip N. Powers admonished scientists to devote careful thought to the postwar institutional structure of science. At the time, Powers, Higinbotham, and David Hawkins were already urging the federation to undertake a study of military support for basic research. In the December 27, 1946, issue of *Science,* Powers wrote, "The important question is whether they [issues regarding the postwar organization of science] will be resolved on the basis of considered opinions of scientists, educators, and others, or whether the whole matter will simply be left to resolve itself or perhaps be left to the Army and the Navy to decide in the way that seems best to them." Although Powers worked for the Scientific Personnel Branch of the Office of Naval Research, he was decidedly ambivalent about the ONR's developing role as primary sponsor for basic research. Like Morrison, he conceded that ONR distributed funds in a manner that recognized "a remarkably clear understanding of the necessity for freedom of scientific research," but he considered the ONR a model for a future National Science Foundation, not a permanent agency for basic research. He concluded with a commentary on the danger of inaction:

> At this point it must be remembered, however, that the easiest thing for Congress to do is nothing. The Army, as well as the Navy, has officially supported a National Science Foundation, and if scientists, educators, and others throughout the country fail to support this view, thereby failing to assume responsibility for encouraging the promotion of science for peaceful purposes, then there is another question to be faced: *Is the primary end of free American science to be the national defense—free because of the necessity of fundamental discoveries leading to novel weapons of war?* [italics in original][74]

Powers agreed with Morrison that military-supported science, no matter how free and open ended it might seem, would inevitably serve war needs, not civilian purposes.

Individual FAS chapters also expressed their concerns about military support of basic research. Announcing its program for 1947, the Association of New York Scientists warned, "We regard military domination of science as inevitable unless a civilian agency is established to carry on this function [basic research]," and it declared its support for a National Science Foundation bill along the lines of the 1946 Kilgore-Magnuson compromise bill.[75] As the debate over the NSF began to heat up again in 1947, the Association of Oak Ridge Engineers and Scientists urged the complete exclusion of military projects from the NSF legislation. Otherwise, "If under present world conditions, the support of military research should be included as a function of the Foundation, there is real danger that military considerations will tend to dominate."[76] The executive committee of the northern California chapter of the FAS also worried that "in the absence of a science research foundation, university, college, and institutional research is forced to depend increasingly on subsidization by military agencies." Although the committee believed the military had provided "generous support and liberal policy," the northern California chapter, like Rabinowitch, Morrison, and Powers, feared such a situation could not last in the long run and urged its members to write their congressional representatives in support of the NSF legislation.[77]

A number of historians of American science have, in recent years, characterized scientists' headlong rush toward acceptance of military funding as ill advised and made without due regard for the consequences.[78] For the elite scientist-administrators, this description may well be true. But they did not represent all scientists. Members of the FAS were hardly so acquiescent. They pushed for a National Science Foundation and a civilian Atomic Energy Commission precisely because they feared they could not accept military funds without serving military ends. Moreover, they believed that not only would military patronage adversely affect pure research, but it would also violate proper democratic principles of government and make the world a more dangerous place. Military support spelled trouble on three fronts—science policy, government, and international relations. Civilian control of atomic energy and the right kind of National Science Foundation, on the other hand, offered hope for a more democratic, publicly responsive, socially conscious science and a safer world.

The original McMahon bill and Kilgore plan constituted much more than self-interested efforts to create autonomous refuges for science. Quite the opposite—liberal and progressive left scientists instead sought to subordinate the autonomy of science to a larger political agenda.

Together, the civilian control and NSF debates were part of an effort to redefine the politics of science and recognize the interrelations between social conditions and scientific research. In seeking to remake science policy, liberal and progressive left scientists also reshaped their own political role. For the first time, scientists sought political power through the rough-and-tumble world of mass-based politics and the legislative process, rather than the White House and executive branch agencies.

Public visibility and an imaginative political prospectus were not without their costs, however. The new politics of science quickly came into conflict with another novel political trend: the rise of the Cold War. The newly perceived dependence of national security on atomic weapons became particularly troublesome for the atomic scientists. They faced internal dissension as well as unwanted attention from the House Committee on Un-American Activities and the Federal Bureau of Investigation. Under the rapidly changing political conditions of postwar America, the scientists' movement soon ran aground upon the shoals of domestic anticommunism.

CHAPTER 2

Fear, Suspicion, and the Surveillance State

The FAS, HUAC, and the FBI, 1945–1948

• • •

As a result of their campaign for civilian and international control of atomic energy, the atomic scientists gained some dangerous enemies. The House Committee on Un-American Activities and the Federal Bureau of Investigation (FBI) quickly seized upon the FAS's political positions as evidence that American scientists posed a menacing threat to the security of the nation. To HUAC and the FBI, the scientists' internationalist agenda symbolized a dangerous naiveté about Soviet intentions, or even worse, an open preference for the Soviet Union over the United States.

In the immediate postwar years, neither HUAC nor the FBI had yet attained the power to intimidate and harass for which they were known and feared by the 1950s. Both already enjoyed substantial visibility and capacity for investigating and plaguing the progressive left, but as of 1946 they lacked the carte blanche soon to be granted by the Cold War political consensus. The House Committee on Un-American Activities was as yet a rather marginal congressional committee of dubious respectability. J. Edgar Hoover, for his part, kept the FBI carefully reined in and restricted the bureau to subtle, yet widespread, abuses of power. The FBI of the 1940s still depended largely on surrogates, sympathetic and well-placed individuals (including members of HUAC) to whom the bureau surreptitiously supplied information, in order to exercise political influence. The FBI's notorious COINTELPRO, the counterintelligence program that employed direct infiltration and harassment of the Communist Party and other political organizations, was not launched until 1956.[1] Even so, in the immediate postwar years, HUAC and the FBI possessed considerable means of quelling dissent, and they brought those means to bear against the atomic scientists.

Internal divisions also plagued the scientists' movement. Uncertain of their next task after the passage of the McMahon bill and the failure in the United Nations to establish international control of atomic energy, the atomic scientists suffered a wrenching split between political moderates and the progressive left. Afraid of what HUAC might do, and distrustful of those scientists who desired a more radical political agenda, the FAS's leaders responded to fears of being labeled procommunist by abandoning the open, public political tactics that had brought the federation modest success in the debate over civilian control of atomic energy.

Suspicion: J. Parnell Thomas, Atomic Energy, and the Politics of Anticommunism, 1946

On June 4, 1946, three days after the Senate passed the McMahon bill and shortly before the House was due to take up the debate, Representative John S. Wood (D.-Ga.), chairman of HUAC, and Ernie Adamson, chief counsel of HUAC, arrived at Oak Ridge to inquire into the operations of two scientists' groups, the Association of Oak Ridge Scientists at Clinton Laboratories and the Oak Ridge Engineers and Scientists. They also wanted information about the activities of local union organizers.[2] The fallout from the visit soon became apparent during House deliberations over civilian control of atomic energy. On July 11 Representative J. Parnell Thomas, a member of the Military Affairs Committee, ranking Republican member of HUAC, and a staunch opponent of civilian control of atomic energy, attempted to kill the McMahon bill in the House Rules Committee. Drawing on the specter of atomic espionage, he argued that the secrets of atomic energy could only be safeguarded as long as the military remained in control. To illustrate the dangers of civilian control, he cited Adamson's preliminary report on "subversive activities" at Oak Ridge.[3]

The "subversive activities" to which Adamson referred were none other than the political efforts of the atomic scientists' movement. Among other things, Adamson observed that the scientists' organizations at Oak Ridge were "devoted to the creation of some form of world government," "very active in support of international civilian control of the manufacture of atomic materials," and "definitely opposed to Army supervision at Oak Ridge." Scientists that he had questioned, he noted, "not only admit communications with persons outside of the United States but in substance say they intend to continue this practice." Adamson also cited the danger posed by Oak Ridge union organizers. Paying special attention to labor's links with scientists, he remarked, "The CIO is now making a desperate effort to unionize all workers of the reservation. . . . The CIO organizer was very militant and seems to be in close contact with the members of the scientific societies." Adamson then claimed that "security officers at Oak Ridge think that the peace and security of the United States is definitely in danger," and he warned, "If jurisdiction and control of the Oak Ridge Reservation passes into civilian hands, the political plans for exploiting the place are well advanced and there would undoubtedly be trouble in the reservation within six months." To head off dire consequences, he recommended that the army retain permanent control of atomic energy.[4]

The atomic scientists immediately denied any implication that their activities were subversive or damaging to U.S. security. Aaron Novick, vice chairman of the executive committee of the Atomic Scientists of Chicago, labeled Adamson's report "naive and unfounded." The scientists' advocacy of civilian control was not subversive; to the complete contrary, their political actions exemplified the deepest tenets of American democracy. Novick declared firmly, "The scientists of this country are not traitors. . . . We have never advocated giving away the secret of the bomb. All that we urged is a form of international control to be placed in the hands of the people and not the military of the United States, because the people are traditional policy-makers of the nation."[5] The following day, July 12, the Association of Oak Ridge Engineers and Scientists wired Senator McMahon to reassure him that Adamson's accusations of subversive activity were "utterly false" and to urge him to redouble his efforts to pass the McMahon bill.[6] The Oak Ridge scientists also sent a telegram to the House Rules Committee charging that Thomas's actions were "typical of the delaying action of those opposing the civilian control of atomic energy." As for any threats to security from the Oak Ridge scientists, the telegram noted that Oak Ridge was safely under the watch of the Manhattan District, which had never found any reason to question the good faith of the scientists' movement.[7]

Thomas failed to keep S.R. 1717, the McMahon bill, bottled up in the Rules Committee and had to carry his fight to the House floor. On July 15, the day before general debate over the atomic energy bill was scheduled to reach the House floor, he announced he was sending a letter to every member of the House. In the letter, Thomas labeled the proposed legislation "the most dangerous bill ever presented to the Congress in the history of the United States." He advised his colleagues to vote against the bill, warning ominously, "A vote in favor of it . . . may be a vote to give a potential aggressor nation the atomic-bomb secret."[8]

In the early postwar years, there were still plenty of members of Congress who were willing to openly contest HUAC's antiradical assaults. In his initial statement on S.R. 1717, Representative Adolph J. Sabath (D.-Ill.) took great care to defend the bill against Thomas's arguments that the nation's security could not be guaranteed if a civilian agency was placed in charge of atomic energy. He attacked Adamson's report as "character assassination by innuendo, by insinuation, by association of ideas." He did not go through all of Adamson's charges point by point, noting, "I have neither time, energy, nor inclination, Mr. Speaker, to point out all the foolish and uninformed statements made by this man Adamson," but he exposed several fallacies in the report. Adamson's comment on the scientists' communications with people from foreign countries, Sabath observed, had been carefully phrased to suggest sinister intent without making any direct allegation that the contacts had been improper. In response to Adamson's contention that security officers at Oak Ridge feared for U.S. security, Sabath cited a direct statement authorized by Colonel David E. Shaw, security officer of the Manhattan District: "I have not said that I consider the security of the Manhattan District in jeopardy as the result of any activity of any current group at Oakridge [*sic*]." Sabath also noted that no connection existed between the American Federation of Labor (AFL) or the Congress of Industrial Organizations (CIO) and the Association of Oak Ridge Engineers and Scientists, and he added that Adamson's "prejudice against all labor organizations impelled him to drag in references to the CIO." In the end, Sabath dismissed Adamson's report as entirely "superficial, meaningless, misleading, and libelous."[9]

Thomas replied to Sabath by attacking S.R. 1717. At first he focused on the bill itself, particularly its provisions for the exchange of scientific information. He denied that the United States could hope to obtain useful knowledge through information exchange, even with allies such as Canada or Great Britain. Sharing information about industrial applications of atomic energy would only "give the atomic-bomb secrets away" and guarantee the loss of the American atomic monopoly.[10]

The next day, Thomas turned to lobbing ad hominem assaults at the McMahon bill. He belittled Edward U. Condon, director of the National Bureau of Standards and scientific adviser to the Senate Special Committee on Atomic Energy, as "an appointee of Henry Wallace." He also implied that Condon's planned 1945 visit to the Soviet Union to attend the 220th anniversary celebration of the Russian Academy of Sciences, canceled at the last minute when his passport was revoked because of his participation in the Manhattan Project, hid a darker purpose. Thomas completely ignored the fact that Condon's passport had been revoked specifically without prejudice to Condon's loyalty or security status. Condon, Thomas contended, provided a typical example of the unreliable types who had written the McMahon bill.[11]

Harold Urey, Nobel Prize–winning nuclear chemist and an active participant in the scientists' movement, also came under fire. Thomas quoted an excerpt from a November 24, 1945, letter to Representative Clare Boothe Luce (R.-Conn.) in which Urey had stated, "I personally would like to see all of the penalty and security violations deleted from this bill." Urey, Thomas said darkly, "is one of those who helped draw up S. 1717 and he wants to see all of the penalty violations taken out of the bill. You can just imagine from this what would happen if this civilian commission ever got the atomic-bomb secret or if Dr. Urey became one of the commissioners." He avoided mentioning Urey's rationale, however, until forced by a fellow member of Congress to read Urey's explanation that laws already on the books provided adequately for security: "I believe that the Espionage Act of the United States is sufficient to cover violations of many security regulations." Nevertheless, Thomas vehemently denounced the McMahon bill as "the creature of impractical idealists." He continued, "I do not say that these one-world-minded persons are unpatriotic. I say that their intense ardor for a better world has blinded them. Their faith in Russia is indicative. And yet, when they advocate a free exchange of atomic secrets with Russian scientists, they completely overlook the fact that all Russian scientists are but tools in a dictatorship of the proletariat. The starry-eyed individuals who molded S. 1717 will strive just as energetically to stack the Atomic Civilian Commission [*sic*], called for under this legislation." Thomas then went on to read Adamson's report on Oak Ridge into the record. He finally concluded that atomic energy could be guarded more safely under military control and contended that communist and communist-front organizations lay behind the support for S.R. 1717.[12] Representative John E. Rankin, a fellow member of HUAC, later echoed Thomas's sentiments and condemned the McMahon bill as "one of the most dangerous pieces of legislation that has ever come before the American Congress."[13]

Thomas continued his opposition to civilian control of atomic energy to the bitter end. As part of the conference committee that reconciled the differences between the Senate and House versions of the McMahon bill, Thomas refused to sign the conference report. He was the only member of the committee to do so.[14]

Thomas's bitter assault on civilian control of atomic energy provided a warning of the shape of things to come. To Thomas, a member of HUAC and a fervent opponent of any view he perceived as procommunist, internationalism and a belief in the nonexistence of atomic secrets constituted subversive ideas that favored Soviet interests. His actions exemplified the way in which anticommunism and postwar fears surrounding the atomic bomb combined to make scientists targets for Cold War political repression. But congressional demagogues like Thomas were hardly the sole source of anticommunist political attacks on scientists. Anticommunism resided at the very center of the scientists' movement, and in the immediate postwar years, it threatened to tear the FAS apart from the inside.

From Progressive Left to Cold War Liberalism: Anticommunism and the FAS

Following HUAC's attack on the atomic scientists, the FAS leadership grew increasingly concerned about the federation's possible vulnerability to red-baiting. Federation of American Scientists chairman William A. Higinbotham, fearful of conservative attacks, took steps to remove any hint of a connection between the scientists' movement and the American left. He condoned the firing of FAS employees with Communist Party connections, and he also began to guard the FAS against ties with organizations labeled left-wing by their political opponents. The federation experienced internal struggles as well, especially between political moderates and more radical scientists within several FAS chapters. By 1948, power was wrested away from the radical factions, and the federation committed itself to a more cautious and limited political agenda.

After the visit of HUAC chairman Wood and HUAC counsel Adamson to Oak Ridge in June 1946, Higinbotham began to worry about the repercussions of the atomic scientists' activities and associations. In a rambling manner and almost frantic tone, he discussed his fears at length in a confidential July 18, 1946, memo to the FAS administrative committee. He did not explain why he thought so, but he felt certain that the FAS and the National Committee on Atomic Information were under close observation. He wrote, "There is no question in my mind but that the FBI is watching us closely. If they ever get anything at all on us they and the Un-American Affairs Committee will go to town."[15] Surveillance, Higinbotham went on to make clear, meant that the FAS had to be careful about its associations and stay above suspicion.

At least two people lost their jobs as a result of the FAS leadership's burgeoning fear of political attacks. In the fall of 1945, the federation fired a secretary thought to be involved with the Communist Party. Higinbotham recalled in his July 1946 memo, "It seemed impossible but we got rid of her anyway, at once. This I believe is the only possible course if there is any real evidence. . . . You can believe me that I am keeping my eyes open. I am sincerely worried because I believe this to be a very real danger."[16] In a more serious move that will be discussed in detail later in the chapter, the FAS successfully helped the National Committee on Atomic Information fire its director, Daniel Melcher, on the shaky allegation that he was following the political line of the Communist Party. In response to Melcher's dismissal, thirteen out of the fourteen members of the NCAI staff resigned.[17]

In addition to conceding the need for internal housecleaning, Higinbotham also worried about the effects of the FAS's connections with other political organizations and the potential repercussions of the speeches scientists gave in various forums. Higinbotham noted, "We could have gotten into trouble before this, but we have been warned in time." In his own experience, he found contacts with "conservative newsmen" crucial in providing such warnings. He described specifically a speech he gave before the Win the Peace Conference in which, alerted by reporters of the group's "bad reputation," he made sure to make explicit "that the scientists did not urge giving the [atomic] secrets to Russia." He also depended on the opinions of Michael Amrine, a journalist who became director of the FAS's public relations efforts in 1946.[18] As we shall see, Amrine was staunchly anticommunist and insistent about rooting out any hint of communist influence in the federation.

Already, then, by mid-1946, in response to conservative politicians' accusations that the atomic scientists' positions on atomic energy were subversive, the FAS had begun to isolate itself and guard its statements carefully rather than risk stigmatization through guilt by association. Internal schisms also plagued the federation. Donald A. Strickland has argued that throughout 1945 and 1946, the scientists' movement consisted of an uneasy alliance between scientists of differing political persuasions and policy goals.[19] With the rise of the Cold War, this coalition began to deteriorate. By 1948, open warfare erupted.

The split paralleled the concurrent rupture in postwar liberalism. On September 12, 1946, Secretary of Commerce Henry A. Wallace delivered a speech at Madison Square Garden that led to Wallace's resignation from the Truman administration and an irreparable rift between liberal anticommunists and the progressive left. Several weeks earlier,

Wallace had come to the conclusion that a lasting peace and reasonably harmonious U.S.-Soviet relations were absolute necessities given the dangers of the nuclear age. Searching for a way out of U.S.-Soviet confrontation to stave off a military buildup and nuclear arms race that would be destabilizing for both sides and doom the further extension of New Deal economic reform, Wallace argued publicly in his New York address for the need to accept balance-of-power politics and Soviet hegemony over Eastern Europe as realities of the postwar world. His views contrasted sharply with Secretary of State James F. Byrnes's increasingly hard-line stance toward the Soviet Union, and his speech created an impression of inconsistency and incompetence in the Truman administration's foreign policy. An infuriated Byrnes demanded that the president either dismiss Wallace or lose the services of his secretary of state. Truman chose to accept Wallace's resignation.[20]

Those liberals who still identified with popular front politics and hoped for continued U.S.-Soviet cooperation in the postwar period responded to the fracas by rallying behind Wallace as the true heir of Rooseveltian internationalism and the standard-bearer of a third-party movement. In December the National Citizens Political Action Committee and the Independent Citizens Committee of the Arts, Sciences, and Professions, both large and influential progressive left organizations, combined to form the Progressive Citizens of America. Only weeks later, anticommunist liberals, consisting of older leftists who regretted their earlier accommodation with the Communist Party (and often, their former membership as well) and younger liberals without strong attachments to the 1930s left, publicly rejected the Wallace movement by forming Americans for Democratic Action (ADA) as an alternative to the Progressive Citizens of America.[21]

The two organizations exemplified fundamentally different visions of liberalism. As noted in chapter 1, the progressive left maintained an optimistic faith in society's capacity for self-reform and hoped for a postwar order based on the expansion of the New Deal and peaceful U.S.-Soviet relations. Although noncommunists within the progressive left coalition often differed with communists, they accepted communists as legitimate participants in American politics and saw no reason not to cooperate with them on common goals. Liberal anticommunists, on the other hand, saw a postwar world filled with danger. They called for hard-headed realism, especially in international relations, and they opposed any cooperation with American communists. In *The Vital Center* (1949), the classic statement of liberal anticommunism, Arthur M. Schlesinger

Jr. categorized the Communist Party as a monolithic conspiracy that sought to subvert every organization in which it participated, and he accused the progressive left of having a "sentimental belief in progress" that made it unprepared to combat the threat of Soviet totalitarianism.[22] On the domestic front, Cold War liberals embraced the New Deal but felt that its reforms were largely complete. Any remaining tasks — the cultivation of economic prosperity, social justice, civil rights — could be accomplished through limited administrative means, such as Keynesian economic management, tripartite negotiations between business, labor, and government, and gradual use of the legal system. Mass political movements were no longer necessary; furthermore, the unbridled passion they represented posed a threat to the stability of the postwar order.[23]

Both types of liberalism contained internal contradictions. The progressive left's idealistic rhetoric about social and economic justice in America was hard to reconcile with its realpolitik acceptance of Soviet repression in Eastern Europe. Liberal anticommunists, on the other hand, with their inflated estimates of Soviet power and their equation of honest left-wing dissent with the mindless acceptance of totalitarianism, abandoned their own credo of rationality and realism.

In the postwar years, the FAS acted out its own version of the schism in postwar liberalism. The Association of New York Scientists (ANYS) experienced a particularly strong split between anticommunists and the progressive left. With more than five hundred members in early 1947, the ANYS was one of the largest and most active FAS chapters.[24] The AEC's decision to build Brookhaven National Laboratory on Long Island further cemented the importance of the ANYS to the atomic scientists' movement, because Brookhaven attracted Higinbotham and other prominent FAS members to the New York area for employment. But factionalism gradually undermined the effectiveness of the New York association. In February 1948, Michael Amrine, one of the liberal anticommunists within the federation, wrote Harold Urey about the political status of the ANYS. By this time, Amrine held a position in public relations at Brookhaven. He told Urey, "My main worry as Public Relations for the scientists was keeping Melba Phillips and her cohorts from wrecking the organization and its reputation." He then castigated Phillips and her colleagues as "an extreme element" that had "virtually nullified and paralyzed the FAMs [FAS] organization." He concluded, "They have in my opinion captured the key local groups and if allowed to continue will do real harm to the general cause and reputation of scientists in general."[25] A month later, Amrine had given up on the FAS. He wrote Urey, "I keep in touch with the Federation and with the New York group of scientists, but regretfully state that I feel they have outlived their usefulness."[26]

Amrine's complaints about Phillips were especially significant because she was a leading member of the FAS. Phillips, an assistant professor of physics at Brooklyn College, was a former student of Oppenheimer's and had worked on the Manhattan Project at the Substitute Alloy Materials Laboratory at Columbia University during the war. In 1946 she served as FAS treasurer and helped staff the FAS Washington office during the debate over the McMahon bill. A year later, she joined the FAS administrative committee. Phillips was also heavily involved in the American Association of Scientific Workers and committed to the political ideals of the progressive left. In targeting Phillips, Amrine revealed a sharp political division within the scientists' movement between liberal anticommunists and popular front liberals.

Cuthbert Daniel, a former member of the Association of Oak Ridge Engineers and Scientists who had moved on to Brookhaven, detailed the situation in the New York association to Urey. An executive council of fifteen members governed the ANYS. Daniel complained that seven members, including Harry Grundfest and Theodor Rosebury, both active in the American Association of Scientific Workers and other progressive left causes, as well as Phillips, "vote the party line on all issues." An eighth executive committee member, Daniel noted, tended to vote with the troublesome seven, giving the left-leaning group a majority on the council. Daniel observed that this group did not have sufficient strength to advance its own agenda despite its working majority on the council, but he felt it prevented the New York association from pursuing other activities. He wrote, "You understand, of course, that a large part of the party line in this organization is not deducible from their activities in orgs [organizations] they control; they simply keep any proposal for action from getting anywhere. Quietism, rather than active support of party projects, is their only hope in FAS."[27]

Those opposed to Phillips, Grundfest, Rosebury, and the other progressive left members of the ANYS were ready to "conduct an open fight" and push them out of power. Daniel informed Urey, "It is our belief that by December, '48, an Exec. Council will be elected with *no* party liners on it. You will, of course, hear much screaming about Red-baiting, witch-hunting, anti-Communist hysteria, and all that, especially from Melba Phillips, during the next six months." Daniel made no effort to hide his animosity or his intent to take control of the ANYS away from what he thought was a communist-dominated faction. He concluded, "You may well receive a note from one Victor Lewinson who is editor of the ANYS Newsletter and a complete fellow traveller. He will tell you that we are splitting the organization by our disruptive and destructive anti-Red hysteria. We are. We are splitting off the 7 members of the Exec. and possibly, judging from the votes on various issues, 25 fellow-travelling members."[28] Urey thoroughly approved Daniel's actions. He replied, "I am very glad to hear of this fight, and I hope that you win it. In fact, I think that it is important that you do, for otherwise this group can completely discredit all of the scientific groups that have worked on this subject."[29]

The battle within the New York association soon came to a head, and just as Daniel had predicted, by early 1949 the ANYS had a new executive council dominated by moderate, anticommunist liberals. The election of the new council failed to alleviate tensions within the ANYS, however. Rather, it exacerbated them. The association's administrative secretary, Gene Weeks, bitterly denounced the new council for having engaged in "a great 'anti-communist' crusade" that in reality was "no more than an election coup of a small well-organized faction that cannot tolerate difference of opinion." He also condemned both factions within the ANYS for putting the association through an excruciating year of "parliamentary bickering" that had constituted "sabotage to ANYS' best welfare." Weeks was so angry that he not only resigned from his post but also renounced his ANYS membership.[30] In 1950, the entire association folded, reportedly because of "policy disputes."[31]

The Association of New York Scientists was not the only FAS chapter divided along Cold War political lines. In his early study of scientists and postwar politics, Harry S. Hall noted that, according to an interview with Thorfin R. Hogness, the FAS cut off funding to both the New York and Cambridge chapters because of communist domination and fellow traveling.[32] Like the ANYS, a significant portion of the membership of the Cambridge Association of Scientists was also involved in the American Association of Scientific Workers and other progressive left organizations. Other FAS chapters experienced Cold War rifts as well. In June 1947, the president of the Association of Monmouth Scientists resigned from both his position and his membership. Although he agreed with the atomic scientists' basic goals, he was upset after the Monmouth association rejected a motion to expel members "proven disloyal to the U.S.A. by any legal, authorized government investigation," and he also objected to "a considerable undercurrent of feeling of sympathy with the views of Henry A. Wallace" that he felt dominated the association.

The following year, the Monmouth association collapsed entirely, apparently as a result of further arguments over the loyalty motion.[33] Early in 1948, the Northern California Association of Scientists (NCAS) clashed over whether or not the association should impose political tests on members of the executive committee and exclude communists or other radicals from holding office. The argument followed allegations by the Tenney Committee, a state-level equivalent of the House Committee on Un-American Activities, that the NCAS was a communist-front organization.[34] In the Association of Los Alamos Scientists, a member of the executive committee resigned for political reasons in March 1948. He questioned the loyalty of another member elected to a high-level position within the association and withdrew to avoid the possibility of being held accountable for any unwise actions the other person might take.[35] Anticommunism, then, was not simply a repressive force imposed from outside. It was an ideology that scientists shared with other Americans. Like American liberals, scientists felt the same pressures and impulses to equate left-leaning politics with disloyalty and dogmatic adherence to the Communist Party and to ferret out radicals within their midst.

Harold Urey's political transformation poses a more detailed and nuanced perspective on domestic anticommunism than that provided by the ugly, partisan sniping in the Association of New York Scientists and other FAS chapters rent by factionalism. In the postwar years, Urey underwent a profound change of heart. He started with an early optimism about the potential of international cooperation and friendly relations between nations but shifted to an unyielding opposition to the Soviet Union and acceptance of Cold War liberalism.

In the 1930s, Urey had been active in various progressive left political organizations run by scientists and other intellectuals, including the University Federation of Democracy and Intellectual Freedom, the American Committee for Democracy and Intellectual Freedom, and the American Association of Scientific Workers.[36] After the war, Urey became an ardent supporter of world government, and he joined the United World Federalists. Urey devoted most of his political energies, however, to the atomic scientists' movement. He spent much of late 1945 and 1946 deeply engaged in the atomic scientists' cause, highlighting the dangers of the atomic age and calling for international control of atomic energy in a variety of forums. As a Nobel Prize winner, he possessed high public visibility, and newspaper and magazine articles constantly quoted him as they followed his pursuit of activism to the exclusion of all other matters. Urey told the *New Yorker* in December 1945,

Harold C. Urey (*center*), with Thorfin R. Hogness (*left*) and Arthur Jaffe (*right*), December 6, 1949, announces that there is "nothing startling" about the revelation that the United States loaned a small amount of uranium to the Soviet Union under Lend-Lease during World War II. (New York World Telegram and Sun Collection, Library of Congress; reproduced by permission of Corbis-Bettmann)

"I've dropped everything to try to carry the message of the bomb's power to the people," and he described a nonstop schedule of traveling around the country giving speeches, preparing congressional testimony, and working on publications.[37]

But by the late 1940s, Urey had grown disillusioned and increasingly wary of organizations that he considered tainted by communist connections. In April 1948, in response to an appeal for funds by the Emergency Committee of Atomic Scientists, Joel H. Hildebrand, professor of chemistry at the University of California at Berkeley, wrote to Urey that he could not provide a donation for the Emergency Committee because its chairman, Albert Einstein, publicly supported Henry Wallace's presidential campaign. Hildebrand added that he had "become sensitive to organizations which are not on their guard against being sucked into the Communist orbit," citing the Independent Citizens' Committee of the Arts, Sciences, and Professions as an example. Disturbed by the specter of scientists eschewing a tone of neutral scientific objectivity in favor of an explicitly political voice, he also complained that the *Bulletin of the Atomic Scientists* devoted too much space to editorializing on "matters of statecraft."[38] Urey replied that he agreed with most of Hildebrand's comments and wanted to distance himself from the scientists' movement: "I am as upset about the matter you discuss as you can possibly be, and as a matter of fact for months have been trying to think of a way to get out of the organization [the Emergency Committee of Atomic Scientists]. I do not know how to do it because my friends say that this would wreck it." As for the scientists' movement, Urey added: "The Atomic Scientists have very little to contribute at present to any of the political problems we are facing."[39] As discussed earlier, Urey agreed with Cuthbert Daniel that the left held too much power in the FAS, and he hoped the anticommunists would be able to establish control.

Beginning in late 1946, Urey gradually severed many of his political ties. Following the lead of Joel Hildebrand, he resigned from the Independent Citizens' Committee in November 1946. For all practical purposes, his wish to leave the Emergency Committee was fulfilled in May 1948, when the committee ended its spring fund-raiser and ceased to function afterward. The following May, it officially closed its operations, and it was legally dissolved in 1951. Urey also lost faith in the United World Federalists. In November 1949, Urey resigned from the group because he felt it was "politically inept" and too willing to appease the Soviet Union.[40]

Urey still believed in world government, but in lieu of its attainment, he felt the United States had to remain on guard because the Soviet Union posed too dangerous and intractable a threat. In response to the Soviet acquisition of the atomic bomb in September 1949, he wrote to Michael Amrine, "We should keep our powder dry, by which I mean, every effort should be made to keep our atomic bomb supply and all other military preparations in readiness."[41] Cooperation with communists, Urey now felt, was not only impossible at present, but always would be. As he wrote to Francis Biddle, national chairman of Americans for Democratic Action and formerly attorney general under President Roosevelt, "We cannot work with Russia, nor with Communist China; never could; never will be able to. Our ideals are too different."[42] By 1951, Urey was a member of the ADA and committed to liberal anticommunism.

The change in Urey's sentiments provides special insight into the political shift that the federation experienced in the early postwar years. Anticommunism within the scientific community should not be caricatured as the misdeed of a few politically reactionary scientists; nor was it just forced on scientists by outsiders. Although fear of red-baiting seems to have motivated Higinbotham's actions, external political repression does not completely explain the split within the federation. For Urey, anticommunism was an honest expression of political belief arrived at through painful consideration. Cuthbert Daniel, who so strenuously opposed the progressive left faction within the Association of New York Scientists, appears for his part to have been an outright political radical who identified himself as such, but he was also a firm anticommunist who rejected the Soviet version of Marxism.[43] Urey, Cuthbert Daniel, and other anticommunist scientists were not uniformly conservative, and they did not have their anticommunist ideology forced on them by HUAC or the FBI. Rather, they established their beliefs through their own understanding of the Cold War international conflict.

The rift within the FAS, however, cannot be considered a case of normal partisan bickering, harsh and bitter but fought out under essentially fair conditions. Documents from the FBI tell, in shadowy detail, another story of informants cultivated and trusts betrayed. They depict the abnormal, distorted social relations that lay underneath the politics of anticommunism. Informants reported regularly to the FBI about the innermost workings of the federation and warned the bureau about individuals they considered suspect. In a few cases, information supplied secretly by the state to the federation swayed the internal operations of what purported to be an independent political organization. In short, FBI documents suggest that the battle within the FAS was not only a factional split—it was a political purge.

The FBI Connection: Methods of Surveillance and Layers of Informing

Before exploring the FBI's role in the FAS's postwar decline, it is first necessary to say a few words about evidence. What follows relies, in large part, on documents obtained from the FBI under the Freedom of Information Act. Records from the Federal Bureau of Investigation are notorious for being difficult, enigmatic sources. Access to them is tightly controlled. Files take years to acquire, and when they are finally released to researchers, they arrive with countless excisions. Names of people are often blacked out on national security or privacy grounds, and paragraphs, pages, or entire documents are frequently removed. Deletions can sometimes be appealed successfully, but only if one possesses patience, time, money, and legal assistance in ample supply. For those who lack one or more of these resources, FBI files are still valuable, but their cryptic nature makes their interpretation problematic. The historian is forced to think creatively about how to use the records as they are, not as one would like them to be. Although censored documents often omit desirable details that historians usually take for granted when sorting through government records to reconstruct a series of events, FBI files still contain a great deal of usable information. In the FBI's files on the Federation of American Scientists, it is often not possible to tell *who* is doing what. It *is* possible, however, to say quite a bit about *what* it is that they are doing.[44]

The materials I have used are incomplete. Thus far, the FBI has released twenty-three of twenty-nine volumes from its main files on the FAS, as well as documents from the Washington Field Office. The materials cover a wide range of time from 1945 to the 1970s, but the bulk of the documents concern the period from 1945 to 1950. Of the more than five thousand pages in these files, the FBI has withheld almost 30 percent in their entirety. Many of the remaining thirty-six hundred pages are heavily censored. Even so, these documents provide the basis for a provocative picture of the FBI's dealings with the federation. When possible, I have filled the gaps in the censored record with informed speculation. In many instances, however, what is hidden under the black ink remains indecipherable, ready to tantalize and vex other scholars in the future.

Like HUAC, the FBI had a pathological attitude when it came to matters of atomic security. Higinbotham's fears of FBI surveillance were well founded. J. Edgar Hoover viewed the FAS's publicly stated agenda as suspicious in and of itself, and as the federation pursued the passage of the McMahon bill throughout late 1945 and on into 1946, the FBI carefully monitored the atomic scientists' activities. As part of this process, the bureau aggressively gathered information on every major FAS chapter. Its agents paid particular attention to the Washington, northern California, New York, and Cambridge associations, using physical surveillance, mechanical surveillance, and confidential informants to probe the federation's political dealings.[45]

Documents originating from the FBI's San Francisco office provide an especially revealing portrait of the FBI's use of physical surveillance. Harry Kimball, the special agent in charge (SAC) in San Francisco, considered the investigation of the Northern California Association of Scientists "one of the 25 most important [cases] in the San Francisco office,"[46] and FBI agents regularly monitored the early political meetings of the Northern California Association of Scientists (NCAS). The FBI records do not always allow one to identify the bureau's methods with absolute certainty. Sometimes the files simply note that a special agent "ascertained" or "learned" of activities; context strongly suggests but does not confirm that the agents acquainted themselves with the scientists' actions through direct observation. In other cases, the use of the term "physical surveillance" is unequivocal. Bureau agents watched scientists enter and leave meetings in public places and private homes, wrote down license plate numbers, identified individuals by sight when they could, and recorded physical descriptions of unfamiliar persons. Concerned with understanding how the association perpetuated itself and its ideas, the FBI monitored the production and distribution of leaflets and other publications put out by the NCAS. At least one enterprising agent attended a meeting of the association, where he dutifully obtained a copy of the NCAS constitution for the FBI's records.[47]

The FBI also employed illegal technical surveillance on the atomic scientists. That the FBI used wiretaps to monitor J. Robert Oppenheimer is already well known. The FBI also maintained technical surveillance on Frank Oppenheimer, Robert's younger brother. The surveillance on Frank had the unanticipated side benefit of allowing the bureau to keep tabs on the NCAS's losing efforts to persuade the conservative American Legion to change its stance on international control of atomic energy during the summer of 1946.[48] The following year, Hoover authorized the San Francisco office to place a wiretap on the telephone at the NCAS's headquarters.[49] Such scrutiny was not limited to the northern California association. Technical surveillance also allowed the bureau to eavesdrop on discussions among New York and Boston area members of the Federation of American Scientists and the American Association of Scientific Workers about a possible merger between the two groups.[50]

The FBI's most common source of information about the scientists' movement, however, was probably confidential informants. The extent to which the FBI relied on informants is difficult to determine precisely. Documents that refer to informants cannot always be taken at their word. As Athan G. Theoharis and John Stuart Cox have discussed in detail, after 1940 Hoover created elaborate filing systems to hide the FBI's use of wiretaps, bugs, mail interceptions, break-ins, and other illegal investigative methods. Because of these filing procedures, regular FBI files that describe "highly confidential" sources or informants may actually refer to illegally obtained information.[51] For example, a brief note from Guy Hottel, special agent in charge at the Washington field office, to Hoover discusses information from "a highly reliable and confidential source."[52] Is this source an informant or was the information acquired illegally? Without corroborating evidence, it is impossible to be certain.

Despite such methodological difficulties, the FBI files on the FAS reveal clearly the importance of informers to the bureau. The records provide relatively detailed information about informants, their cultivation, the clandestine nature of their relationship with the bureau, and, in a few cases, their identities. The bureau received at least a portion of its intelligence on each FAS chapter from outside contacts. By August 1947, the FBI had developed informants within the New York, Chicago, Cambridge, northern California, Pasadena, and Dayton chapters of the FAS, and it had plans to groom confidential sources in the federation's other affiliates.[53] Informants could be anyone, from a worried taxi driver who found political materials left behind in his cab to a congressional staffer who, in connection with his official duties, attended an FAS meeting.[54] They could also be scientists and FAS members. Confidential informants delivered to the FBI opinions about individuals, reports of activities at meetings, and copies of documents, including documents that were clearly not meant for distribution. Some developed a long-term relationship with the FBI and established regular meetings with agents. Informing could also become a two-way relationship, in which the FBI or other security-oriented agencies provided information to FAS members for political purposes.

It should be emphasized that there are types and degrees of informing, and they do not all carry the same implications for betrayal of trust and abuse of power by the state. Not all persons who provided information to the FBI were informers in the traditional sense of the word. As Sigmund Diamond notes, "The FBI had several levels of informants, ranging from those who furnished information 'during a single interview' to those 'from whom information was regularly received under an expressed assurance of confidentiality.' "[55] The former were usually persons who, in agreeing to be interviewed by the FBI about a particular individual or organization, assumed they were cooperating with the legitimate policing functions of the bureau. They simply answered questions; they did not volunteer privileged information or hand over documents. As loyalty and security clearance investigations proliferated after the creation of the federal loyalty program and the establishment of the AEC in 1947, this least invasive type of informing became increasingly common. By the late 1940s, it was not unusual for FBI agents to take information gleaned from interviews about individual applicants for government clearances and apply it toward unrelated investigations of organizations. For example, one informant who commented on an individual's activities in the Federation of American Scientists in response to a regular security clearance inquiry not only assisted the FBI in its statutory responsibility to engage in clearance investigations but also unwittingly aided the bureau's efforts to monitor liberal and left-wing political organizations.[56] The true informers, on the other hand, were those persons who pursued a much more intimate relationship with the bureau and actively collaborated with the security apparatus of the Cold War state.

Throughout the spring and summer of 1946, as the FAS and the National Committee on Atomic Information lobbied for the McMahon bill, the FBI used informants to determine, and if necessary, limit, the extent of what it suspected was communist influence within the two organizations. On March 7, 1946, FBI assistant director D. Milton Ladd reported to Hoover about "highly confidential information" relayed by an informant the previous day. This informant, Ladd noted, had previously supplied the bureau with information about the FAS. Earlier FBI documents show that he (for the sake of convenience, I will use masculine pronouns in the absence of specific reason to do otherwise) had already given the FBI extensive information about the National Committee on Atomic Information, including an FAS document labeled "Not for release in any form."[57] Now the informer proposed to develop a new contact for the FBI. Ladd described at length the informer's concerns about the atomic scientists' activities and his acquaintance's eagerness to assist the bureau. Ladd wrote:

> According to [name deleted, he] has been for a number of months convincing [name deleted] of the tremendous danger involved in the activities and political orientation of a large number of scientists presently the top men in the atomic research field, including such men as Julius Robert Oppenheimer, Frank Oppenheimer, and a number of other scientists who are interesting themselves in the atomic energy problem, including Dr. Harold Urey and Dr. Edward U. Condon, present Head of the Bureau of Standards. [Name deleted] stated that [he] has now completely convinced [name deleted] that the situation has rapidly passed from the dangerous to the [unreadable] and that [name deleted] is extremely anxious to assist the Bureau on a [highly] confidential basis in furnishing any information which may be of interest [illegible—probably regarding or concerning] the Federation of Atomic Scientists or matters related thereto. [Name deleted] specifically stated that [name deleted] would be more than willing to furnish [illegible—probably the bureau] confidentially with any of the records of the Federation of Atomic Scientists for copying purposes, or with any other information [he] could procure bearing on the activities of this group and the scientists connected therewith.[58]

STRICTLY ~~CONFIDENTIAL~~

March 7, 1946

The Director

D. M. Ladd

Time of Interview: 3:30 P. M. 3-6-46

FEDERATION OF ATOMIC SCIENTISTS
INTERNAL SECURITY - R

…SIFIED BY 9803 [redacted]
ON 10/22/93 #340, 198; 339,335
#377, 338; 339,333; 339,1337; 330,196
#343,494

At the above time, [redacted] appeared at the Bureau and furnished the following highly confidential information to [redacted] [redacted] concerning the above group and related matters.

[redacted] stated that with regard to the previous information which [redacted] had furnished the Bureau concerning the Federation of Atomic Scientists, which it is noted was included in my memorandum to you dated March 4, 1946, in the above case, [redacted] wanted to refer the Bureau to [redacted]

According to [redacted] has been for a number of months convincing [redacted] of the tremendous danger involved in the activities and political orientation of a large number of scientists presently the top men in the atomic research field, including such men as Julius Robert Oppenheimer, Frank Oppenheimer, and a number of other scientists who are interesting themselves in the atomic energy problem, including Dr. Harold Urey and Dr. Edward U. Condon, present Head of the Bureau of Standards. [redacted] stated that [redacted] has now completely convinced [redacted] that the situation has rapidly passed from the dangerous to the [illegible] and that [redacted] is extremely anxious to assist the Bureau on a [illegible] confidential basis in furnishing any information which may be of interest [illegible] the Federation of Atomic Scientists or matters related thereto. [redacted] specifically stated that [redacted] would be more than willing to furnish [illegible] confidentially with any of the records of the Federation of Atomic [illegible] for copying purposes, or with any other information [redacted] could procure [illegible] the activities of this group and the scientists connected therewith. [illegible] be noted, of course, that in addition to the above case, considerable information can be furnished by [redacted] regarding the present active investigation in the Cinrad Case since many of the individuals whose names appear in the Cinrad Investigation are interested in the Federation of Atomic Scientists.

RECORDED

[redacted] specifically pointed out, in addition, that [redacted] is now convinced of the necessity for an educational campaign among scientists and the public generally regarding the inherent dangers in the present proposal for [illegible] atomic energy, including specifically the proposed McMahon Bill which, you will recall, was recently briefed for you. Further, [redacted] stated [redacted] would be glad to discreetly further the viewpoint of the Bureau [illegible] connection with atomic energy, with particular regard to such matters as the

ORIGINAL COPY FILED IN 100-343044-

The challenges of working with FBI files: A typical selection from the file on the FAS. (Federal Bureau of Investigation)

Based on the original informant's fears, a second individual was now apparently convinced of the potential threat posed by the atomic scientists, and he was willing to provide FAS records to the FBI without the approval of the federation.

Both informers knew they were playing a dangerous game. Their actions required betrayals of trust that, if exposed, would bring condemnation on them, terminate their relationship with the federation, and destroy their ability to gather further information. In his regular meetings with the FBI, the original informant repeatedly emphasized that "under no circumstances should [he] be revealed as a source of any information turned over to the Federal Bureau of Investigation."[59] Elaborate measures had to be taken to avoid discovery. In his March 7 memo, Ladd related to Hoover the mechanisms being established to preserve the new informant's anonymity:

> According to [name deleted, name deleted] is anxious to be of any possible help [he] can and will be very glad to have a Bureau representative contact [him] periodically in confidence concerning the above case and related matters. It was suggested to [name deleted] and [he] agreed, that in view of the political orientation of the number of scientists connected with the Federation of Atomic Scientists, it might not be wise for [name deleted] to make any number of formal visits to the Bureau. In view of this, [name deleted] suggested that a Bureau representative come to [line deleted] March 9, 1946, and at that time [name deleted] will introduce the Bureau representative to [name deleted] and assist in effectuating these arrangements.[60]

Ladd felt that this new contact would yield information of "extreme value" to the FBI and stressed that it was "strongly advisable that the Bureau effectuate this relationship with [name deleted]."[61]

On March 11, Ladd informed Hoover of their new contact's initial report: "[Name deleted] was extremely cooperative during this lengthy interview [from 10:20 A.M. to 12:45 P.M.] and furnished a considerable amount of information, including the names of approximately 25 individuals who have aroused his suspicion in connection with agitation and pressure activities aimed at influencing the control of atomic energy and the dissemination of technical and other information concerning it." The new informant also provided FBI agents with copies of NCAI and FAS publications.[62] The original informant continued to report to the bureau as well, relating particulars about visitors to the NCAI and FAS offices, the internal workings of the atomic scientists' lobby, and his suspicions about Communist Party connections on the part of various individuals.[63]

Who were these informers? Were they scientists? It is difficult to tell, but in this particular instance, I suspect they were not. The NCAI and the FAS Washington office had adjoining office space, and these informants' reports show a lack of intimacy with the scientific community that suggests they were probably more closely affiliated with the National Committee on Atomic Information. But other FBI records, although less detailed regarding the process by which informants were recruited, provide substantial, in some cases unequivocal, evidence that the bureau's informers included FAS members.

The FBI's deletions usually make the identification of informants as scientists a matter of intelligent conjecture. For example, a January 28, 1947, report by the FBI's Washington, D.C., office contains an informant's detailed comments about communist infiltration in FAS chapters across the country. The informant noted:

> Cambridge — seems OK
> New York — main characters are [name deleted] and [name deleted]
> Philadelphia — This group was started by the Philadelphia Chapter of American Association of Scientific Workers. I haven't been in close enough touch to know much about them. They haven't been very active.
> Rocket Research — this group was scattered in the spring.
> Washington — new group — have been in close touch and nothing suspicious.
> Pittsburgh — I know a lot of the boys in this who are OK but they work a lot with [name deleted] who is a [*sic*] more suspect.
> Cornell — This summer it seems to have heard the "line." I didn't have time to find out where it came from. It didn't come from [name deleted] in fact he doesn't have it at all as far as I can tell.
> Rochester — Don't know about this one.
> Dayton — Don't know about this one.
> Oakridge — The two groups here have combined and are very active. They are overwhelmingly for world government.
> Chicago — Mostly they are anti-Communist, don't think there is any trouble here.
> Northern California — Active group, they have a number of guys who love organization. They haven't indicated any particular bias in anything they have done although I think they aren't inclined to be very careful.
> Cal. Tech — This group has apparently disbanded. I have several reports leading me to believe that the trouble is confusion raised by Communists. They were quite taken with the Hollywood Writers Mobilization.
> Los Alamos — Not very active.[64]

Who could have felt competent to discuss the character of the different FAS chapters so extensively? The commentator's familiar tone, his easy acquaintance with members such as "the boys" in the Pittsburgh chapter, and his ability to remark freely on so many FAS chapters strongly suggest that he was a scientist actively involved in the federation. Other informant-based reports on various FAS chapters also indicate an intimacy with their inner workings that implicates scientists as sources of the FBI's information.[65]

Occasionally, some FBI documents contain enough clues to make guesswork unnecessary and provide undeniable evidence that the FBI's informants within the federation included scientists. In June 1947, a member of the FAS's northern California chapter who was also apparently a person "engaged in scientific work" began to furnish information about the NCAS to the FBI because he felt that "the professed opinion of the NCAS is not the true opinion of the majority of its members and, on too many occasions, has appeared to be an opinion directly parallel to the Communist Party line."[66] Several months later, another scientist involved in the FAS visited the FBI's Washington, D.C., office for an hour and a half. During that time he talked about the federation, expressed his "full sympathy and agreement with the Bureau's work," and put forth his willingness "to help the FBI in every way possible." The FBI agent who spoke with the scientist found him "a very intelligent individual, but perhaps naive on matters outside the realm of science." Whoever the scientist was, he made a positive impression and seemed like a potentially valuable informant. Guy Hottel, SAC Washington, thought the scientist should be "recontacted in the near future" and that he was a good candidate for "a possible confidential national defense informant."[67]

Some of the informants within the FAS came from leadership positions within the federation. As mentioned earlier, an executive committee member of the Association of Los Alamos Scientists resigned because of his suspicions about another high-ranking member's loyalty. He immediately reported those suspicions to the FBI.[68] One of the informants within the Association of New York Scientists, identified in FBI documents as "Confidential Informant T1, of known reliability," was an employee of the AEC and "an active member of the Executive Council of A.N.Y.S." T1 apprised the bureau of the 1949 campaign for the ANYS executive council, the election loss of "a 'pro-Communist minority' " led by Harry Grundfest, Theodor Rosebury, and Melba Phillips, and the subsequent demise of the ANYS.[69]

At one time, such revelations about informing by trusted insiders would have been shocking. But in the current day and age it should not be surprising that the FBI's informants on the Federation of American Scientists included scientists and high-ranking federation members. Rather, it would be naive to assume that the FBI's sources consisted solely of an array of secretaries, janitors, and other outsiders and that scientists stayed away from the netherworld of informing. There are already too many well-known instances of prominent scientists who turned informer during the 1940s and 1950s to try to deny the extent of informing within the atomic scientists' movement. As Sigmund Diamond has shown, James B. Conant routinely provided the FBI and Central Intelligence Agency (CIA) with information about Harvard University faculty members and their political affiliations during his tenure as president of the university.[70] During World War II, J. Robert Oppenheimer reported his students' political associations to Manhattan Project security officers. He later served as a cooperative witness during HUAC's 1949 investigation of alleged wartime espionage at the Berkeley Radiation Laboratory, where he again described his students' Communist Party connections, to the chagrin of other physicists. Hans Bethe, Victor F. Weisskopf, and Edward U. Condon sharply berated Oppenheimer for his testimony.[71] Physicist Samuel Goudsmit, scientific director of the wartime ALSOS project that evaluated the status of the German atomic bomb effort, regularly passed information about his meetings with other scientists and the status of scientific projects overseas to Walter F. Colby, director of intelligence at the AEC, and also to the CIA in the postwar years. A longtime lover of codes, criminology, and mysteries, Goudsmit became enamored of intelligence work during the war and hoped to build a postwar career in intelligence.[72] Goudsmit would later serve on the FAS's Scientists' Committee on Loyalty Problems, where he would help determine the extent of the FAS's assistance to scientists denied security clearances. Such informing by these scientists violated notions of trust, openness, and autonomy that are assumed necessary for the functioning of a democratic polity. But the violation of these norms was far from uncommon in the early Cold War years.

Had the transactions between the atomic scientists' movement and the FBI been limited to the transfer of information from the former to the latter, they might have had little immediate effect on the FAS and the National Committee on Atomic Information. But the relationship between the FBI and the FAS went in two directions. Some FAS members received information from the FBI and other internal security organizations, and they used that information to dismiss individuals and perpetuate a moderate political identity for the federation.

From Partisan Politics to Political Purge

When NCAI director Daniel Melcher was fired in July 1946, the leaders of the FAS and the NCAI denied that his dismissal had any connection to his politics. Competence, not communism, was the issue. Melcher had worked in the Treasury Department, where he had overseen a highly successful effort to sell war bonds before becoming director of the NCAI in February 1946. When hired, he seemed like someone who could direct the education and mobilization of the general public for the atomic scientists' movement. However, according to the official explanation for his firing, his leadership style created problems. Too often, the National Committee on Atomic Information had released statements without the requisite prior approval of the federation, and despite Melcher's successful fund-raising and organizing efforts, his relationship with the scientists and the NCAI executive committee had become antagonistic. Ralph MacDonald, head of the National Education Association's division of higher education and chairman of the NCAI, attributed Melcher's removal entirely to personal instability. MacDonald explained to the renowned physicist Albert Einstein, chairman of the Emergency Committee of Atomic Scientists, that the NCAI director had developed a "delusion of persecution" and become "at least temporarily upset mentally."[73]

If Melcher felt persecuted, however, he had good reason. Although the FAS and Melcher did have some genuine disagreements about the relationship between the federation and the NCAI and the appropriate extent of oversight by the former over the latter, these differences were not the cause of Melcher's firing. Rather, despite FAS leaders' claims to the contrary, red-baiting played a central role.

Nothing in the records of the NCAI hints of incompetence on Melcher's part before June 1946. Throughout the first half of 1946, the NCAI effectively produced and distributed pamphlets and other printed information, provided names of speakers, and performed other services to help communities launch discussions about atomic energy. Melcher's relationships with the atomic scientists seemed cordial and friendly, spiked occasionally with some high-spirited, but amicable, intellectual debates. The executive committee of the NCAI also seemed pleased with Melcher's work. On June 4, it awarded Melcher a 50 percent pay raise to compensate him for the financial loss he had taken in leaving the Treasury Department to work for the NCAI, a gesture that hardly suggested a lack of confidence in his abilities.[74]

This atmosphere of congenial collegiality changed completely after the publication of the June 17, 1946, issue of *Atomic Information*, the NCAI's semimonthly newsletter. The leaders of the FAS were alarmed by an article about the British physicist Alan Nunn May's recent conviction for atomic espionage, and they feared *Atomic Information* was beginning to follow the line of the Communist Party.[75] On June 20, Michael Amrine commented on what he viewed as a dangerous pattern of biased reporting in *Atomic Information*. He carefully noted that its writings did not by themselves prove Communist Party influence, but he thought them "strange appearing in a bulletin which was supposed to be dedicated to impartial sound atomic information." Amrine was especially disturbed by an item that featured extensive quotations from Frédéric Joliot-Curie, who was both a leading nuclear physicist and a prominent member of the French Communist Party. Joliot-Curie, Amrine complained, was a man "given to wild speculation" and "an avowed Communist." Amrine found the coverage of the British Association of Scientific Workers' opinion about the Alan Nunn May spy case equally alarming. He observed, "Many Britishers look on the B.A.S.W. as representing the communist line pure and simple." No other scientific organization had protested May's sentence. Since the FAS had not commented on the case, Amrine felt it "extremely strange" that the NCAI should devote any attention to it. To Amrine, these and other stories indicated that *Atomic Information* was treading a dangerous path: "In the hands of the wrong person, such as a representative of the Un-American Committee, these [writings] could be worked into an extremely damaging attack on the National Committee for Atomic Information. They are strange, and to me they indicate a trend which could be quite alarming if allowed to continue unchecked."[76]

Amrine understood correctly how HUAC worked and what it might do with *Atomic Information*. But he was not simply worried about how HUAC might misconstrue *Atomic Information*; he himself clearly feared that the NCAI had taken a political turn for the worse. His accusations, however, were overblown. The accounts from Joliot-Curie and the British Association of Scientific Workers appeared on a page devoted to international news about scientific freedom. They were legitimate news, and they were narrated in a factual, neutral tone. *Atomic Information* simply reported what Joliot-Curie and the British association said; it did not endorse their positions. By comparison, the publication of Leo Szilard's "Letter to Stalin," which appeared in the *Bulletin of the Atomic Scientists* the following year, was much more daring and open to misinterpretation. Eugene Rabinowitch conceded that Szilard's letter, a desperate appeal to Stalin to speak to the American people and reverse the deterioration in U.S.-Soviet relations that had occurred since the end of World War II, might lead to criticisms of the atomic scientists for "pro-Soviet leaning," or at the very least, naiveté, but the *Bulletin* suffered no reprisals.[77] Like the

Bulletin, Atomic Information offered a moderate liberal-left political outlook that viewed stable and cooperative U.S.-Soviet relations as necessary for the maintenance of peace in the nuclear age. If *Atomic Information* was on its way to becoming a scientific version of the *Daily Worker*, it had a very long distance to go.

Melcher met with Higinbotham and Amrine to discuss the FAS's concerns that the atomic scientists' movement would be misrepresented in *Atomic Information*. Melcher thought they reached a mutual understanding that the NCAI would check its public statements with the FAS for scientific accuracy but not political content. Amrine, however, was livid. He angrily wrote to Gardner Jackson, a staffer with the National Farmers Union (NFU) and a member of the NCAI Executive Board, that Melcher wanted to take over the NCAI and "sabotage" it. Jackson, no stranger to red-baiting, joined Amrine in pushing for Melcher's ouster. Meanwhile, in an interesting twist, Jackson's own political fortunes were shifting. A 1930s agrarian radical turned virulently anticommunist by the mid-1940s, Jackson would lose his position in the National Farmers Union at about the same time that the NCAI fired Daniel Melcher. Following his successful introduction of an anticommunist resolution at the 1946 convention of the National Farmers Union, he was dismissed from the NFU by its president, James Patton, a liberal unwilling to stomach the politics of anticommunism.[78]

The dispute over Melcher came to a head in July. For three weeks, the federation and the executive committee of the NCAI debated over what to do about Melcher. During that time, certain individuals from the NCAI and the FAS called on the Justice Department for advice and assistance in making the case for Melcher's removal. Two visits were paid to the Justice Department. In the first, according to an FBI informant, someone from the NCAI met with Assistant Attorney General Theron Caudle and "complained to him of the alleged Communist infiltration of the National Committee on Atomic Information."[79] Then on July 26, two representatives from the FAS, almost certainly Higinbotham and Joseph Rush, treasurer of the FAS and formerly a physicist at Oak Ridge, spoke to Attorney General Tom Clark.[80] Again, an informer's report relates what happened: "[Name deleted] reported to the Attorney General that he was fearful that the communists were obtaining positions in his organization for the purpose of taking over. Mr. Clark is reported to have stated if [name(s) deleted] would give him the names of the suspects he would have them checked *and turn the information over to the Federation of American Scientists for their confidential use* [italics mine]."[81] Any such information would, of course, have come from the FBI. Melcher was fired from the directorship of the NCAI two days later. On July 29, an informant reported news of Melcher's dismissal and the subsequent resignations of all but one of the NCAI's staff members to the FBI. The next day, the informant let the FBI know that a reporter from the *Washington Times-Herald* had heard there were political grounds for Melcher's firing and wanted verification before running the story. The informant, anxious to keep the story out of the press either because the incident would embarrass the federation and the NCAI or possibly out of desire to hide the NCAI's and FAS's appeals to the Justice Department and the FBI, lied and told the reporter that his impression was "entirely untrue." Melcher had been dismissed, he said, because "the NCAI intended to expand and needed a man of higher qualifications and a bigger name to head the organization. . . . Any other resignations were entirely voluntary and had nothing to do with any alleged communistic activities."[82]

Melcher did not know of the Justice Department and FBI's involvement, but he did know who his opponents were. In early July, as Melcher grew increasingly aware of his precarious position, he brooded over his critics and excoriated their reasoning and motives. In an angrily drafted response to Amrine and Jackson's attacks, he accused them of "intellectual bankruptcy." He paid extra attention to Jackson, who, he wrote, was engaging in a "malicious attack" that showed he was "neither sober, nor sane." A slightly more temperate letter written to the NCAI executive committee two days later complained that the Reverend E. A. Conway, NCAI treasurer, followed "the 'Holy War against Russia' line of the Catholic Church." In this letter, Melcher also contended that Conway and Jackson were themselves former communists "now obsessed with the idea that impartiality means uncritically opposing anything that the Russians are for." As we shall soon see, Conway was the FBI's foremost informer against Melcher. Melcher seemed unclear about Higinbotham's motives, but he noted, "Higinbotham goes along with all of them—though not, I'm sure, with the support of members of the federation." A month later, after his firing, Melcher wrote to the NCAI executive committee and accused Jackson, Conway, Higinbotham, and Amrine of being "in on the original red-baiting."[83]

Higinbotham and Rush ultimately decided not to take up the attorney general's offer of information. The dismissal of Melcher and mass resignation of his staff, Higinbotham informed Clark, allowed the NCAI "to start with a clean slate in this organization" and rendered the Justice Department's assistance unnecessary, at least for the time being. Higinbotham added that he thought the federation now had control of its political situation: "I feel that we have a pretty good idea whom to watch among the scientists and at this time it does not seem to us that any of them are in a position where they can do any harm."[84] The Federation of American Scientists, Higinbotham implied, understood the necessity for anticommunist action and was prepared to prevent the progressive left from gaining too strong an influence within the federation.

Had Higinbotham and Rush gone through with Clark's offer, it is very unlikely that the attorney general would have had new information to provide with respect to Daniel Melcher. When the FBI first began to investigate the NCAI in March 1946, it held no record on Melcher and no evidence that he had any past communist affiliations. Virtually everything the FBI knew about Melcher's political beliefs and activities dated from the second half of June 1946 or later. Most of what the bureau knew came from three sources. The first was a concerned taxi driver who had turned in a briefcase containing NCAI documents and memos that Melcher had accidentally left behind after a late night cab ride on July 10. The second was an informant within the NCAI (or possibly multiple informants—the censored nature of FBI documents does not preclude the possibility) who supplied opinions and additional documents to the bureau.[85] This informant also passed information from a third source to the FBI.

A July 13, 1946, memo from Guy Hottel to J. Edgar Hoover unequivocally identifies the second of the bureau's sources on Melcher as the Reverend E. A. Conway. According to the memo, Conway had already provided the FBI's Washington office with information about the NCAI. Furthermore, his high-level position and usefulness as a regular source apparently led the FBI to designate Conway a confidential informant, a status that warranted granting Conway the added protection of a symbol number to be used rather than his name in FBI reports.[86]

Over the next several weeks, Conway continued to report his own observations to the FBI and to supply the bureau with copies of Melcher's letters and other internal NCAI correspondence. He also solicited from a member of the FAS (presumably someone of relatively high rank within the federation) a lengthy typewritten and signed statement about Melcher's political opinions and likely communist sympathies, which he gave to the FBI on July 23. This FAS member became the FBI's third source of information regarding Melcher. Did the FAS member know that Conway planned to pass his statement to the FBI? The FBI files do not provide a direct answer to this question, but some sketchy circumstantial evidence suggests that he may well have known Conway's intent. The FBI did feel that it could keep the original copy of the statement, an indication that the bureau did not consider its acquisition of the FAS member's report particularly surreptitious. In addition, the fact that the statement's author wrote such a formal document, almost as if he were supplying a signed deposition, hints that he knew Conway would take it to a higher authority that shared his anticommunist leanings.[87]

Federal Bureau of Investigation memoranda written two years later shed more light on this incident. In late March 1948, an investigator from HUAC approached a present or former FAS member in the New York area. According to FBI documents, HUAC wanted to know about "the alleged subversive activities of former employees of the Federation of American Scientists" and hoped to obtain the federation's internal correspondence about its efforts to expel suspected communists from the atomic scientists' movement. The FAS member immediately consulted the FBI's New York office about how to stonewall HUAC in favor of furthering his cooperation with the FBI. Edward Scheidt, New York SAC, noted that this FAS member had previously "been used as a confidential source of information by the New York Office" and had "proved to be of considerable value." In the past, the FAS member had been a confidant of the bureau's Washington office as well.[88]

In the course of consulting with the FBI, the FAS member made reference to Michael Amrine's June 1946 memo on *Atomic Information* and the unknown FAS member's July 22, 1946, statement about Melcher that E. A. Conway had given to the FBI. Guy Hottel, SAC Washington, checked the FBI's old files for copies of these 1946 documents and the means by which the FBI had acquired them. Apparently, Conway's source within the federation had knowingly developed his relationship with the FBI. Hottel informed Hoover, "Special Agent [name deleted] stated that he had met [name deleted—the FAS member] through [name deleted—Conway] but that [the FAS member] was desirous of furnishing information through [Conway] as an intermediary for reasons of discretion and security. The information thus furnished by [Conway] was given with the understanding that the identities of both [Conway] and [the FAS member] be treated with the strictest confidence."[89] Conway, then, was not the only significant informer against Daniel Melcher.

For their part, Higinbotham and Rush, through their dealings with Attorney General Clark, appeared willing to accept clandestine aid from the FBI if necessary. The prospect of backing from the Justice Department may have given the FAS's leaders the confidence to dismiss Melcher and take a tough stance against communism. At the very least, given Higinbotham's fears of FBI surveillance and red-baiting, a cooperative attitude toward the Justice Department and the FBI provided assurance that the federation was able and willing to clean its own house without government intervention.

Melcher was not the only person adversely affected by informing. The FBI doggedly cultivated informants throughout the chapters of the FAS, and in some cases, the federation clearly used privileged information from the FBI or other security agencies to restrict its political associations and remove individuals. In November 1945, as the atomic scientists launched their lobbying operations in Washington, D.C., they initially took up office space supplied by the Independent Citizens' Committee of the Arts, Sciences, and Professions. Then a Chicago scientist apparently consulted with the Security Office of the Manhattan Project's Metallurgical Laboratory about "the nature and background" of the Independent Citizens' Committee. The scientists subsequently vacated the office space "because the Federation of Atomic Scientists had learned the nature of the ICCASP."[90] The Chicago scientists also consulted with Metlab security about individuals connected with the federation: "It is understood according to the informant that [name deleted] submitted the name of [name deleted] to the Security Office of Metlab *after consultation with the executive committee of the Atomic Scientists of Chicago* [italics mine]."[91] Although an individual scientist, most likely Metallurgical Laboratory chemist Farrington Daniels, took the initiative in contacting security officials, the ruling committee of the Atomic Scientists of Chicago sanctioned his actions.[92] Four lines deleted from the FBI records then seem to have suggested, correctly or incorrectly, that the person under consideration was a member of the Communist Party. The record continued, "[Name deleted,] the informant advises, consulted telephonically with [name deleted] whom he stated he considered to be level headed in such matters. The informant advises that [name deleted] was desirous of handling the matter in such a way as to avoid publicity because he felt that if it became known that a Communist Party member had succeeded in obtaining employment with the Federation of Atomic Scientists, this would do damage to the efforts of the atomic scientists." The individual under discussion was probably the secretary that Higinbotham referred to in his July 18, 1946, memo.[93] She was quietly fired without further ado.

Months later, the FBI felt relatively reassured that the Atomic Scientists of Chicago could find and root out communist influence within the atomic scientists' movement on its own. Bureau informants, probably either from Metlab security or from within the Atomic Scientists of Chicago, reported that the Chicago scientists now seemed able to protect the federation against the dangers of communism. As a December 19, 1947, report from the FBI's Chicago office noted, "[Name deleted] advises that the Atomic Scientists of Chicago have been particularly aware of the necessity of keeping Communist influence out of the organization. [Name deleted] says they are aware that in the past there have been instances where it appeared that the organization has cooperated with groups having possible Communist influence. [Name deleted] states, however, that [name deleted] believes that the organization is now mature enough to recognize the Communist viewpoint and avoid it."[94] As part of their political education, the Chicago scientists had learned self-monitoring and self-censorship as means of survival in the Cold War years.

Informers' motives varied. Some appear to have contacted the FBI out of a real sense of patriotism. Genuinely anticommunist, they went to the bureau as a way to take firm measures against what they believed was a serious threat to the political identity of the atomic scientists' movement. Others seemed motivated by fear and cooperated with the FBI to gain protection from the threat of red-baiting. Sometimes the desire for protection was personal. For example, an FBI agent noted that an informant who resigned from the American Association of Scientific Workers came to the bureau "to clear the record" and "indicate his desires [*sic*] to cooperate with the FBI."[95] In other cases, informers acted to secure their political organizations against anticommunist assaults. As mentioned earlier, one informer from the FAS approached the FBI specifically to consult about how to head off HUAC's 1948 attempt to expose the firings that had taken place two years earlier. The quiet dismissal of suspected communists from the FAS, the informer stressed, had taken place with the FBI's approval. As an FBI agent noted to Hoover, the informer "said the matter was one which was handled very discreetly, without publicity, and was guarded as a secret, to avoid any possible publicity. [Name deleted] said that Agents of the F.B.I. in Washington had urged him not to discuss this matter generally outside of the Federation." The informant emphasized his past cooperation with the FBI and explained "that [he] wanted to cooperate with the Bureau as much as possible and did not feel inclined to go over this matter with the representative of the House Committee on un-American [*sic*] Activities."[96]

The classic caricature of the informer conjures up the image of the rat, the stoolie, the underhanded individual who informs purely for the sake of greed, personal advancement, or pure deviousness. These cartoonish labels cannot be used easily to dismiss the informers within the FAS. The informers' intentions were not always the most noble. Selling out individuals to preserve an organization from external attack does not generally sit well with most Americans' sense of ethical conduct. Within the anticommunist mentality that many Americans came to share as the postwar years progressed, however, informing became a moral gray area. Informing remained a distasteful task, but according to the dictates of Cold War ideology, it was also a necessary means of combating a mortal danger.[97]

Nonetheless, informers' actions cannot be exonerated. Informing corrupted the internal politics of the federation. The firings of Melcher and the secretary had serious implications for the FAS. Had the factional fighting and red-baiting within the federation remained purely an internal matter, the firings could be considered part of a normal political conflict, ugly and distasteful yet based on legitimate means of political debate and decision making about the FAS's future. Instead, leaders of the FAS proved willing to bring in the power of the state to bear on the decisions of the federation, thereby compromising its independence and autonomy. By inviting the state to intervene, the federation's leaders elevated political infighting to a political purge.

Did the purge extend beyond the firings of these two individual nonscientists? No evidence has emerged that information from the FBI helped determine the outcome of partisan disputes within the Association of New York Scientists and other FAS chapters, but the expurgated nature of FBI files leaves the possibility open. Even if the bureau did not intervene directly in these factional struggles, by providing the federation with information about its members' beliefs and associations, it still exercised a subtle influence that compromised the FAS's independence. Knowledge of FBI surveillance led the FAS's leaders to avoid carefully too audacious or radical a political agenda. The act of informing itself, regardless of whether or not it resulted in direct intervention by the FBI, almost certainly affected the political calculations that went into the factional struggles. By cultivating the good graces of the FBI, informers knew they had a valuable ally that they could fall back on, if necessary. With this assurance of the state's support, they could feel confident as they pursued open combat against the progressive left. Although the repressive power of the state did not solely determine the outcome of the ideological conflicts within the FAS, it did help to tip the political balance away from the progressive left and in favor of liberal anticommunists during the early Cold War years.

The FBI Perspective: Fact Gathering and the Uses of Information

J. Edgar Hoover always claimed that the FBI was a neutral, nonpartisan fact-gathering agency that merely reported information it received; it did not evaluate the veracity of that information. The notion that the FBI provided competent, responsible, professional law enforcement was still widespread in the late 1940s. Even Harvard astronomer Harlow Shapley, a staunch civil libertarian and a firm believer in progressive left politics, once referred to "the recognized efficiency of the Federal Bureau of Investigation in the protection of people against real dangers."[98] Shapley hoped to make the case for the abolition of HUAC by suggesting that the FBI's law enforcement activities made HUAC superfluous. He either did not know or felt it politically astute not to mention that the FBI was keeping close tabs on his own political activities.

Fact gathering was anything but politically neutral. The type of information the FBI collected and the language FBI agents employed in their reports illuminate clearly the bureau's underlying Cold War political assumptions.[99] Political biases pervaded the bureau's paperwork on the FAS. One typical description referred to the federation's statements about the February 1946 Canadian spy cases as "propaganda publications."[100] Many FBI reports targeted FAS chapters' regular activities in support of civilian and international control of atomic energy as in and of themselves causes for concern. For example, an April 25, 1946, document cast a suspicious eye on the FAS's use of normal political strategies. From the FBI's perspective, a letter-writing campaign organized by the federation looked like "pressure tactics" rather than an expression of popular will: "It would appear that the volume of letters received by the Senate Committee is more of a testimonial to the efforts of such organizations among those they influence rather than voluntary expression on the part of the people generally regarding the subject of atomic energy."[101] Other documents described the attempts of the Northern California Association of Scientists to persuade the local American Legion chapter to reverse its position on international control of atomic energy. It was a futile effort, since the American Legion staunchly backed the Cold War political consensus and provided the FBI with a prime source of informants to boot. Nevertheless, in one report on the northern California association's efforts, Hoover penned the notation, "A shocking commentation on Communist penetration."[102] Repeatedly, the FBI implied that there was something antidemocratic and conspiratorial about normal political lobbying conducted by the federation.

The FBI's collecting of information was legitimate only insofar as it related directly to the bureau's fulfillment of its statutory duties. But, as is well known, Hoover's FBI tried to use its voluminous files for political gains. In addition to releasing information that eventually reached the FAS, the FBI also tried to exploit its records about scientists in other arenas. Throughout Truman's presidency, Hoover attempted to expand his influence in the White House by passing political intelligence to Truman's political associates. George E. Allen, director of the Reconstruction Finance Corporation, and White House aide Harry Vaughn, both personal friends of Truman, regularly received urgent missives from Hoover about supposed communist-sponsored activities. In the case of scientists, Hoover took careful note of the atomic scientists' campaigns for civilian and international control of atomic energy and related matters, and he reported them to the White House as issues requiring immediate attention.

On May 29, 1946, Hoover wrote to Allen about a subject that he thought would be of interest to the president. According to "a source believed to be reliable," Hoover reported, there was "an enormous Soviet espionage ring" in Washington, D.C. Undersecretary of State Dean Acheson, former assistant secretary of war John J. McCloy, Secretary of Commerce Henry A. Wallace, and Edward U. Condon, among others, were alleged to be involved. James R. Newman, counsel to the Special Senate Committee on Atomic Energy, served as the alleged leader of the spy racket. The primary evidence for the existence of this atomic espionage ring, however, was nothing more than the regular activity surrounding atomic energy legislation. Hoover wrote, "The informant has drawn the conclusion that the entire setup of the McMahon Committee to investigate and recommend legislation on atomic energy and its use is a scheme to make available information concerning the atomic bomb and atomic energy, and that it all amounts to Soviet espionage in this country directed toward the obtaining for the Soviet Union the knowledge possessed by the United States concerning atomic energy and specifically the atomic bomb."[103] Hoover's informant apparently assumed that any exchange of information regarding atomic energy constituted atomic espionage and that the supporters of civilian control of atomic energy aimed to give away American secrets. Starting from these axioms, the informant constructed a theory of subversive conspiracy.

J. Edgar Hoover, 1948. (New York World Telegram and Sun Collection, Library of Congress)

Later information supplied by Hoover was less imaginative but similar in its political content. Early in 1947, Hoover alerted the White House to the tactics that Cambridge scientists were planning to employ to support the confirmation of David Lilienthal as chairman of the AEC. Hoover reported, "According to a highly confidential source, Wendell Furry advised Norman Levinson recently that the national officers of the Federation of Atomic Scientists [*sic*] had agreed that no action should be taken to strike back at legislative or police efforts to control scientists until David Lilienthal had been confirmed." Only after Lilienthal's confirmation would the FAS "lead a countercampaign." Furry and Levinson, Hoover noted (correctly, in their cases), had been reported to be Communist Party members as recently as 1945. Hoover also related that Harlow Shapley, acting in his capacity as a member of the Progressive Citizens of America, had advised the Progressive Citizens of America and the FAS that efforts to support Lilienthal publicly would backfire. According to FBI sources, Shapley took the position that "no action should be taken because all scientists were regarded by Congressmen as 'Left Wingers' and that any protest by Liberals or scientists would defeat Lilienthal's appointment."[104]

Shapley was also the subject of several other 1947 letters from Hoover to the Truman White House. A March 29 letter reported further actions by Shapley in connection with Lilienthal's confirmation.[105] On February 6 Hoover wrote to George E. Allen about Shapley's support for the National Science Foundation. Shapley's NSF proposal, the FBI director noted, called "for the establishment by Congress of a central national laboratory of scientific research to be paid for by Federal funds but to be free of Federal control." Hoover added, "All developments produced at the foundation would be free of 'Secret' classification." At a strategy session with Harvard University sociologist Talcott Parsons, Harvard geologist Kirtley Mather, and MIT physicist Victor Weisskopf, Hoover observed, Shapley had promoted the creation of a national science committee, formed from preexisting scientists' organizations, to "employ a subtle attack so that the public would not know what organizations or individuals were lobbying for the establishment of this foundation, but rather they must make it appear that the demand is universal and that it was being made by all persons in all walks of life." Shapley, Hoover observed, rejected cooperation with Frank B. Jewett and Detlev W. Bronk, chairman of the National Research Council, on the grounds that they were "undercover representatives of the National Association of Manufacturers." In conclusion, Hoover noted that in the past Shapley had been reported to the FBI as "exhibiting pro-Russian tendencies," and his methods displayed "a tactical scheme very similar to the general operational procedures of the Communist Party in an effort to reach his goal."[106]

Once again, the FBI director distorted ordinary political lobbying into subversion. Scientists viewed the NSF as a way to guarantee postwar funding for nonsecret basic research that would be free from the pressures of military and industrial support. To Hoover, these goals smacked of subversive intent. Like HUAC chairman J. Parnell Thomas, Hoover believed advocacy of the open exchange of scientific information to be an indication of disloyalty and communist influence. Furthermore, he automatically considered any positions or actions taken by individuals whom he deemed communist sympathizers suspicious, worthy of surveillance, and meriting notification of the White House, regardless of the ideological content or national security implications of those positions. The strategizing of Shapley and others over the promotion of the NSF legislation was no different from the maneuvers that any political group would consider. But because the FBI thought of Shapley as a communist sympathizer, it interpreted and represented all of his actions as possible manifestations of the Communist Party's program.

The scrutiny Hoover devoted to Shapley and other scientists reflected common attitudes toward the Communist Party and the progressive left. The anticommunist mentality simply did not distinguish between normal political activities conducted by the progressive left and subversive subterfuge. Anticommunists, both conservative and liberal, viewed the Communist Party as a monolithic conspiracy. Part of the fault lay with the Communist Party itself, for the party's leaders were all too subordinate to Moscow. Recent evidence from Soviet archives also indicates that the party, at times, served as a conduit for espionage. At the same time, however, the rank-and-file members who represented the "public world of American communism" were seldom the Soviet-controlled automatons that popular culture made them out to be.[107] For example, the communist-led United Electrical, Radio, and Machine Workers of America, the third-largest CIO union, earned workers' support not by employing underhanded, undemocratic means but through strong, able leadership and the negotiation of substantial material benefits for its members.[108] The FBI, congressional conservatives, and other promoters of anticommunism, however, rarely regarded American communists from a nuanced perspective, and they tended to suspect subversive intent behind any political program that garnered communist support, be it labor unions or the National Science Foundation.

The FBI and its defenders claimed that the FBI merely gathered facts for the use of other federal agencies. Other government bodies determined how those facts were used. As several scholars have demonstrated, however, the FBI was more than a simple gatherer and supplier of data. Under Hoover's leadership, it actively sought to mobilize public opinion toward an anticommunist consensus. At the very least, Hoover sent constant reports to the White House of espionage threats to head off charges of laxity on the part of the bureau should any of the reports later turn out to be true. But Hoover also viewed FBI files as a far more potent resource. Despite his public disclaimers about the quality of the FBI's information, he used data from FBI files to enhance the authority of the bureau. When confronted with criticisms that FBI reports were compilations of unsubstantiated rumors and allegations, Hoover retreated to the position that they were undigested reports from informants and not meant to be taken on blind faith. Through his channels to the White House, however, Hoover touted them as examples of the importance of the FBI's initiatives. On December 23, 1946, for example, Hoover reported to George E. Allen, in a note he wanted to be passed on to the president, that FBI personnel investigations for the AEC had uncovered "derogatory information" about individuals ranging "from subversive activities to murder." Hoover informed Allen, "This is being brought to your attention to show the great need for investigations of this type, especially in connection with such an important problem as atomic energy." The FBI also regularly channeled political intelligence to receptive journalists, members of Congress, and other persons whom Hoover trusted to use the information to attack and discredit the liberal-left.[109]

This is not to underestimate the dangers of the postwar world. As historian John Earl Haynes and others have recently begun to document, Soviet espionage was real, and Soviet intelligence organizations aggressively sought information about America's nuclear program. But conceding the reality of atomic espionage does not lend credibility to the allegations of HUAC, the FBI, and other would-be protectors of the "secret of the atom" or justify their actions. The problem with Hoover's FBI was that it was more interested in the suppression of unpopular political views than in the investigation of specific criminal acts, and it had little ability or desire to differentiate between real and imaginary threats.[110] As a political and cultural phenomenon, Cold War anticommunism was less about legitimate security requirements than it was about domestic political needs. In targeting liberal and left-wing organizations and individuals, employing illegal surveillance methods against them, and encouraging practices of informing that distorted normal relationships of trust between individuals, the FBI did more to enhance the power of J. Edgar Hoover than it did to ensure the security of the nation.

Immediately after World War II, the atomic scientists' movement emerged as a vigorous political campaign, appealing to scientists and nonscientists across the political spectrum. Less than two years later, the FAS went on the decline. Its turn of fortune was not simply, as Alice Kimball Smith's standard account suggests, a matter of waning political interest and combat fatigue after the failure to establish international control of atomic energy.[111] As we have seen, the FAS's deterioration also came from the internecine battles within the federation over the rapidly evolving politics of anticommunism. As the FBI watched, the federation's membership dropped. By late 1949, the FAS had only fifteen hundred members, down from a high of more than three thousand. The Cambridge and Los Alamos associations, in addition to several of the smaller chapters, folded completely.[112] On April 21, 1950, the FBI closed its active investigation of the federation in light of the FAS's shrinking membership and the FBI's conclusion that "investigation failed to reveal any indication of Communist control of the organization on a national scale, although there had been indications of Communist infiltration and influence in some local organizations and that some of the local officers were known to be very pro-communist or pro-Soviet."[113]

The scientists' movement did not succumb completely to liberal anticommunism, however. Although Harold Urey abandoned the progressive left for ADA-style liberalism in the late 1940s, within the vital center, he was an iconoclast. Throughout the 1950s he continued to speak out against Cold War limits on political freedom, in contrast to the willingness of the ADA and other liberal organizations to rein in the First Amendment in the name of national security. Urey also sharply condemned the judicial system's treatment of Julius and Ethel Rosenberg, to the dismay of the American Committee for Cultural Freedom.[114] Similarly, the FAS's politics did not fit neatly within vital center liberalism. Despite the firings of staff members and the intense divisiveness within the Association of New York Scientists and other FAS chapters, the federation did not initiate a total purge of popular front liberals. For example, Cornell physicist Philip Morrison, a member of the Communist Party from 1936 to 1942 and one of the most active scientists involved in progressive left political organizations after the war, served on the FAS administrative committee and delivered valued advice to the federation throughout the early Cold War years.[115] Nor was the FAS's break with progressive left organizations complete. At times, the federation continued to meet with groups such as the American Association of Scientific Workers and the National Council of the Arts, Sciences, and Professions to discuss the threat of HUAC to civil liberties and other matters of common interest. Despite reservations, Higinbotham even participated in the 1949 Cultural and Scientific Conference for World Peace, a meeting widely denounced by anticommunists, liberal and conservative, as a procommunist gathering.[116]

Nevertheless, red-baiting had a stifling effect on the FAS. After the firing of Daniel Melcher, the NCAI limped along for another year. The FBI continued to monitor the NCAI until the committee's final dissolution at the end of 1948 and then closed its books on the organization.[117] With the death of the NCAI, the federation lost its most direct path to a mass political base. Meanwhile, throughout 1946 and 1947, the atomic scientists continued to be dogged by rumors that the FAS was next on HUAC's list of targets.[118] In the following years, the political style of the scientists' movement changed markedly. The Federation of American Scientists traded its early spontaneity for caution, its emotionally charged and driven tone for one of staid neutrality. It decried the doctrine of guilt by association, but at the same time it succumbed to that doctrine, avoiding relations with suspect organizations and refusing to defend directly individuals under attack as loyalty or security risks.

When faced with reports from scientists of growing difficulties with the security clearance system, the FAS turned inward. Although scientists continued to speak out publicly against the excesses involved in denials of clearance, they relied more and more heavily on a strategy of quiet diplomacy, working behind the scenes with the AEC to promote changes. In the short run, this strategy actually proved fairly successful within its limited goal of procedural reform. But in the long run, as domestic anticommunism grew in strength and the AEC came under increasing fire, scientists were poorly prepared to combat the relentless political and cultural drive in America to stamp out opinions that challenged the orthodoxy of the Cold War political consensus.

CHAPTER 3

Individual Encounters I

Scientists and AEC Security, 1946–1948

• • •

Denials of security clearances plagued scientists long before the 1954 Oppenheimer case. Loyalty and security investigations of scientists began during World War II, and with the deterioration of U.S.-Soviet relations afterward, they continued with growing intensity in the postwar years. Before 1939, civil service rules established in the late nineteenth century prohibited the federal government from questioning individuals about their political and religious opinions and affiliations as a condition for employment. Limited consideration of employees' political associations began with the 1939 Hatch Act, which barred members of organizations deemed communist, Nazi, or fascist from government employment. The Smith Act, passed in 1940, forbade teaching and advocating the overthrow of the government by force or violence or belonging to any group considered to do so. The alliance between the United States and the Soviet Union defused the initial impact of the Hatch and Smith acts on radicals and the progressive left during World War II, however, and neither act was enforced regularly. Wartime security measures, though, did allow the military wide latitude in dealing with personnel whose employment the armed forces judged as contrary to national security interest.[1]

President Truman's federal loyalty program, announced in March 1947 nine days after the Truman Doctrine speech, surpassed earlier loyalty investigations in scope and scale. The program required FBI investigations of all government employees, allowing an unprecedented expansion of the FBI's investigative powers. More important, it legitimated the concept of guilt by association by employing the vague standard of "sympathetic association" with any organization or group of persons deemed by the attorney general to be totalitarian, fascist, communist, or subversive as grounds for dismissal. Under the loyalty program, federal workers could lose their jobs because of their political beliefs and associations regardless of whether or not they worked in sensitive positions that might affect national security.

Personnel security programs that covered individuals with access to classified information also focused on trying to identify ideological risks, persons who might betray state secrets for the sake of ideology and a higher loyalty to another nation, namely the Soviet Union. The history of wartime espionage suggests that a few such persons did indeed exist. The problem was whether it was possible to identify such individuals before the fact and without unduly restricting the political freedom of all persons subject to loyalty tests. During the Cold War, political tests inevitably edged into political repression. The security clearance system assumed that there existed a certain ideological type or profile that predisposed persons to commit espionage or other crimes of subversion. In practice, loyalty was a vague, ill-defined concept, and all too often the search for ideological risks equated potential disloyalty with liberal-left political action.

The loyalty program and personnel security system affected thousands of scientists. By 1950, almost sixty thousand scientists and engineers, more than one out of every ten in the country, were federal employees who therefore had to undergo loyalty clearance; those engaged in classified research had to acquire security clearances as well. Industrial or university scientists working on secret military projects also had to have security clearances. The number of such scientists is difficult to determine, but surely it ran into the tens of thousands. In mid-1949, the *New York Times* published a report about delays in industrial research caused by the security system. According to government estimates, anywhere from twenty thousand to fifty thousand scientists, engineers, and technicians in industry alone were either not working because they were waiting to be cleared or working with only temporary security clearances.[2] In the early postwar years, then, loyalty and security clearances and the political tests that went with them became a fact of life for many American scientists.

This chapter and the next examine five early postwar episodes involving six scientists—Eugene Rabinowitch, John and Hildred Blewett, Robert H. Vought, Harlow Shapley, and Edward U. Condon—who suffered varying degrees of government-sanctioned scrutiny and harassment because of their political beliefs and actions. The response of the scientific community to civil liberties issues affecting scientists will be dealt with more systematically in chapters 5 and 6. My purpose here and in the following chapter is to concentrate on individuals as a way to explore the diverse situations and dilemmas faced by scientists who found themselves politically suspect during the early Cold War period. Rather than attempt a comprehensive survey of loyalty-security cases, I have chosen to focus on these five cases because they cover a broad range of experiences and analytical issues. They feature scientists at varying stages of their careers and of differing levels of prestige, from Vought, a recent Ph.D., to Shapley and Condon, elder statesmen of science with brilliant records of past research. Levels of political activity among these scientists also varied, from relatively limited political involvement on the part of Vought and the Blewetts to Rabinowitch's fervent commitment to promoting nuclear arms control and Shapley's leading role in the progressive left. In addition, the cases illustrate the institutional complexity of postwar science. The six scientists' places of employment included industry, Atomic Energy Commission laboratories, federal agencies, and universities. Their political beliefs were evaluated and judged by different governmental bodies, including the army, the AEC, and the House Committee on Un-American Activities. The case of Edward U. Condon, unusual in its bureaucratic and political convolutions, came under the purview of three separate government enterprises: HUAC, the AEC security clearance system, and the federal loyalty program. Four of the five incidents involved security clearance; the fifth, Harlow Shapley's 1946 encounter with HUAC, constituted neither a security nor a loyalty clearance case. What tied all of these individuals together was that their political beliefs, associations, and activities became cause for unwanted attention from the federal government. To that extent, their experiences matched those of thousands of nonscientists who also faced political repression in the Cold War years. Scientists, however, were somewhat set apart because of the position of the atomic bomb in American politics and culture. Whether in the backdrop or at the fore, the postwar preoccupation with preserving the "secret of the atom" touched all six of these scientists as well as many others.

This chapter focuses on the postwar security clearance difficulties of Rabinowitch, the Blewetts, and Vought. Their cases are probably representative of most scientists who were ever denied clearance. They involved lesser known scientists and occurred without public scrutiny. This relative isolation allowed these scientists to pursue efforts to obtain security clearance free from public pressure and possible press headlines about the threat of atomic espionage, but it also left them with fewer resources to apply pressure on the AEC (or in Rabinowitch's case, the army) to reach a favorable decision. They negotiated with the AEC largely on their own, sometimes appealing to the FAS as a last resort but even then obtaining minimal publicity.

Shapley and Condon's encounters with HUAC are discussed in chapter 4. Unlike the AEC, HUAC specialized in staging public spectacles with sensational news headlines and confrontational hearings. In its attack on Condon, the committee both capitalized on and perpetuated postwar fears of atomic espionage. Although in actuality HUAC inflicted its inquiries upon relatively few scientists, the committee had an impact well beyond the boundaries of its hearing rooms, and it reinforced general public anxieties about atom spies and the potential loss of the U.S. atomic monopoly.

Whether scientists faced HUAC, the AEC, or other government agencies, fundamental procedural rights and political freedoms were at stake. In many cases, scientists denied clearance were not informed of the charges against them and had no clear way to contest those charges. Evidence of unreliability was often based on FBI interviews with anonymous sources, and even when granted hearings, scientists lacked the right to cross-examine FBI informants. Moreover, the charges themselves often lacked clear relevance to the protection of national security. Instead, they cast liberal political beliefs and actions as threats to the state.

Rabinowitch, the Blewetts, Vought, Shapley, and Condon all had to cope with the peculiar Cold War logic that support of any political position held by the American Communist Party or labeled dangerous to vaguely defined American interests, or association with any organization believed to have communist members or communist-supported goals, implied one might be a threat to the national security of the United States. Within the worldview of domestic anticommunism, support for labor unions, civil rights, international control of atomic energy, and scientific cooperation between the United States and the Soviet Union constituted grounds for suspicion, and these were among the political positions that the federal loyalty-security apparatus and HUAC used as indicators of potential or actual disloyalty. For scientists, their perceived access to atomic secrets, real or imagined, meant they were subject to extra close attention from the national security state. The pressure to yield to the ideology of anticommunism, to watch one's associations and guard one's opinions, was strong. In the following three security clearance cases, Eugene Rabinowitch, John and Hildred Blewett, and Robert H. Vought faced confused and arbitrary procedures and questions about their loyalty with the knowledge that their jobs and careers were at stake. Their experiences provide telling examples of the shock and anguish felt by scientists whose loyalty the state challenged, and their stories testify to the limited means lone individuals had to contest judgments handed down from the loyalty-security system.

Eugene Rabinowitch

In April 1946 biophysicist Eugene Rabinowitch, cofounder and coeditor of the *Bulletin of the Atomic Scientists* and an active participant in the Atomic Scientists of Chicago's efforts to establish civilian and international control of atomic energy, accepted a job offer to become head of the chemistry department of Clinton Laboratories at Oak Ridge. A few months later, however, as he prepared to leave his previous appointment at the Massachusetts Institute of Technology, the army suddenly refused to grant him security clearance. The refusal came despite his wartime work on the Manhattan Project and the army's prior indication that he would be cleared for the new position.

Before the passage of the McMahon Act and the establishment of the AEC, the army continued to administer the Manhattan Project and atomic energy research. Even in the AEC's early years, the new agency continued to follow the army's improvised wartime procedures and use army personnel.[3] Thus it is appropriate to consider Eugene Rabinowitch's 1946 encounter with the army's clearance procedures comparable to later incidents involving AEC clearance.

Readily obtainable information about Rabinowitch's case is scant, and the reasons for the army's position remain unclear. Rabinowitch was originally from Russia, but since he had been cleared previously for the Manhattan Project, his birthplace probably does not explain the army's decision against him. Rabinowitch was born on August 26, 1901, in St. Petersburg. His father was a lawyer, his mother a musician, and he later described his family as a "typical Russian emancipated Jewish intellectual family in relatively easy circumstances." Circumstances became much more difficult, however, after October 1917. He was elated by the February revolution and the fall of the czarist regime, but once Lenin came to power, the situation rapidly unraveled. He recalled decades later, "I remember vividly the day the communists came to power and the wave of indignation and protest which went through the whole democratic intelligentsia when the communists started to trample the freedoms newly acquired six months earlier." Desperate food shortages in St. Petersburg followed in the winter of 1917–18, and Rabinowitch, his parents, and his older brother headed for German-occupied Kiev with the hope that the communist government would soon fall and they could return to St. Petersburg. Instead, by 1919 civil war broke out in the Ukraine. The family remained in Kiev for some time, but early in 1921 they decided to risk the difficult trip to Germany. First they went to Minsk, where they were delayed for weeks amid fighting between Red Army and Polish troops. Then they reached Warsaw, where Rabinowitch's older brother died of an illness. Finally they arrived in Berlin, which would be Rabinowitch's home for the next twelve years.[4]

Rabinowitch studied inorganic chemistry at the University of Berlin and received a doctorate in 1926. He also supplemented his family's income by writing for newspapers, and for a time he served as the Berlin correspondent for a liberal Russian paper. After graduating, he spent two years at the Kaiser Wilhelm Institute as a research associate, and then he devoted four years to Göttingen working with James Franck. The malignant resurgence of anti-Semitism, however, made remaining in Germany a grim prospect. Franck contacted Niels Bohr and arranged a fellowship for Rabinowitch in Denmark. In 1933, shortly after Hitler came to power, Rabinowitch and his wife went to Copenhagen, where Rabinowitch spent several months at the Institute of Theoretical Physics. Then they went to England, where he worked as a research associate at University College, London, for the next four years. England did not offer the possibility of long-term advancement to foreign scientists, so Rabinowitch and his young family immigrated to the United States in 1938, where he became a research associate at MIT's solar energy research project. In 1942, he joined the Manhattan Project and went to work at the Metallurgical Laboratory at the University of Chicago. After the war, he briefly returned to MIT before deciding to go to Clinton Laboratories. Then in July 1946, he found his new position endangered when the army turned him down for security clearance.[5]

Rabinowitch's life was one of constant changes. Not only was he forced to move from one country to the next, but he was also peripatetic in his research interests. Although his laboratory training at the University of Berlin emphasized chemistry, most of his course work focused on physics. His dissertation research on volatile hydrides consisted purely of inorganic chemistry, but at the Kaiser Wilhelm Institute and Göttingen he gravitated toward physical chemistry. Then in London he turned to research on chlorophyll, which generated a lifelong interest in photosynthesis. With this research, he moved in the direction of biology and biophysics.

Yet another change took place between Rabinowitch's service on the

Manhattan Project and his decision to accept employment in the postwar atomic energy project. He became politically active. In a series of seminars held throughout the spring of 1945, Rabinowitch and other Metallurgical Laboratory scientists began to discuss their reservations about using the atomic bomb on Japan. In June, they organized a committee on Social and Political Implications that quickly put out the well-known Franck Report. The report was named after the committee's chairman, James Franck, Rabinowitch's mentor back in his Göttingen days. Drafted by Rabinowitch, the Franck Report recommended the atomic bomb be used on Japan only with prior warning and on an uninhabited target. A surprise attack on a city would not only be immoral, but it would also damage the reputation of the United States on the world stage, breed mistrust on the part of other nations, especially the Soviet Union, weaken prospects for arms control after the war, and increase the probability of a postwar nuclear arms race. On June 11, in an effort to influence official decision making, Franck traveled to Washington, D.C., where he met with Arthur H. Compton, a member of the Scientific Panel. The panel, consisting of Compton, Ernest O. Lawrence, Enrico Fermi, and J. Robert Oppenheimer, would soon issue an opinion to the high-level Interim Committee regarding use of the bomb. Franck and Compton also delivered a copy of the Franck report to Interim Committee chairman Henry L. Stimson. Unbeknownst to Franck and the other Chicago scientists, at the end of May, the Interim Committee had already recommended to Truman that the atomic bomb, when ready, be used on Japan without any special consideration. In mid-June, the Scientific Panel also backed deployment of the bomb against enemy targets without prior warning.[6]

After the atomic bombings of Hiroshima and Nagasaki, Rabinowitch and others continued their organizing as part of the postwar atomic scientists' movement. In September Rabinowitch helped form the Atomic Scientists of Chicago (ASC), and in December he and H. H. Goldsmith became founding editors of the *Bulletin of the Atomic Scientists*. He did not travel to Washington to lobby for the McMahon bill, but from Chicago he participated in efforts to coordinate the atomic scientists' discussions and activities at the different Manhattan Project sites. His most important role was as a publicist, and, as discussed in chapter 1, he wrote eloquently in the *Bulletin of the Atomic Scientists* and other publications about the interrelationship between civilian control of atomic energy, scientific freedom, international control of atomic energy, and the preservation of peace in the postwar world.[7]

When denied clearance to work at Oak Ridge, Rabinowitch immediately concluded that his participation in the atomic scientists' movement was responsible. He told Farrington Daniels, a former Metallurgical Laboratory colleague now working on reactor design at Oak Ridge, "You probably heard that the Army has refused my clearance for Oak Ridge. Since this was a change of mind from 3 months ago when Monsanto was given [the] green light to make me an offer, I have little doubt that it is the result of my ASC activities in general and of the Bulletin editorship in particular."[8]

Rabinowitch condemned the army's actions. There was, he informed Daniels, "something wrong with a system where a man can be 'blacklisted' without having been given a hearing," and he stressed that during his tenure on the Manhattan Project he was "as careful as anyone . . . to protect classified information." He added that in promoting the atomic scientists' objectives he was "more anxious than most to preserve the ASC and the Bulletin from any influence of conscious or unconscious followers of the 'party line.'" Nevertheless, he lamented, given the current political atmosphere, "in the eyes of some, being identified with the fight for international control of atomic energy" was by itself "a sign of subversive activities."[9]

The army's decision posed problems for Rabinowitch's professional life, as he had already resigned from MIT, and he had ended employment negotiations with the University of Illinois at Urbana-Champaign after deciding to accept the Oak Ridge position. He also feared that the denial of clearance might affect his chances for employment if he reopened negotiations with Illinois or with other universities, and he asked Daniels for whatever support he could provide. Daniels replied that he was "greatly distressed" by what Rabinowitch told him and that he would make inquiries at Oak Ridge regarding Rabinowitch's situation.[10]

Little other information about Rabinowitch's clearance difficulties has emerged, but even the minimal account in his letter to Daniels is telling. Rabinowitch's uncharacteristic comment to Daniels about the care he took to keep the Atomic Scientists of Chicago and the *Bulletin of the Atomic Scientists* away from "the 'party line'" says much about the pressures of the clearance system. Rabinowitch believed ardently in the right to free speech and freedom of political association. Throughout the 1940s and 1950s he wrote forceful editorials that denounced the government's drive for absolute security as a road to totalitarianism, and he defended the rights of communists to air unpopular opinions.[11] Faced with a serious threat to his career and livelihood, however, he felt

Eugene Rabinowitch, early 1950s. (James Franck Papers, University of Chicago Library)

compelled to provide a statement of his own diligence in ensuring the propriety of his political activities. At the same time, he could only guess at the army's reasons for denying him clearance, but he could not be absolutely sure. In a system in which charges were vague or unknown and procedures arbitrary, there existed powerful incentives for persons to make statements in which they emphasized their own political orthodoxy and vigilance.

Rabinowitch reapproached the University of Illinois at Urbana-Champaign and accepted a position as professor of botany and biophysics. Only days later, he was informed that the army's denial of security clearance had been an error, and he was free to work at Oak Ridge if he wished. Instead, he went to the University of Illinois and never regretted his decision.[12] Perhaps he realized that employment at Oak Ridge might compromise his own political freedom, not through direct repression but through self-censorship. In a 1951 editorial he wrote, "The first step toward totalitarianism is to impress the majority of the people that their best interests are served by keeping their mouths shut. The fear of unfavorable public reaction, worry for professional security and advancement which could be jeopardized by participation in public discussion, undermine the mechanism by which public opinion is formed in a democracy."[13] In writing this editorial, he may have been speaking through personal experience.

At some level, Rabinowitch probably realized the inconsistency between his faith in scientific freedom and international cooperation and his initial willingness to work under tightly controlled security conditions at Oak Ridge. Given his political commitments, it is somewhat surprising that Rabinowitch would have wanted to go to Oak Ridge in the first place.

He later recalled, "I really did not want to stay with atomic energy research and in Oak Ridge with all its security and all its many implications," but he added that the offer there was more attractive than elsewhere (presumably in terms of salary).[14] At the same time, he perhaps foresaw the difficulties of continuing his political pursuits while working at an installation dedicated to secret research. After discussing his clearance problems and his decision to go to Illinois in a 1964 interview, he concluded, "And that's how my double existence since 1947 began," a double life of photosynthesis research and political engagement.[15]

In the decades after the war, Rabinowitch's political involvement matched, and often overshadowed, his scientific research. He continued to edit and write for the *Bulletin of the Atomic Scientists*. In the early 1950s, he brought scientists from different nations together to discuss in an informal, candid setting the prospects for international control and the peaceful application of atomic energy. In 1957 he helped launch the Pugwash conferences, at which scientists from the United States, the Soviet Union, and other nations on both sides of the Iron Curtain met to discuss the technical and political aspects of nuclear arms control and the proliferation of other weapons that endangered world security. Pugwash embodied Rabinowitch's faith in scientific cooperation as a means of expanding international understanding, and Pugwash discussions eventually contributed to the Nuclear Test Ban Treaty in 1963 and the Strategic Arms Limitation Talks in the 1970s. Until his death in 1973, Rabinowitch remained a creative force behind attempts to employ science as a model for challenging conventional assumptions about the arms race, U.S.-Soviet confrontation, and the Cold War.[16]

John and Hildred Blewett

In September 1946, John P. and M. Hildred Blewett, husband and wife and both physicists, were offered positions at the AEC's newly established Brookhaven National Laboratory. Excited by the prospect of being at the forefront of accelerator development, they decided to leave their positions at the General Electric Company (GE), where they had both worked for several years. They purchased a house on Long Island in late January 1947 and moved from Schenectady to begin their new jobs. When the Blewetts arrived at Brookhaven, however, they discovered they had not yet received security clearance.

Brookhaven, to a far greater extent than Los Alamos and other AEC laboratories, was dedicated to basic research in nuclear physics. Most research there was unclassified, and the Blewetts had no desire to work on classified projects. Nor did the Atomic Energy Act of 1946 require security clearances for all employees. Under the law, only employees who needed restricted data had to obtain clearance. Nevertheless, Brookhaven and other AEC installations had an unofficial policy that required clearance of all scientific personnel, on the grounds that scientists working on unclassified projects might in the future require access to classified information.[17]

Like Rabinowitch, the Blewetts were émigrés, although they came from much closer quarters. Both were born in Toronto, Ontario, John on April 12, 1910, and Hildred on May 28, 1911. They met in college and married in 1936. Hildred earned her undergraduate degree from the University of Toronto in 1935. She then spent a year at the University of Rochester and, later, several years at Cornell University. In 1942 she took up work in applied physics at the General Engineering division of GE. John received his bachelor's and master's degrees in 1932, also from the University of Toronto. He then entered Princeton University, where he earned a Ph.D. in 1936. From there he went on to Cambridge University, England, where he conducted research in nuclear physics under Marc Oliphant, Ernest Rutherford's "right-hand man." Then he returned to the United States to work at GE's Research Laboratory. During the war, he was "side-tracked into various forms of electrical engineering [and] vacuum tube development," and he worked on radar as part of a collaborative project between the MIT Radiation Laboratory and GE. Hildred was also involved in radar research, working on an underwater object locator for GE. They became U.S. citizens in 1943. After the war, John desired a switch from vacuum tubes and tackled accelerator development at GE before he and Hildred left for Brookhaven in January 1947. Eager to pursue fresh research interests, they instead faced a frustrating vigil in their newly purchased home as they waited to receive security clearance.[18]

Brookhaven director Philip Morse discussed the Blewetts' situation with AEC general manager Carroll L. Wilson in mid-January. Wilson believed the case could be resolved quickly. Morse assured the Blewetts that the matter would be rectified within a month, but the couple was understandably distressed. They later told AEC commissioner Robert F. Bacher, "We were very much upset at what appeared to be an implication of some sort of disloyalty on our part, an accusation which is completely unjustified and which no one had ever made before." John Blewett later recalled that he found the situation to be especially mystifying because he had held Top Secret clearance during the war, and he and Hildred had both worked as outside consultants for Brookhaven in the months before their move to Long Island. On Morse's advice, the Blewetts agreed not to publicize their plight, on the grounds that doing so might prejudice their case and cause embarrassment to the laboratory. Trusting Morse's assurances, they began what they thought would be a relatively brief wait.[19]

February came and went without any sign of progress. Then, on March 3, two FBI agents paid the Blewetts a visit. The agents questioned them for an hour or so about their political affiliations, several of their friends and acquaintances, their union-related activities at General Electric, and their attitudes toward "communism, international control of atomic energy, and the appropriate treatment for spies and saboteurs."[20]

The Blewetts' political engagements were fairly modest, but they were sympathetic to the ideals of the progressive left. John later recalled being exposed to "a lot of fairly radical thinking" at the University of Toronto during the harshest years of the depression. As an undergraduate, he participated in the Student League of Social Reconstruction, one of a variety of left-wing political organizations on campus. As a graduate student, however, he and the other young physicists at Princeton were preoccupied with "trying desperately to make a name for themselves" so that they would be able to find employment during lean economic times. Scientific work left no time for politics.[21] But the Blewetts' political instincts remained unchanged. After the war, they joined the Independent Citizens' Committee of the Arts, Sciences, and Professions. Although they were "never active" in the Independent Citizens' Committee, they agreed with its basic domestic and foreign policy objectives.[22]

Until the FBI interview, the Blewetts had had no indication of the obstacles standing in the way of their being cleared. They had not received a formal statement of charges or the opportunity for a hearing. Based on the questions in the FBI interview, the Blewetts believed their associations with Israel Halperin (he was a good friend and John's college roommate) and their support for the union at GE were the sources of their troubles. Halperin, a professor at Queen's University in Canada, had been arrested during the Canadian spy scare along with twenty-five other individuals. The arrests implied little in the way of guilt, for the defendants were refused access to counsel, jailed without charges, and denied other basic rights. Halperin was tried and acquitted in March 1947, "speedily," according to the *New York Times*. Both of the Blewetts believed Halperin was innocent and felt vindicated by his acquittal. Years later, John vividly recalled the FBI agents' interest in Halperin: "The thing I remember most is they said, 'What do you think of your friend Halperin now?' And I said, 'I don't think he's guilty.' "[23]

The Blewetts had also openly backed the United Electrical, Radio, and Machine Workers of America (UE) at General Electric. They assisted the efforts of the Office Workers Organizing Committee to unionize white-collar office workers, and they sided with the UE during the 1946 strike. Part of the massive nationwide strike wave of 1946, the UE strike closed every GE and Westinghouse plant in the country. John helped bring in entertainers such as Zero Mostel to perform for the workers, attended a few union meetings, and marched on the picket line. "That didn't endear me with the management too much," he recalled. John nonetheless viewed his union involvement as relatively limited. He definitely approved of the UE and its goals, but it "was a sympathetic sort of thing," not a consuming passion or major time commitment.[24]

The UE was not just any union, however—it was the most prominent left-led union in the CIO. The FBI pressed the Blewetts on the subject of communist influence in the union, but the two physicists remained "convinced that its processes were never other than open and democratic." As Ronald W. Schatz has persuasively demonstrated, communists in the leadership of the UE won rank-and-file support not through underhanded, subversive means but by helping to build a strong union with the ability to meet workers' needs. In the political context of Cold War America, however, anticommunism not only led to the decline of the UE but also threatened individuals like the Blewetts, nonmembers of the union who supported it from the sidelines.[25]

The FBI interview failed to produce a decision from the AEC, and the Blewetts grew increasingly impatient. Mixed signals from Morse worsened their predicament. On March 4, Morse suggested they seek work elsewhere. Physicist and Nobel laureate I. I. Rabi heard of their difficulties and offered John a job at the Columbia Radiation Laboratory. When John told Morse of Rabi's offer, however, Morse expressed the hope that the Blewetts would stay and be a part of Brookhaven's high-energy accelerator program. They decided they wanted to remain, and John turned down Rabi's offer. But they also knew that they faced an unstable future—"unemployed here, yet in no position to look for jobs elsewhere since we have no explanation for our failure to go to work at Brookhaven."[26] If they ultimately had to find other work, not all potential employers promised to be as understanding as Rabi.

Morse also kept insisting that the Blewetts keep their troubles secret and avoid publicity. They "leaned over backward in this regard" by staying away from scientific meetings and not even telling friends of their difficulties.[27] John later termed Morse's attitude "fairly obnoxious" and expressed the feeling that their case would have been resolved more quickly if they had appealed immediately for help outside the AEC.[28] At the time, however, the Blewetts chose to follow the Brookhaven director's advice and rely on his assistance. Morse continued to consult with Carroll Wilson, but no news was forthcoming. Norman F. Ramsey, head of the physics department at Brookhaven and a good friend of the Blewetts, made overtures to AEC director of research James B. Fisk and AEC commissioner Robert F. Bacher but failed to attain a quick resolution. With Ramsey and Morse's approval, the Blewetts decided to write directly to Bacher and tell him at length about their "absurd situation."[29] Ramsey agreed with the Blewetts that their continued silence was of no benefit, and the Blewetts began to consider the possibility of publicity. Sometime in June or early July, they contacted the FAS, and William Higinbotham began to pressure the AEC.

Bacher promised to take up the Blewetts' case with the security division of the AEC, and he hoped to "obtain a speedy answer" for them.[30] The AEC commissioners discussed the Blewetts' situation on June 5. David E. Lilienthal, chairman of the AEC, was upset by what he learned, and in his diary he wrote of the lack of due process in clearance procedures, the shadowy nature of evidence based on FBI reports, and the need for the AEC to establish better personnel security procedures. Lilienthal observed with dismay, "But, as things now stand, the fact that ten years ago a scientist contributed to the defense of the Scottsboro boys, or believes in collective bargaining or the international control of atomic energy—such things as these are solemnly reported and regarded as 'derogatory information.'" Recalling the attacks on him during his confirmation as AEC chairman and uneasy about his own role in the AEC, he added, "Well, this process makes me sick at the stomach—a lot more when I find myself part of it than when it is operated against me."[31]

Despite Lilienthal's personal feelings, the AEC commissioners failed to reach a decision about the Blewetts at the June 5 meeting and decided to seek more information, including statements from the Blewetts themselves. On June 18, the commissioners authorized the appointment of an ad hoc review committee to make a recommendation in the Blewetts' case. A man named Desmond chaired the committee, with Admiral Sidney G. Souers, first director of the Central Intelligence Group and an AEC consultant, and a man named Lansdale (probably John Lansdale, formerly General Groves's chief of security for the Manhattan Project) as the other committee members.[32]

The Blewetts heard nothing about the discussions in Washington. In July they received the opportunity for a hearing, for which they later credited Higinbotham's efforts. The hearing was a strange affair, however. On the morning of July 9, at 6:30 A.M., they received a phone call from an AEC representative who told them they could report that day to Washington for a hearing. At great inconvenience, John made the trip at his own expense, and he went before a shadowy three-person committee that identified itself "only to the extent of stating it was not part of the AEC organization."[33] The committee members did identify themselves by name, and in retrospect it is clear that the hearing took place before the ad hoc review panel. Why the panel declined to provide its institutional identity remains a mystery.[34]

The hearing covered more or less the same ground that the FBI had trod back in March. Toward the end, in what was an all too familiar refrain, John was told to expect a decision in a week to ten days. Several more weeks passed. At the AEC's July 29 meeting, the ad hoc panel recommended to the commissioners that the Blewetts be cleared. In a four-to-one decision, with Lewis Strauss dissenting, the AEC decided to grant them limited clearance. The Blewetts, however, received no immediate notice of this decision and remained unable to commence work at Brookhaven.[35]

The Blewetts again appealed to the FAS. On August 10, Robert E. Marshak, chairman of the FAS council, wrote to David Lilienthal and the other AEC commissioners on behalf of the council to request that the Blewetts' situation be resolved and the AEC's clearance policy clarified. Marshak conceded "the need for continued security measures, especially in the field of military applications of atomic energy," but he emphasized that the FAS was "very disturbed about the unnecessary infringement of civil rights inherent in present procedures." The federation had received reports of cases in which no reason for denial of clearance had been given and scientists were forced to wait months for decisions. The Blewetts' situation constituted "a most flagrant case in point." Marshak warned Lilienthal that stories like that of the Blewetts impaired relations between scientists and the AEC: "The scientists of the country will have confidence in the Commission, and will be willing to work in its laboratories, only if their civil rights are safeguarded."[36]

In addition to requesting a resolution of the Blewetts' case, Marshak appealed for the establishment of "a general clearance policy . . . which would obviate all such cases." He suggested several steps the AEC might take, including publication of the criteria it used in making clearance decisions, timely resolution of cases, and financial compensation for individuals awaiting decisions. Marshak also recommended that individuals denied clearance have the right to know the charges against them, the right to a hearing with the observation of due process, and the establishment of appeals procedures. At the end of the letter, Marshak took care to reassure Lilienthal that the FAS was on the side of the AEC. He noted that the FAS "fought more vigorously than any other single group of citizens for a civilian atomic energy commission" and stated, "We are vitally interested in your success." He simply wanted to remind Lilienthal that while the AEC needed to ensure public confidence in its ability to protect the nation's security, "it also depends crucially on the measure of confidence which it enjoys among the scientists and engineers whom it employs."[37]

On August 21, the Mohawk Association of Scientists and Engineers, the FAS chapter based in Schenectady, sent a letter to the AEC similar in content to Marshak's August 10 letter. The president of the association, Herbert C. Pollack, emphasized that the Blewetts had suffered considerable "financial loss," "even more serious professional loss," and "severe psychological strain," and he recommended procedural revisions along the lines of those suggested by the FAS council.[38] Unknown to the Mohawk chapter, by the time it wrote its letter, the Blewetts' case had finally been resolved. On August 14, after seven months of waiting during which they were reduced to "living on charity," the Blewetts received permission to work at Brookhaven uncleared, without access to secret information or restricted areas.[39]

John and Hildred went on to successful careers in physics. Both made important contributions to the cosmotron, the first big accelerator at Brookhaven. John worked on accelerator development at Brookhaven for the next thirty years; by the early 1960s he was in charge of advanced accelerator design there. Hildred left Brookhaven in 1964 to become a senior physicist at Argonne National Laboratory, and in 1969 she joined the European Organization for Nuclear Research as a senior physicist.[40]

Like Rabinowitch, the Blewetts encountered a vague and arbitrary security system. They were initially required to obtain security clearance, even though they had no interest in classified research. When faced with delays, they had no practical means of discovering why they had not been cleared. There was no official procedure for them to follow; they could only consult Morse and Ramsey for advice on how to proceed. After their first FBI interview, they were left to guess that their acquaintance with Halperin and their support of the UE were the causes of their problems with AEC security, but they received no statement of charges and had no clear right to a hearing. Lacking direct contacts with the top of the AEC hierarchy, they had to depend on others to apply pressure on the AEC. Only after a long, drawn-out process mediated by Morse and Ramsey (and perhaps aided by the FAS as well, although whether or not the federation's intervention had any influence in their case is unclear) were they allowed to work at Brookhaven.

The Blewetts' case had important implications for general clearance policy, for it provided a precedent for the employment of senior personnel without security clearance. The AEC's decision in their favor seemed to indicate a willingness to discard the commission's unofficial clearance requirement for all scientific personnel and insist on a limited definition of security clearance, one that would restrict considerations of employee loyalty and reliability only to those employees who required access to information with national security implications. However, this boundary that the AEC was beginning to draw was far from clear-cut. The establishment of President Truman's loyalty program in March 1947 created a completely opposite precedent, one that established loyalty investigations for all executive branch employees, regardless of the nature of their work. In its first two years of operation, the loyalty program investigated more than two million federal employees in nonsensitive positions.

Furthermore, within the AEC there were significant professional reasons to obtain clearance in order to take full advantage of the commission's resources. Philip Morse once noted to a colleague the difficulty of working at Brookhaven without security clearance:

> We have a great number of scientists working here who are not interested in having access to classified information, particularly in the biological and medical fields. We have found, however, for those scientists working here that it is somewhat embarrassing if they do not have full clearance. Once in a while, they wish to consult with Dr. Borst about the reactor or to go to one of the few classified laboratories here at the site. This is easy if they have clearance and is difficult if they do not. A number of the scientists came here with the same feeling that you have, that they did not want to have anything to do with clearance. We honored these desires and did not ask for full clearance, at first, for them. Most of them have since agreed that they would like to get cleared to spare themselves some of the slight complications which I mentioned above.[41]

Morse's comment indicates that working without clearance at Brookhaven sometimes required a willingness to accept a certain amount of social marginalization, as well as hindrances in carrying out research. As Stuart W. Leslie has discussed in his study of military research at MIT and Stanford, classified research created a new set of social relations for American scientists. Access to secret information and the defense-related social network that went with having clearance greatly facilitated professional advancement.[42] Under such circumstances, few scientists could easily refuse to obtain security clearance when doing unclassified research at institutions with secret projects. The option to work without being cleared meant relatively little when social and professional pressures made clearance a virtual necessity. Despite these pressures, the Blewetts were able to build a place for themselves at Brookhaven without obtaining full security clearance. Robert H. Vought, a young physicist at GE, would not be as fortunate.

Robert H. Vought

In July 1946 Robert H. Vought, a twenty-six-year-old physicist who had recently received his doctorate from the University of Pennsylvania, accepted a position at the Research Laboratory of the General Electric Company in Schenectady, New York, beginning in September. Vought saw his job opportunity as an excellent chance to do basic research. The offer, he later recalled, implied he was free to undertake research "in any field in which I was interested," with an added implication that GE hoped he would choose to work on projects sponsored by the AEC. With this possibility in mind, Vought applied for security clearance upon his arrival at GE on September 3.[43] Five weeks later, the personnel director at GE informed Vought that he could not be granted a security clearance.

There was little in Vought's background to suggest why he could not be cleared. Vought was born on January 30, 1920, in Ridgway, Pennsylvania, a town of six thousand at the southeastern edge of the Allegheny National Forest. His was a "traditional, small town life." As a boy he attended public school, where he excelled in his studies. In 1937 he was salutatorian at his high school graduation, and he earned scholarships that, combined with a variety of odd jobs, allowed him to attend Allegheny College. There he was interested in science but not initially intent on a scientific career. His father had once been a schoolteacher, and Vought's first thought was to become a high school science teacher. In his first year at Allegheny, however, he did well in his physics course, and his professor, Charles Wilbur Ufford, inspired him to become a physicist. He graduated from Allegheny in 1941, the only physics major in his class.[44]

In the fall of 1941, Vought went on to graduate school at the University of Pennsylvania. He became a student of William E. Stephens and initially worked on a joint project with the medical school in which he was responsible for producing methane enriched in ^{13}C for use as a radioactive tracer.[45] With the bombing of Pearl Harbor, however, the Physics Department quickly shifted to work on radar for the National Defense Research Committee (NDRC). The University of Pennsylvania's contribution to the radar project involved the application of silicon as a semiconductor detector. Ultrapure silicon was required for this work, and Vought determined the dissociation energies of silicon tetrachloride and its dissociation products to provide necessary data for the production of ultrapure silicon. The research also served as Vought's doctoral dissertation.[46]

Although the nation was at war and scientists were heavily involved in military programs, the radar project lacked the day-to-day tensions between scientists and the military that characterized the Manhattan Project. Vought later recalled, "We were aware that this was ultimately for military use, but we never saw any military people at all. In fact, the big shots who came to visit us were from the Radiation Lab at MIT . . . but there was no evidence that this was being run by the military—it wasn't run by the military, it was NDRC." The military's presence was minimal, and security requirements were not strict. While John Blewett acquired Top Secret clearance for the MIT-GE collaboration, Vought did not recollect ever having received any kind of formal security clearance to work on radar.

Although the 1930s and 1940s were years of adversity and rapid change in American society, politics never became a major preoccupation for Vought. During his college years, the physical isolation of Allegheny supplied a buffer against the tumultuous events of the outside world. There were few radios on the campus, and Vought later remembered only one professor who seemed attuned to national and international affairs. In graduate school, however, he became moderately active. In 1943, he and several other physics students joined the United People's Action Committee (UPAC), a local Philadelphia civil rights organization.[47] When the scientists' movement formed after the war, Vought joined the local FAS chapter, the Association of Philadelphia Scientists, but he devoted relatively little effort to the debate over civilian control of atomic energy.

Vought received his doctorate in June 1946, and he had employment offers from Westinghouse, General Electric, and the University of Chicago. Decades later Vought no longer recalled his reasons for rejecting the offer from Chicago, but he expressed some regret over not having accepted: "I can think of three clear offers I had . . . and one maybe is the one I should have taken. Harold Urey offered me a chance to come with him, and that would have been exciting. . . . I think probably if I had done that I would have stayed pretty much in academic life." For whatever reason, he narrowed his decision down to a choice between Westinghouse and GE. The research opportunities at both companies were comparable. A little tired of having spent his entire life in Pennsylvania and attracted to the prospect of living near the Adirondack Mountains, Vought took the position at GE.

One of the reasons that John Blewett decided to leave GE in 1946 was that he thought several major retirements, including that of the director of the Research Laboratory, meant that GE's better days had passed. The new director, C. Guy Suits, "was a much lesser man than the previous directors."[48] But to Vought, GE still seemed like an exciting place for basic research. He later recalled that scientists at the Research Laboratory "were pretty free to do what they thought they should do rather than what the company thought they should do. And a lot of them did at that time. . . . There were some remarkable people who just had bright ideas that spawned something that was worthwhile. It was a pretty good place to do research." Only weeks after his arrival, however, Vought's application for security clearance was rejected.

Vought's inability to obtain clearance placed him in a difficult position, for like Brookhaven, the Research Laboratory required clearance for all scientific personnel. The laboratory actually had few classified projects at the time, but management wanted maximum freedom to transfer personnel to classified work if necessary.[49] As at Brookhaven, clearance also functioned as a way of fostering maximum discussion among scientists, whether or not their own research was secret. The clearance requirement was originally unofficial policy, but it became mandatory for employment at the laboratory on November 4, 1946, two months after Vought's arrival at GE. Vought was not fired immediately, but he realized that noncompliance with the new policy meant his employment might be terminated at any time.[50]

Vought immediately tried to obtain a clear statement of the reasons why his clearance application had been rejected. On October 14, he spoke with Lieutenant Vernon K. Schumann, the army official at the

Area Engineer's Office responsible for clearances, but Schumann refused to provide any specific information. Vought later wrote to Robert F. Bacher, "He [Schumann] told me some things that were *not* responsible for my not being cleared, and then lectured briefly on the evils of organizations in reference to this organization I had belonged to in Philadelphia. However, he would not commit himself to the statement that this was the reason for refusing my clearance." Vought added, "[Schumann] said my record indicated no reason to doubt my personal loyalty."[51] In his letter to Bacher, Vought did not name the organization mentioned by Schumann, but he knew it was the United Peoples' Action Committee.

On Schumann's advice, on October 16 Vought asked laboratory director Suits to request a reexamination of the clearance decision. On October 28, Suits informed him that he had once more been denied clearance. Again, Vought was not given a specific statement of the charges against him. Vought later wrote Bacher that Suits told him nothing other than "the only apparent evidence was that some of the members of this organization to which I had belonged were Communists." Vought added, "Again there was no reason to doubt my personal loyalty." Suits also pressured Vought to transfer to another GE position outside the Research Laboratory, where clearance would not be required.[52]

Vought spoke to Schumann again on November 5. Vought noted that he now dealt with "Mr. Schumann," who had "changed to civilian clothes coincidentally [*sic*] with the Atomic Energy Program"; his comment suggests the continuity between the army and initial AEC security procedures. This time Schumann confirmed Vought's strong suspicion that his membership in the United People's Action Committee from 1943 to late 1945 or early 1946 constituted the sole reason for his being denied clearance. Vought noted that Schumann seemed sympathetic toward him and suggested that if he discussed the nature of his UPAC membership and activities in writing, Schumann would "do all he could."[53]

Later that day, Vought wrote a letter to Schumann describing UPAC. He reported that the organization's purpose "was to work for equality of opportunity for all people regardless of race or color." Vought assured Schumann that UPAC's goals were "in accord with those of churches and other humanitarian organizations" and not "in any way subversive or Communistic." Typical activities included writing letters and telegrams to local politicians to protest against discrimination, promotion of fair employment practices legislation, support for the enforcement of state laws prohibiting segregation in public places, recommendations to the mayor for "a city-wide educational program on racial tolerance," and the formulation of a plan to provide low-income public housing.[54]

Throughout the postwar period, civil rights activism raised a red flag for agencies responsible for loyalty and security decisions. Despite the fact that the Communist Party was one of many diverse organizations committed to the ideal of racial equality, those individuals staunchly opposed to efforts to guarantee equal rights to African Americans viewed advocacy of civil rights as an indicator of communist influence. Racial intolerance went hand in hand with antiradical nativism. The House Committee on Un-American Activities, the FBI, and the executive branch agencies that administered the federal loyalty program all tended to equate the promotion of civil rights with communist subversion. The FBI was especially suspicious of civil rights organizations. Personally racist, J. Edgar Hoover viewed civil rights groups as little more than easy targets for communist infiltration.[55]

Vought went to great lengths to assure Schumann that UPAC had no communist connections, or at least none of which he was aware. Vought protested that even if he were now to accept rumors of communist members in UPAC as true, "I didn't know it when I was in the organization. I had no way of knowing it for no one there ever preached, or indicated belief in, Communism." Vought added, "The organization was not only not Communistic, but it wasn't political, except insofar as politics affects every phase of life, and, of course, issues affecting minority groups are frequently made political issues." In anticipation of possible counterarguments, Vought also told Schumann, "I am not so naive that I don't know that such organizations can be used as fronts for subversive activities." From his personal participation in UPAC, however, he concluded that it operated in an open, democratic manner and could not possibly have been involved in any secret subversive efforts. Finally, Vought argued that since he no longer had any connection with UPAC, his past membership was irrelevant to any consideration of his suitability for security clearance.[56] Like Rabinowitch, Vought felt compelled to demonstrate his awareness of the danger of communist influence in his political associations. In pleading his case, he implicitly accepted the security system's use of associations as an indicator of individual loyalty and avoided directly challenging his accusers' inference that liberal political action provided evidence of potential danger to national security.

With the arrival of the new year, Vought still had not obtained clearance, and he did not have the benefit of a sympathetic superior to aid him in negotiating with the AEC. Suits pressured him to find another job within GE. Vought refused, explaining that he was in the middle of an experiment, another job within GE "would just be losing time as far as physics is concerned," and finding a new position outside GE would be difficult because of "the non-clearance stigma." Vought, like Rabinowitch and the Blewetts, feared his clearance situation rendered him unable to find other employment. He also liked his new job and preferred to buy time in the hope that the AEC would come through with a favorable decision. Vought reached an agreement with Suits to stay at the laboratory until May 1, in order to finish his experiment and settle his case.[57]

Toward the end of January, Vought wrote to Robert F. Bacher and explained his problems. He noted that he had the options of seeking assistance from a former senator and the Association of Cambridge Scientists, but, he told Bacher, "I have preferred not to use these methods unless necessary because I did not want to appear to be trying to bring too much pressure to bear on something that might take care of itself."[58] As in the case of the Blewetts, Vought preferred to exhaust internal administrative channels before turning to advocates outside GE for help.

The May 1 deadline came and went, and Vought's situation remained unresolved. In February Vought had received word from Bacher that the AEC had begun an investigation and the FBI was interviewing his friends and acquaintances. In anticipation of an imminent decision, GE allowed him to remain in the Research Laboratory. In early June he learned from Schumann that the FBI had completed its investigation and the AEC had commenced deliberations over his case.[59] By this time, the FAS had learned of Vought's predicament, but no major effort was undertaken to assist him. Vought expressed his willingness to help the FAS or the American Civil Liberties Union (ACLU) with efforts aimed at clarifying general clearance policy, but he felt ambivalent about the possibility of publicity in his own case. He told Robert S. Rochlin, secretary of the FAS Committee on Secrecy and Clearance, "I have put my trust in Bacher by writing him the enclosed [January 23] letter, and I don't want him to think that I'm trying to bring pressure to bear on him." Vought wanted no publicity that Bacher "might interpret in that way." Somewhat incongruously he added, "I would of course favor any publicity which might help him in his efforts to establish a more reasonable policy." Vought hoped the FAS or the ACLU would make use of his case as they saw fit to address general problems with government clearance policies, but at the same time he did not wish to risk offending Bacher.[60]

On August 22, Vought again wrote to Bacher. He had heard from an unnamed source that, based on the FBI investigation, the AEC had recommended he be cleared but then some sort of objection had led to further FBI interviews. He plaintively asked Bacher to inform him of his status. He told the AEC commissioner, "It is now nearly a year since my clearance proceedings started, during which time I didn't know but that the next day I would have to look for a new position." He said he understood that the AEC's personnel investigations took "an enormous amount of work," but in his own case, he could see no reason for further delays. At the end of his letter he pleaded, "What necessitated rechecking the investigation? How much does Dr. Suits' saying that it is no longer urgent postpone final consideration of the case? When may I expect a decision?"[61]

In October Suits advised Vought to seek other employment. Vought began to look for another position, but as he had feared earlier, his clearance situation made finding a new job difficult. In January 1948 Argonne National Laboratory offered him a position but then withdrew the invitation after learning of his failure to obtain security clearance at GE.[62]

Meanwhile, Vought's case only grew more confused. On November 24 William E. Stephens, Vought's graduate adviser, wrote to Bacher on Vought's behalf. Vought learned from Stephens that Bacher was surprised his case still had not been resolved and thought the AEC might have mistakenly believed Vought did not require clearance because his work at GE was unclassified. At the December meeting of the American Physical Society, Vought spoke with Bacher directly, but he learned little. Bacher said there was " 'something peculiar' " about the treatment of his case but refused to elaborate further.[63]

On March 9, 1948, Vought wrote to FAS executive secretary R. L. Meier. He related the events of the past nine months and asked Meier to use whatever contacts the FAS had with the AEC to expedite the processing of his case. Meier soon reported back that, according to confidential sources, Vought had "past associations" that meant he faced three alternatives. He could try to appeal the AEC's findings regarding his associations, he could attempt to remain at GE doing unclassified research, or he could look for a new position at a university or another industrial laboratory.[64]

Meier's cryptic message suggested that the AEC had objections to "past associations" other than Vought's participation in the United People's Action Committee. On May 11, a local AEC official again informed Vought that his application for clearance had been denied. The AEC still failed to provide a specific statement of charges, but this time Vought learned that the AEC based its refusal to grant him clearance on his membership in four organizations. Vought was unfamiliar with three of the four mentioned.[65]

Throughout the summer of 1948, Vought attempted to obtain a statement of charges and a hearing from the AEC. The Mohawk Association of Scientists and Engineers tried to help by appealing directly to Carroll L. Wilson. In a July 24 letter, Harold A. Gauper, secretary of the Mohawk association, stressed that the Research Laboratory had given Vought a strict cutoff of September 10 to either obtain clearance or leave the laboratory. The deadline meant Vought's plight had reached a crisis point: "*If Dr. Vought's clearance status is not determined on a fair basis by that date, the clearance request will automatically lapse and he will be branded as uncleared for the rest of his career* [italics in original]. This is a matter of urgency if a possible grave personal injustice is to be avoided."[66]

By late summer, Vought gave up his struggle to obtain security clearance and began to search for an academic position. He wrote in a letter to the FAS Committee on Secrecy and Clearance, published in the September 1948 issue of the *Bulletin of the Atomic Scientists*, "Frankly, I'm quite discouraged and am ready to give up. I'm looking for a university position now . . . [ellipsis in original] I hope I never hear the word clearance again. This is probably running away from my problem, but I've wasted enough time here, and I've got to get out and do some physics."[67] In September, Vought joined the faculty of nearby Union College as an assistant professor of physics.

Vought's exciting job opportunity at GE had degenerated into an exasperating, almost Kafkaesque nightmare. The charges against him were unclear. He was unable to elicit useful information directly from the AEC and had to rely on secondhand reports from other scientists for news about his case. What charges he knew of were trivial, except within the worldview of Cold War anticommunism, but he had no formal way to contest them. As a young physicist, he also had few informal channels to exploit. His own superior, Suits, was unsympathetic. He did manage to obtain information and advice from his thesis adviser and the FAS, but as a junior physicist with little standing, he could not apply much pressure on the AEC for a favorable decision. The Blewetts were able to remain at Brookhaven because of Morse and Ramsey's support and pressure from the FAS. Vought had no backing from GE management, and he was forced to leave.

Vought's ordeal did not end with his move to Union College. On September 15, he finally received a statement of charges and notice that he had been granted a hearing from the AEC. The commission's charges centered on Vought's membership in the United People's Action Committee, purported to be a communist-infiltrated organization that "allegedly followed the communist party line"; his associations with UPAC members alleged to be members of the Communist Party (including a fellow graduate student in the University of Pennsylvania Physics Department); his presence at a UPAC meeting where the scheduled speaker was reportedly a communist; and his membership in the American Association of Scientific Workers, an organization whose leadership included "persons who are either communists or communist sympathizers."[68] In addition, a September 23, 1948, FBI memorandum cited Vought's membership in the Association of Philadelphia Scientists, a chapter of the FAS that the bureau considered "under some communist influence." The AEC, however, apparently discarded Vought's involvement in the Philadelphia association from its official list of charges. The agent who authored the FBI memo also doubted Vought's earlier account of his affiliation with UPAC. Although "some evidence" suggested that Vought had joined UPAC "primarily for humanitarian reasons," the agent believed it was "obvious . . . that he could not have been unaware of the extensive Communist connections of his associates."[69]

The charges sounded relatively serious, but at the hearing itself they proved almost irrelevant. As the hearing proceeded, it became apparent that the decision to grant clearance would not be based on a searching investigation of specific matters of contention. Rather, the ruling in Vought's case would rest on an amorphous evaluation of political attitudes and personal character.

The hearing took place on October 4, 1948, at the Knolls Atomic Power Laboratory. It was a full-day event, with a break for lunch. Present were the three members of the local AEC security board, chairman Cooper B. Rhodes, William A. Erickson, and W. R. Kanne, a physicist at Knolls Atomic Power Laboratory; Vought and his counsel, Research Laboratory physicist Lewi Tonks; William D. Denson, an AEC observer; and, throughout the day, several character witnesses. Surprisingly, it did not occur to Vought to seek a lawyer for the hearing. Indeed, he originally planned to attend the hearing alone, but Tonks suggested that he might want some assistance and offered to serve as his counsel. Vought was grateful for Tonks's help and later commented, "I don't know what I would have done without him." Although Tonks was not a lawyer, he proved an effective advocate throughout the hearing. Vought recalled that Tonks had a talent for argumentation and rhetoric: "He was outspoken. He would challenge people, and he was very effective in his manner of speaking. . . . Some people questioned whether I was being smart to have him as my counsel, because he could antagonize people [chuckles], very rightfully, I felt, many times. . . . Lewi was a very well-informed guy. . . . I don't think he antagonized the board more than he helped."[70] Throughout the hearing, Tonks took great care to clarify testimony that otherwise could be misunderstood to reflect negatively on Vought, and he took every opportunity to highlight remarks that redounded to Vought's favor.[71]

The hearing was a somewhat odd, surrealistic affair. Vought later recalled that he had a terrible cold the day of the hearing, and the medication he was taking had "the effect of making you sort of not feel too real." He laughed and continued, "And then the hearing itself is the sort of thing that would make you not feel too real, and so it was not a very pleasant day."[72] The board stated the charges at the outset, but they were not the primary focus of the hearing. Vought averred that he was not and had never been a member of the Communist Party, he described his associations with the individuals in question as limited, he emphasized that the activities of the United People's Action Committee were concerned solely with race relations, and he testified that neither communism, Russia, nor political philosophy in general had ever been topics for discussion at UPAC meetings. As for his membership in the American Association of Scientific Workers, Vought's sole connection to the organization consisted of having donated a dollar.[73] Other than Vought's testimony, however, no attempt to evaluate the accuracy of the charges occurred. Indeed, since Vought and his counsel had no access to the FBI reports that provided the allegations and no right to cross-examine witnesses who supplied the FBI with information, it was impossible for them to engage in an incisive, detailed analysis of the allegations. In his closing summary, Tonks called attention to the general problem of accusations based on inaccurate recollections and hearsay, and he noted that neither UPAC nor the AASW appeared on the attorney general's list of subversive organizations, which, despite its own major flaws, provided the one source of legal authority to designate domestic groups as threats to American interests.[74] But Tonks possessed no means of challenging the content of the charges directly. Under the circumstances, a trained lawyer could have done no better.

Instead, the hearing ranged far beyond the charges themselves and became a general inquiry into Vought's character and political outlook. Shortly after the hearing opened and the allegations were read, Chairman Rhodes invited Vought to offer "any evidence bearing on your character, associations and loyalty." After giving his initial responses to the charges, Vought delivered a general summary of his life history in order to indicate that he led "a normal life, normal American life, with no indications of instability or insecurity."[75] Then he presented a number of affidavits from members of the Physics Department at the University of Pennsylvania, personal friends, his adviser at Allegheny College, leaders of the Boy Scouts from his childhood in Ridgway, and even his supervisor in the dining halls at Allegheny. The affidavits all attested to Vought's good character, reliability, carefulness of judgment, and unquestionable loyalty to the United States.[76] In 1993, when I asked Dr. Vought about his use of affidavits and my impression that the hearing was not about the charges but about assessing his character, he replied, "I suppose I brought some of that [the affidavits] in because, what else do I have to offer, other than denying what they have accused me of. . . . What else did I have? Except I'm a good guy?"[77]

In the afternoon session, Vought presented seven character witnesses. Most of them were fellow physicists from the Schenectady area, and because they spoke late in the afternoon, the board did not question them at length. Their testimony covered much the same ground as the affidavits. William E. Stephens traveled from Philadelphia to testify for his former student, and as the first witness, he was questioned extensively. At Vought's request, Stephens described his general relationship with Vought, his opinion of Vought's political attitudes, and his views about the AASW and the Association of Philadelphia Scientists. Then Chairman Rhodes questioned Stephens about his own background, as well as his political and professional affiliations. In response to an earlier comment by Stephens that he did not believe one of Vought's fellow graduate students to be a communist, Rhodes asked his opinion on how one would identify a communist. At one point, Rhodes also asked Stephens what he knew about Vought's views on the exchange of scientific information between the United States and the Soviet Union. The questions about the identification of communists and Vought's opinions about information exchange suggested the way in which the legitimate problem of protecting classified information was filtered through a vision of the world in which scientists might unwittingly release vital information, through either communist infiltration or commitment to internationalism.

A third of Stephens's testimony centered on an incident that had nothing to do with Vought. In response to Rhodes's query about publications by Vought, Stephens mentioned his former student's participation in a series of seminars on nuclear fission held at the University of Pennsylvania soon after the atomic bombings of Hiroshima and Nagasaki.

The seminar papers, Stephens noted, were supposed to be put out by McGraw-Hill, but the publishing company refused to distribute them without clearance from the Manhattan District, even though the papers were all based on publicly available information. Rhodes and security board member W. R. Kanne both became preoccupied with the details of the incident, and their fixation further illustrates the postwar obsession with the protection of "atomic secrets." Stephens objected that Vought had nothing to do with the publication process, and he did not see the relevance of the topic to Vought's clearance. Rhodes replied lamely, "Perhaps what now doesn't seem relevant to you may seem relevant to Dr. Vought as the hearing progresses."[78] A few minutes later, when Vought returned to the stand, the board also asked him about the publication of the seminars. The incident became so conspicuous that at the end of the hearing Tonks objected, "Although the question of the publication of this book was not listed as one of the difficulties in the way of clearance, it seems to have been given a good deal of prominence here at the proceedings, and it just seems that if it had that importance in the determination in Dr. Vought's case that there might have been some forewarning about that matter."[79]

The board also brought up other topics at random. Rhodes asked Vought about his relationship with several members of the United People's Action Committee who were not named in the charges, simply because their names had come up during the hearing. Rhodes also inquired into Vought's willingness to bear arms for his country in wartime, asked whether Vought had ever subscribed to the *New Masses* (which he had) or engaged in any other research about the nature of communism, and questioned Vought on his attitudes toward military security requirements.[80] Vought later recalled that he felt the hearing board members were "quite fair" but their questions were sometimes "a little obtuse."[81] Although legal scholar Walter Gellhorn considered hearings "a means of enabling the Commission to arrive at a just and discerning conclusion," Vought's own hearing risked becoming a haphazard fishing expedition.[82] While Rhodes and the other board members showed no overt bias against Vought, their questioning lacked clear direction. Although the AEC had specified a limited set of allegations, the unrestricted range of subjects covered at the hearing indicated that any type of information could be used for or against Vought in reaching a decision in his case.

Four months after the hearing, Vought received a brief letter from John E. Gingrich, director of AEC security. It read in full: "The Deputy General Manager of the Atomic Energy Commission on January 28,

1949 directed me to notify you that the hearing and review of your case resulted in the finding that your employment on work of a classified nature would not have endangered the common defense and security."[83] Twenty-eight months after he had first been denied clearance, Vought had finally been vindicated.

Vought stayed at Union College for eight years, where he taught physics and performed research under government contracts. In 1956, he returned to General Electric as a solid-state physicist and stayed until his retirement in 1985. Although military contracts funded much of his research, he never again had difficulty obtaining security clearance.

In *Naming Names*, Victor Navasky wrote about the moral dilemmas of actors, writers, and directors who became informers during HUAC's investigations of Hollywood in the 1940s and 1950s. Faced with the prospect of having their careers ended unless they provided the names of Hollywood colleagues who had been Communist Party members, and surrounded by peer pressure and convenient rationalizations to name names, they betrayed the confidences of friends and colleagues. The House Committee on Un-American Activities and the movie studios created and institutionalized a ritual in which those who were under suspicion could clear themselves and prove their loyalty only by becoming informers. If they refused to cooperate, they were blacklisted. On the other hand, if informers named names, they were rewarded with the ability to continue their careers, but they were also forced to cope with anguished doubts about the morality of their actions.[84]

The AEC's investigations of scientists raise similar issues about coercion and the effects of questioning individuals' loyalty. The AEC's security system was not as coercive as Hollywood's. The AEC did not ask persons under investigation to name names. Nor did it establish formal mechanisms to make informing a way to regain political purity. Nevertheless, the security clearance system did compel people to feel and act in ways that undermined their sense of self-assurance and dignity. In addition to the threat of unemployment, the act of questioning individuals' loyalty and trustworthiness by itself proved intimidating and painful for the persons involved.

Scientists' written appeals to rectify their clearance problems consisted largely of descriptions of events and did not often articulate emotional states. Fears about loss of employment and damage to careers were common—Rabinowitch, Vought, and the Blewetts all expressed such concerns—but other emotional responses were seldom discussed at length. When they did mention other feelings, however, they talked about stunned surprise and distress. In their letter to Robert F. Bacher, John and Hildred Blewett wrote that they were "very much upset" at the questioning of their loyalty. In a case discussed in the *Bulletin of the Atomic Scientists*, one industrial researcher described his "complete shock" at being denied clearance, and a young chemist at Los Alamos described his "shock," "personal discomfort," and "great personal embarrassment" due to his failure to obtain clearance.[85] Rabinowitch expressed anger at being "blacklisted" rather than distress. Even so, his description to Farrington Daniels about the steps he took to preserve the security of the Manhattan Project and prevent undue communist influence in the *Bulletin of the Atomic Scientists* suggests the defensive, threatened position in which a person denied clearance was placed.

It does not require much imagination to suggest that such sentiments were widespread among scientists who experienced clearance difficulties. Robert H. Vought did not describe his feelings in his correspondence with Bacher, but he faced a particularly precarious situation. He was a young scientist, doctorate recently in hand. He did not have the support of his laboratory superior. The forces arrayed against him were vague and anonymous. He had difficulty discovering the AEC's charges against him, and he had no clear procedure to follow to appeal the AEC's decision. Although he was able to consult a few individuals for advice, he was largely isolated in his efforts to negotiate with the AEC. Similar themes resonate in the Blewetts' case. At least they had the support of their laboratory superior, but like Vought, they had to improvise their appeals to the AEC in the context of vague and nonexistent procedures. For many months, they also functioned literally in isolation, remaining at home and concealing their troubles from friends and colleagues on the grounds that to discuss their problems openly with others might damage their chances of being cleared. John Blewett later recounted with some bitterness that although they knew many people at Brookhaven, during their eight month wait, only Norman Ramsey and his wife came to visit them.[86] Although Hildred later viewed their suffering as a minor inconvenience compared with those who lost their careers, the ordeal took its toll. Hildred was ill throughout the spring of 1947 and was unable to begin work immediately when the Blewetts finally received clearance.[87] It does not require much insight to conjecture that the stress of their long wait contributed to her illness. In my interview with John Blewett, the physicist clearly preferred to discuss his scientific accomplishments rather than recall the painful period of waiting. He was glad that I wanted to write about his and Hildred's political difficulties, but he did not enjoy talking about them. He admitted, "You know, after I got to the lab [at Brookhaven], I tried to concentrate on my scientific [work], and as a result I think I've forgotten. A lot more stuff is in the notes than I remember." Forgetting the details was only natural—John was attempting to recall events that had taken place almost a half century earlier. But it was also apparent that John had tried to forget.[88]

When denied clearance, some individuals adopted an uncomfortably deferential posture. Both Rabinowitch and Vought, uncertain of how to proceed in appealing their cases, provided unsolicited statements about the steps they had taken to ensure the propriety of their political activities. Rather than defend their right to freedom of political association, they implicitly conceded to authorities the power to use liberal and left-wing political opinions as an indicator of potential disloyalty. A hearing was a particularly dangerous place to take a stand against unjust procedures. As discussed in Vought's case, hearings functioned more as a test of character than a way to examine charges closely. To become combative was to risk failing the test.[89]

In some cases, scientists were actively discouraged from protesting their treatment too strongly. One young researcher who failed to obtain clearance wrote to Lilienthal that he felt "deep indignation" and "helpless," and he had been done "a profound injustice." He declared openly that he had "political sympathies left of center," but so did many other "non-Communist American liberals." He refused to concede that there was nothing of value to be found in the Soviet Union. He wrote, "Although I find totalitarianism repulsive, whether of the left or the right, I have nevertheless felt that the Soviet Union has valuable contributions to make to humanity," and he expressed his support for U.S.-Soviet scientific cooperation as a way to encourage the rise of a more liberal regime in the Soviet Union. His interest in the Soviet Union, he insisted, in no way made him disloyal. He declared, "But at no time have I demonstrated that my fundamental loyalty does not lie with my own country and that I have any other feeling than that the United States has great gifts to make to mankind and deserves my whole-hearted; even if critical devotion."[90] The researcher sent a copy of the letter to the FAS, and R. L. Meier read it with dismay. Meier told him, "Unfortunately, if you had been a communist sympathizer, and I know you well enough to understand that this would be completely out of character, the tone and content of the letter would have been the same as that you used." Security clearance, Meier wrote, was an area "where protest creates more suspicion than sympathy."[91] To oppose injustice openly was to endanger one's own chance of acquiring clearance. Isolated individuals with few sources of support could not afford to dispute the security system's fundamental assumptions about the nature of loyalty. It was far safer to play by the rules than to defend the appropriateness and necessity of diverse political opinions in a democratic culture.

As of the late 1940s, Rabinowitch, the Blewetts, and Vought were all lesser-known scientists. Their peers respected their work within their specialties, but none of them held leadership positions within the scientific profession's honorary societies, prominent research awards, or the other trappings that accompany the highest professional status in science. As scientists outside the top echelon, they had few strings to pull and depended heavily on the help of friends and colleagues to pursue their cases. They had difficulty obtaining descriptions of the charges against them, much less information about how to proceed and contest those charges. Faced with uncertain situations when their careers were in danger, they could not bear the risk of challenging the fundamental legitimacy of a system that made unorthodox political views grounds for denial of security clearance and, potentially, livelihood as well. Reluctant to seek publicity lest they alienate representatives of the security system, they instead quietly used whatever contacts they could establish to clarify their circumstances.

For well-known scientists faced with political attacks, the dynamics of their cases were very different. As Rabinowitch, the Blewetts, and Vought worked behind the scenes to address their difficulties with the army and the AEC, Harlow Shapley and Edward Condon confronted HUAC out in the open. Their experiences were not necessarily easier to endure, but, as we shall see, their high visibility and status at least provided them protections and avenues unavailable to other, less prominent scientists. Both would survive their ordeals with careers and reputations intact, but the scientific community as a whole would not be able to escape the long-term implications of congressional pressure.

CHAPTER 4

Individual Encounters II

Scientists and HUAC, 1946–1948

• • •

The three cases in the previous chapter dealt with early AEC or postwar Manhattan Project security clearance procedures. They involved lesser-known scientists who depended primarily on informal administrative channels, isolated from publicity, to untangle their problems. The following two cases emerged from investigations by the House Committee on Un-American Activities, and they proceeded quite differently. They were public spectacles that featured two prominent, outspoken scientists, Harlow Shapley and Edward U. Condon. Because both were well known, they were able to meet HUAC head-on without fears for their professional standing. In their confrontations with postwar anticommunism, Shapley and Condon brandished their individualism and held steadfast to their political ideals. Self-confident in their convictions and protected by their status within the scientific profession, Shapley and Condon felt able to criticize the rise of the Cold War publicly and to condemn the abuses perpetrated in the name of national security throughout the 1940s and 1950s. Both weathered years of on-and-off political attacks for their forthright candor, as anticommunists tried to exact as high a price as they could for open political discussion and dissent in the postwar years.

A Brief Confrontation: Harlow Shapley and HUAC, 1946

On November 4, 1946, Harvard astronomer Harlow Shapley was subpoenaed to appear before the House Committee on Un-American Activities and to produce the records of four organizations, the Political Action Committee of the Congress of Industrial Organizations (CIO-PAC), the National Citizens Political Action Committee, the Joint Anti-Fascist Refugee Committee, and the Independent Citizens Committee of the Arts, Sciences, and Professions. A closed HUAC hearing took place in Washington, D.C., on November 14. Representative John E. Rankin was the sole committee member present. On the face of it, the hearing appeared to have nothing to do with science, Shapley himself, or even supposed communist infiltration in the cultural and political institutions of the United States. Instead, it involved the recent Massachusetts congressional race between incumbent Joseph Martin, a friend of Rankin's as well as a powerful Republican House member, and Democratic challenger Martha Sharp. The committee had somehow obtained a copy of a letter from Shapley to Sharp that seemed to imply Sharp wished to obtain political contributions from left-wing organizations without undue publicity. Sharp wanted mainly to avoid alienating conservative voters, but Rankin hoped to prove that Sharp had received unacknowledged, and therefore illegal, campaign contributions from the National Citizens Political Action Committee. Rankin also wanted to know of any support for Sharp from the CIO-PAC, the Joint Anti-Fascist Refugee Committee, and the Independent Citizens Committee.[1]

Sharp lost the race, so the hearing did not suggest an effort to contest the election results. In part, the hearing reflected HUAC members' predilection for embarking on haphazard fishing expeditions, but it also had larger political implications. Of the four groups, the CIO-PAC, the National Citizens Political Action Committee, and the forerunner of the Independent Citizens Committee, the Independent Voters Committee of the Arts and Sciences, had been powerful progressive left political organizations and had played a critical role in Roosevelt's 1944 reelection campaign. The CIO withdrew from the popular front coalition in the fall of 1946, but the National Citizens Political Action Committee and the Independent Citizens Committee remained active. Although Rankin's immediate interest in Shapley originated from the Massachusetts congressional race, his targeting of these political action groups exemplified the more general decline of the New Deal liberal-left coalition and mounting assaults on the progressive left from a conservative alliance of Republicans and southern Democrats.[2]

Shapley's hearing ended quickly and in a confused uproar. According to the single-page transcript made by Nelly Thomas, Shapley's stenographer, after Shapley was sworn in and asked to identify himself, HUAC counsel Ernie Adamson asked him if he was "affiliated in any way directly or indirectly with the Joint Anti-Fascist Refugee Committee." Shapley responded that he had a statement he wished to read. At this point the transcript ends with the notation that Rankin asked both Thomas H. Eliot, Shapley's lawyer, and Nelly Thomas to leave the room, "and when Dr. Shapley asked that they remain, Rankin called for the guards."[3]

Exactly what happened next is a matter of some conjecture. Shapley either read or attempted to read a statement in which he contended the committee had no right to launch investigations into campaign financing. As for his own political affiliations, Shapley's statement noted that he belonged to neither the National Citizens Political Action Committee nor the CIO-PAC, he was only a sponsor of the Joint Anti-Fascist Refugee Committee, and his position as Massachusetts chairman of the Independent Citizens Committee did not give him the authority to produce for HUAC the records of the national organization. At some point, Shapley and Rankin launched into a fierce argument over whether or not Shapley's statement would be entered into the record unedited. Then, according to Shapley, Rankin grabbed the statement from his hands. Rankin and Adamson, on the other hand, claimed Shapley had given them the statement, only to snatch it back. Whatever actually happened, the following is clear: both Shapley and Rankin were enraged, and the hearing ended abruptly with Rankin threatening to have Shapley cited for contempt of Congress.[4]

No contempt citation ensued, and HUAC never called Shapley before the committee again. But the skirmish between Shapley and Rankin was only the beginning of a series of politically motivated attacks that struck Shapley and his acquaintances. Outspoken and active in progressive left politics since the 1930s, Shapley had to endure the politics of Cold War anticommunism throughout the 1940s and 1950s.

Harlow Shapley was one of the foremost astronomers of his generation, known for his work on variable stars and globular clusters. He was born on November 2, 1885, on his family's farm five miles from Nashville, Missouri. Lack of family funds meant that his early education was sporadic, but both his mother and older sister valued the acquisition of knowledge and encouraged Harlow, and his brothers Horace and John, to strive for higher learning. Shapley received an elementary education at a local schoolhouse that he augmented with home study and a brief business course in Pittsburg, Kansas. At the age of sixteen, he went to work as a newspaper reporter in Chanute, Kansas, and he reported for papers in Kansas and Missouri for the next several years. Seeking further learning, in 1905 he enrolled in a Missouri preparatory school. Then in 1907 he entered the University of Missouri, where he intended to major in journalism. But the university's journalism school had not yet opened, and he happened upon astronomy instead. Shapley earned an undergraduate degree with high honors in mathematics and physics in 1910 and a master's degree in 1911.[5]

Next, Shapley turned to graduate study at Princeton University, where he made the first of several major discoveries. Working with Henry Norris Russell on variable stars, he wrote a dissertation on the characteristics of eclipsing binaries. As a sideline to his dissertation research, he found that the Cepheid variables were not binary stars, as previously thought, but single pulsating stars. Fellow astronomer Otto Struve later described Shapley's dissertation as "the most significant single contribution toward our understanding of the physical characteristics of very close double stars"; Shapley's Harvard colleague Bart J. Bok simply labeled it "a classic in the field."[6]

After receiving his Ph.D. from Princeton in 1914, Shapley became an astronomer at Mount Wilson Observatory. He began to study variable stars in globular clusters, and this work led to his most famous discovery, when in 1918 he demonstrated that the Milky Way galaxy was much larger than previously thought and that the earth's solar system was located at the periphery of the Milky Way, rather than near the center. Shapley's findings had both scientific and philosophical implications. Some contemporary observers, as well as Shapley himself, likened the discovery to Copernicus's formulation of the heliocentric solar system and observed that Shapley's work further removed the human race from the center of creation.

In 1921 Shapley moved to the Harvard College Observatory, where he became director only six months after his arrival. Under his leadership, the observatory became a major center of astronomical research throughout the 1920s and 1930s. During this period, Shapley began to acquire the honors that mark prominence in science. He was elected to the National Academy of Sciences in 1924, and he received the academy's Draper Medal in 1926. His other honors included the Rumford Medal of the American Academy of Arts and Sciences in 1933 and the Gold Medal of the Royal Astronomical Society in 1934. In the 1940s he served as president of several important scientific societies: the American Academy of Arts and Sciences from 1939 to 1944, the American Astronomical Society from 1943 to 1946, and the honorary scientific research society Sigma Xi from 1943 to 1947.

In the 1930s, Shapley began to spend more and more time on public affairs. As Peter Kuznick has shown, the depression years were a period of increased social awareness for many American scientists.[7] Politically active scientists during the 1930s concerned themselves primarily with issues related specifically to science, especially the need for scientists to bear greater responsibility for the results of scientific research. But some scientists moved beyond the immediate interests of science and promoted efforts to oppose fascism overseas and protect civil liberties at home in the United States. Shapley was a leading member of the latter group. In the early 1930s, his first political efforts concerned aid to foreign academics. As a member of the Emergency Committee in Aid of Displaced German Scholars, he helped place refugee scholars at academic institutions in the United States. By the late 1930s, Shapley had expanded the scope of his involvement and joined the wider antifascist movement. In 1938 Shapley, Harold C. Urey, and Arthur H. Compton attempted to meet with President Roosevelt to discuss the possibility of U.S. support for the Spanish Loyalists. As a sponsor of the American Committee on Democracy and Intellectual Freedom, an organization formed in 1939 by anthropologist Franz Boas and composed primarily of scientists and social scientists, Shapley participated in various progressive left campaigns. He was active in the Joint Anti-Fascist Refugee Committee, which sought to aid refugees from the Spanish civil war, and the Independent Voters Committee of the Arts and Sciences. Soon a leader in the progressive left, Shapley traveled to the White House in 1944 as part of a delegation from the Independent Voters Committee that delivered a statement in support of Roosevelt's reelection to the president himself.[8]

After the war, Shapley continued to work diligently for a variety of political causes. A firm internationalist, he vigorously advocated scientific cooperation between nations, and he strived to ensure that science would be a priority of the United Nations Educational, Scientific and Cultural Organization (UNESCO). Although not a central figure in the FAS, throughout 1945 and 1946 he worked for civilian control of atomic energy as a member of both the federation and the Independent Citizens Committee of the Arts, Sciences, and Professions. He also helped to lead postwar efforts to establish a National Science Foundation in line with Senator Kilgore's principles of public accountability and social responsibility.

Although Shapley was an important figure in progressive left politics, his main ideological commitment was not to any particular economic system but to internationalism, individual rights, and the general improvement of the human condition. Like other internationalist scientists, Shapley thought that the international cooperation found in science could serve as a model for relations between nations. As he told a journalist in 1949, "We see every day how co-operation across national boundaries helps to advance the sciences. It's natural to think that it would be a good idea to apply the same principle to human relations in general."[9]

Shapley pinned his hopes for postwar peace on a decline of nationalism and an increased dedication by individual nations to global problems. In an October 1946 *New York Times* interview during the fall meeting of the National Academy of Sciences (NAS), Shapley appealed to scientists to answer to a higher cause and "increase the importance of their world citizenship over their local loyalties."[10] Two months later, in his inaugural presidential address to the American Association for the Advancement of Science, he repeated this appeal and spoke at length about the need for international cooperation and friendship. He emphasized the hope that "positive friendship for civilization, expeditiously organized and steadily maintained," could replace balance-of-power politics, and he asked for "persistent systematic friendship and tolerance, more correspondence across the borders, more travel across the boundaries of nations, more collaboration across the national political lines, until finally the boundaries become worn thin by so much international traffic."[11]

As relations between the United States and the Soviet Union deteriorated throughout 1946 and 1947, Shapley sharply criticized American foreign policy and the rise of nationalism. In September 1947, he attacked the "travesty and hypocrisy" of American policy toward Greece for its failure to address "the tyranny" that persisted in that country, and he advocated efforts to change American policy by placing "social idealists and . . . students of human rights into political office and at policy-making levels."[12] His political language, with its emphasis on an optimistic faith in human nature, embraced the tenets of popular front liberalism. In a 1948 article, he wrote fervently against nationalism, arguing that feelings of selfish nationalism "curse international relations." Ordinary human values, he contended, suggested much better alternatives: "Love of family and of friends breeds into the individual a natural human generosity, that propaganda cannot wholly erase — a willingness, in a pleasant social order, to live by decent rules, to compete and contest with sportsmanship and toleration." This natural human decency, however, was too often subverted by "super-heated nationalism," "the cancerous greed of political and financial groups," and the "evil phrases" of "National Aspirations and Manifest Destiny."[13]

Shapley also assailed efforts to discourage or limit the expression of dissenting political opinions. As chairman of an October 1947 conference on cultural freedom and civil liberties sponsored by the Progressive

Harlow Shapley (*rear left*) as co-chairman of the Progressive Citizens of America, with Jo Davidson (*rear right*), Henry Wallace (*front left*), and Elliott Roosevelt (*front right*), March 31, 1947, at Madison Square Garden. (New York World Telegram and Sun Collection, Library of Congress)

Citizens of America (PCA), Shapley hailed the gathering for its recognition of "the dangers of the hysteria threatening our country."[14] In other forums, he singled out the actions of HUAC as particularly reprehensible. When the Hollywood Ten were cited for contempt of Congress, Shapley, in his capacity as chairman of the Arts, Sciences, and Professions Council of the PCA, warned the president of the Motion Picture Association of America that "to yield to hysteria by establishing blacklists and purges would be a betrayal of the trust of the American people." In early 1948, Shapley also signed on as an initiating sponsor of the Committee of One Thousand, an organization that sought to abolish HUAC.[15]

Throughout the postwar years, Shapley helped to lead the progressive left. He held prominent positions in the Independent Citizens Committee and, later, the Progressive Citizens of America. As a high-ranking member of the PCA, Shapley, along with other PCA members, shared the stage with Henry Wallace before a sellout crowd at Madison Square Garden in April 1947. Shapley supported Wallace's 1948 presidential campaign, although he later wrote that he was disappointed by Communist Party influence at the Progressive Party convention.

Although certainly to the left by American standards, Shapley denied that he was a radical, and he was especially offended by accusations that he supported the Soviet government. Well aware of the purges of Soviet astronomers in the 1930s, he had no sympathy for totalitarianism in any form. But he refused to capitulate to Cold War political orthodoxy. Instead he identified himself as "a fellow traveler with all the minorities in turn when their constitutional rights are invaded by a thoughtless majority." He explained unapologetically, "I have a sense of obligation to defend American principles and constitutional policies, even if my speaking out is sometimes embarrassing to friends who prefer, even in these dangerous days, to be quiet on civic issues and resign themselves to hope, or resign themselves to fate."[16] Above all else, a deep concern for human freedom and dignity inspired Shapley's politics.

Rankin's brief attempt to question Shapley focused on the Martin-Sharp race and avoided emphasizing Shapley's other political engagements. Nevertheless, the incident leads one to consider the repercussions of progressive left political activity during the postwar decade, both for the individual in question and for his or her acquaintances. Although Rankin's threat to cite Shapley for contempt failed for lack of congressional support and Shapley never testified before HUAC again, the committee cited and criticized him in its publications. The FBI also kept a watchful eye on Shapley, and FBI investigations of other persons paid careful attention to their relationships with him. Although Shapley did not suffer from any legal reprisals or loss of employment, he was susceptible to public vilification throughout the 1940s and 1950s.

Review of the Scientific and Cultural Conference for World Peace, a 1949 report by HUAC, identified Shapley a number of times as a member of groups the committee considered subversive and listed him as "affiliated with from 11 to 20 Communist-front organizations." Shapley shared this designation with, among others, Albert Einstein. Similar references to Shapley appeared in HUAC's *Report on the Communist "Peace" Offensive*. Published in 1951, the *Report on the Communist "Peace" Offensive* consisted largely of excerpts from previous HUAC reports and contained little new information.[17]

The committee listed Shapley's political affiliations in accordance with a common HUAC technique: the indiscriminate and blatant application of the concept of guilt by association. The House Committee on Un-American Activities regularly identified left-wing organizations that had a few communist members, or had members who belonged to other organizations that were allegedly dominated by communists, as communist-controlled and dangerous. The committee made little distinction between liberal and left organizations that might attract communist support and the American Communist Party itself, and it invoked the doctrine of guilt by association to the extreme of identifying individual or group support of any political position that also happened to be held by the Communist Party as evidence of communist sympathy and subversive potential. The committee's reports were also, as Cornell legal scholar Robert K. Carr gently put it in 1952, "poorly organized and badly written," and they were filled with exaggerated and inflammatory statements about the threats various organizations posed to American values and institutions.[18]

The committee's treatment of scientists in the *Review of the Scientific and Cultural Conference for World Peace* provides multiple examples of HUAC's flawed and deliberately incendiary reasoning. According to HUAC's account of the conference, Communist Party member Richard Boyer advocated civil disobedience as a form of political protest and declared that it was "the duty of Americans to defy an American Government intent on imperialist war." The report suggested that Boyer's statements were particularly ominous given that Harlow Shapley, William A. Higinbotham, Walter Orr Roberts, Philip Morrison, Victor Weisskopf, Oswald Veblen, and Albert Einstein were in the audience. Their mere presence, the report implied, indicated that the scientists shared the speaker's views. Insinuations aside, the committee lacked even a firm grasp of the facts. Veblen sponsored the conference but was absent from the gathering; Einstein and Weisskopf probably did not provide sponsorship, and it is doubtful that either attended.[19] In addition, HUAC incorrectly identified all the scientists as "atomic scientists." Several of the names were misspelled, Roberts's first name was given as "William," and Shapley, Roberts, Veblen, and Einstein had had no part in wartime or postwar atomic research. But the committee, playing off postwar fears of atomic espionage, carelessly applied the label "atomic scientists" and suggested, "If the Communists could succeed by playing upon the notorious political naiveté of physical scientists, in inciting scientists to a 'strike' against their own Government, or sabotage, it would be a real achievement for the Soviet fatherland. They would like nothing better than a repetition in the United States of the cases of the Canadian atomic scientists . . . who divulged atomic secrets to the Soviet Military Intelligence."[20] To the committee, civil disobedience on the part of communists had to mean sabotage or espionage, and the attendance of scientists at an open, public meeting where a Communist Party member spoke meant a high likelihood that those scientists planned to commit disloyal acts against the United States government.

The House Committee on Un-American Activities was not alone in attacking the gathering. The Cultural and Scientific Conference for World Peace (HUAC reversed the order of the words "scientific" and "cultural") provoked enormous controversy. The National Council of the Arts, Sciences, and Professions, chaired by Shapley, organized the meeting as a way of bringing together citizens from around the world to find ways of improving U.S.-Soviet relations. Conference organizers especially wanted participants from the Soviet Union and Eastern bloc nations. The State Department admitted delegates from the Eastern bloc because they came as official representatives of their governments, but it refused to grant visas to other foreign delegates alleged to have communist ties. Hundreds of anticommunist protesters, demonstrating in the name of religious freedom, picketed the conference site. The meeting also became a battleground between Cold War liberals and the progressive left. Liberal anticommunists, led by Sidney Hook, staged an alternative assembly, where they spoke about intellectual freedom and savaged the "communist-inspired Waldorf conference."[21]

Cold War liberals pilloried the Waldorf conference on the ground that its participants took too little account of political and intellectual repression in the Soviet Union. They also contended that the meeting itself made a mockery of free speech and discussion when Eastern bloc delegates obviously could not speak candidly but were compelled to follow their governments' line. Conference organizers replied that there had been plenty of free discussion and disagreement, and the Soviet delegates had taken some tough comments and questions about Lysenkoism (the Soviet crusade against Mendelian genetics), arbitrary arrests in the Soviet Union, the stifling of scientific and artistic freedom in Russia, and partial Soviet responsibility for the deterioration of U.S.-Soviet relations.[22]

At the same time, as Shapley pointed out in an earlier letter to a fellow organizer, the primary purpose of the conference was to transcend the divisions of the Cold War. He wrote, "We live in a glass house. The admission of Jews to Medical Schools [a reference to the Jewish quota, common in American universities, that placed an upper limit on the admission of Jews] and the demotions of Russian geneticists are not too distantly related. Our Conference for World Peace will be, I hope, a *peace* conference and not a war conference; a bid for understanding and cooperation and survival, not a further incitation to hate. . . . Twice I have spoken on the radio against Lysenkoism; but I decline to be bullied by the hysteria into arguing that there is something unique in this Russian attempt to dictate on the subject of natural laws."[23] When pressed, popular front liberals like Shapley did not deny the unsavory realities of life in the Soviet Union. Where the progressive left parted from liberal anticommunism was in its demand that Americans focus attention on repression within the United States. More important, the progressive left also insisted that some form of U.S.-Soviet cooperation was not only possible but necessary for the sake of world peace.[24] Such goals were hardly evidence of the subversive conspiracy implied by HUAC. But to antiradical nativists, the objectives of the Waldorf conference constituted grounds for suspicion. For HUAC, the FBI, and liberal anticommunists, advocacy of anything other than a hard-line stance toward the Soviet Union smacked of capitulation to totalitarianism.

The House Committee on Un-American Activities was not the only government organization concerned about Shapley's endeavors. The FBI monitored the Harvard astronomer and his political affairs closely. I have not examined Shapley's FBI file, but, as discussed in chapter 2, letters from Hoover to the White House show that the FBI took keen interest in Shapley's support for the confirmation of David Lilienthal as AEC chairman and his efforts to establish a National Science Foundation. Under the dictates of guilt by association, the FBI also considered connections to Shapley highly relevant when investigating other individuals. In 1948 Harvard astronomer Donald H. Menzel applied for a security clearance as part of an application for an AEC consultantship at Los Alamos. As noted by Sigmund Diamond, the FBI was especially interested in Menzel's relationship to Shapley and Walter Orr Roberts, his colleagues at Harvard College Observatory. Diamond also notes that when Harvard sociologist Talcott Parsons was the subject of a loyalty investigation in 1953 and 1954, his connection to Shapley again tantalized the FBI.[25]

Shapley also attracted the attention of the infamous Senator Joseph McCarthy (R.-Wis.). In his notorious February 9, 1950, speech in Wheeling, West Virginia, McCarthy specifically named Shapley, along with John Stewart Service, Gustavo Duran, and Mary Jane Keeney, as among 205 State Department employees who were members of the Communist Party. McCarthy repeated Shapley's name two days later in Reno, Nevada, although by that time McCarthy's list of State Department subversives had shrunk to "57 card-carrying members." In Senate hearings a month later, McCarthy continued to cite Shapley's name, despite the fact that Shapley had never been either a State Department employee or a Communist Party member.[26] Shapley sustained no damage from the incident, but three years later his name was invoked in Senate confirmation hearings over Harvard president James B. Conant's appointment as United States high commissioner for Germany. Senator Robert A. Taft (R.-Ohio), Senate majority leader and an uneasy ally of McCarthy, pointedly questioned Conant about Shapley, a Harvard faculty member. Taft claimed that during Conant's tenure Shapley had "attended every Communist meeting, important meeting, . . . in this country."[27] Shapley wrote Taft to protest that the senator was "much misinformed." The 1930s purges of Russian astronomers, Shapley explained, meant that "naturally no American astronomer has been pro-communism since that time, however pro-humanity we may be."[28]

Shapley suffered relatively little from his brushes with HUAC and other branches of the national security state. As a tenured professor and eminent scientist, his livelihood was safe. A contempt citation in 1946 might have done him harm, but Rankin's unsavory conduct and lack of congressional support spared him. The House Committee on Un-American Activities could have launched direct inquiries into Shapley's actions, but a House rule that kept Rankin off the committee after 1948 probably freed Shapley from further scrutiny. Moreover, Shapley was an astronomer, not an atomic scientist, and the absence of any national security implications in his scientific work further diminished his appeal as a political target.

Shapley's confrontations with Cold War anticommunism might seem like a long series of minor annoyances, but they point to serious consequences. Even for liberals and radicals who escaped material sanctions, such as loss of employment, political activism carried at the very least the risk of being publicly vilified. Shapley was a strong-minded and resilient person who refused to feel threatened, but for individuals who were less confident or without job security, intimidation could easily result. Furthermore, underneath the public attacks lay a system in which an individual's friends and acquaintances might also find their loyalty questioned. As far as the FBI was concerned, Donald Menzel's and Talcott Parsons's acquaintance with Shapley constituted a black mark in their respective efforts to obtain security and loyalty clearance. Shapley's name also functioned as a political symbol in James B. Conant's confirmation debate and served as a means to question the political purity of both Conant and Harvard University. In Cold War America, the actions of a person guilty of no more than outspokenness and unorthodox political views could have unforeseen detrimental effects on friends, colleagues, acquaintances, and institutions.

Shapley himself refused to allow his beliefs to be reduced to Cold War terms or to give in to red-baiters' tactic of guilt by association. He once explained passionately to the editor of an Illinois newspaper in which he had been maligned:

> I have joined a number of citizen movements to protest against violation of civil liberties and human rights. I shall continue to do so, and if it happens that extreme radicals or extreme reactionaries also support these actions, I shall not back out in a cowardly fashion. Undoubtedly Communists have walked on Broadway, and I shall not avoid the street for that reason; undoubtedly Communists have shopped at Macy's, and the Daily Worker (which I never see) is distributed into Illinois — but that does not blaspheme the store or State for me. Probably Communists have attended meetings over which I have presided; and I know that some reactionary Republicans have attended these meetings. . . . I know a few Communists abroad. . . . And I know a few extreme fellow-travelers who are not Communists. . . . I am not afraid of them, or of Dixiecrats. Both are a legitimate part of the wide and healthy political spectrum in America.[29]

Rather than deny his associations to try to prove his own political purity, Shapley insisted that the value of ideas could not be judged on the basis of who held them. Harlow Shapley defined his political beliefs not by the character of his associations but by the strength of his convictions.

HUAC and the Condon Case, 1947–1948

Although Shapley's 1946 confrontation with Rankin did not involve atomic energy, J. Parnell Thomas did not lose sight of atomic espionage as a potential investigatory mission for HUAC. When the Republicans gained control of the House of Representatives after the 1946 elections, Thomas became HUAC chairman. In January 1947, he announced an eight-point program for future investigations by the committee, including "investigation of those groups and movements which are trying to dissipate our atomic bomb knowledge for the benefit of a foreign power."[30]

The committee held no hearings related to atomic energy in 1947, but in the time that had passed since the fight over the McMahon bill, Thomas had not forgotten about E. U. Condon. In March 1947, Thomas continued the attacks he had begun during the McMahon bill debate by launching several trial balloons directed at Condon. On March 23, the *Washington Times-Herald* published a front-page story headlined, "Condon Duped into Sponsoring Commie-Front Outfit's Dinner." The article stated that Condon had allowed his name to be listed as a sponsor of a dinner of the Southern Conference for Human Welfare, an alleged communist front under investigation by HUAC. Two days later the *Times-Herald* announced that Condon's loyalty was being probed because of his membership in the American-Soviet Science Society, an organization formed during World War II that promoted the exchange of scientific information between the United States and the Soviet Union.[31] Thomas provided the background information for both *Times-Herald* accounts.

Soon Thomas began to attack Condon more openly. In the June 1947 issue of *American Magazine*, the congressman pointed to Condon's membership in the American-Soviet Science Society as evidence of a Soviet plot to acquire information about atomic energy.[32] The following month, on July 17, the *Times-Herald* declared in a page-one headline, "House Unit to Quiz Dr. Condon on Reds' A-Bomb 'Know-How.'" The article was an exercise in innuendo. It mentioned that according to Thomas, HUAC considered Condon qualified to discuss Soviet atomic research because of his "contacts with Russian scientists and pro-Communist sympathizers in this country." The *Times-Herald* also listed six reasons why HUAC planned to subpoena Condon, including his American-Soviet Science Society membership, his friendship with Harlow Shapley, and, in a particularly egregious example of insinuation, the fact that "Condon was constantly checked by military authorities for security reasons at the same time, but not as frequently as Dr. Frank Oppenheimer, card-carrying member of the Communist party, whose brother, Dr. J. Robert Oppenheimer, led the team of physicists who exploded the first atom bomb in the Western desert."[33] Speaking before the House on July 22, Representative Chet Holifield (D.-Calif.) defended Condon point by point against the suspicions reported by the *Times-Herald*. He concluded by stressing Condon's eminence as a scientist and challenging "rumor-mongering character assassins" to "put up or shut up."[34]

Where the articles were leading did not become fully apparent until March 1, 1948. On that day, speaking to the press from a bed in Walter Reed Hospital, where he was recovering from gastrointestinal hemorrhages, Thomas issued a special subcommittee preliminary report to the House Committee on Un-American Activities. The subcommittee, consisting of Thomas, Representative Richard B. Vail (R.-Ill.), and Representative John S. Wood, accused Condon of being a security risk. Exploiting postwar fears about the atomic bomb and the need for preservation of its supposed secret, the report contended, "From the evidence at hand, it appears that Dr. Condon is one of the weakest links in our atomic security."[35]

In choosing to attack Condon, HUAC picked a prominent target. Condon was a highly regarded physicist, recognized for his contributions to quantum theory in the 1920s. By the 1940s, he was an elder statesman of American physics. Condon was elected to the National Academy of Sciences in 1944, and he served as president of the American Physical Society in 1946. In addition to recognizing his professional achievements, the physics community also esteemed Condon for his cheerful demeanor, sharp wit, and love of telling humorous stories, tales that usually poked fun at those he considered foolish or short sighted.[36]

Condon was born on March 2, 1902, in Alamogordo, New Mexico. Decades later he liked to claim that a suspicious senator wanted to know how he had come to be born so near the location of the Trinity explosion, the first test of the atomic bomb. His parents separated when he was young, and in his early years he led a somewhat nomadic life with his mother, moving from town to town in the western United States. He identified with his western upbringing, and later in life his friendly, boisterous attitude seemed in marked contrast to the more ascetic character of his East Coast colleagues. Philip Morse, one of his students at Princeton, later recalled that Condon's "western vocabulary, the proletarian outlook, the rough-edged kindliness" seemed out of place among "the eastern establishment manners that were then the Princeton norm."[37]

Condon and his mother eventually settled in the San Francisco Bay area, where Condon attended junior high and high school. As a youth, he was interested in science, but on graduating from high school, he spent three years as a journalist before going to college. As a reporter for the conservative *Oakland Enquirer* in 1919, he wrote what he recalled were "lurid and sensational" stories about the Communist Labor Party of California. The stories resulted in an indictment against the party, and Condon was brought in to testify against its members during their trial. He found the experience disillusioning, and he turned to science.[38]

Condon earned both his undergraduate and graduate degrees from the University of California at Berkeley. He then studied at Göttingen and Munich on a National Research Council fellowship after receiving his Ph.D. in 1926. Before leaving for Göttingen, Condon accepted an assistant professorship at Berkeley, but he changed his mind upon his return to the United States in 1927. Feeling a need for exposure to the eastern intellectual establishment, he instead lectured in physics at Columbia University before becoming assistant professor of physics at Princeton University in 1928. He departed from Princeton in 1929 to become professor of physics at the University of Minnesota, but he returned in 1930 and remained at Princeton for the next seven years as an associate professor. In 1937, he left Princeton permanently and became associate director of the research laboratories of the Westinghouse Electric and Manufacturing Company. There he established a research program in nuclear physics, solid-state physics, and mass spectroscopy. After June 1940, Condon took charge of Westinghouse's microwave radar project, which soon led him to wartime physics research.

Condon, unlike Shapley, does not appear to have been particularly active during the heady political atmosphere of the 1930s. He served as a local organizer in the 1939 Lincoln's Birthday Committee for Democracy and Intellectual Freedom, the forerunner of the American Committee for Democracy and Intellectual Freedom, but there is no evidence indicating Condon played a central role in the organization.[39] He may have attended meetings and signed petitions during the 1930s, but he was not a leading figure as were Franz Boas, Harlow Shapley, and others.

In the fall of 1940, Condon began to labor full-time on military projects. As a consultant to the National Defense Research Committee he helped to organize the Radiation Laboratory at MIT, the wartime venture responsible for the development of radar. Then he turned to work on the atomic bomb. In 1943 Condon spent six weeks as associate director of Los Alamos under J. Robert Oppenheimer. There he quickly earned the animosity of General Leslie Groves after several arguments about security regulations. His resignation letter listed his opposition to the policy of compartmentalization as one of his main reasons for leaving. Like many of the Manhattan Project scientists, Condon considered compartmentalization, the idea that scientific information on the atomic bomb should only be shared with those considered to have a demonstrable need for it, an impediment to research. He also found other conditions, such as the censorship of mail and telephone calls and the possibility of complete militarization of Los Alamos, "extreme" and "morbidly depressing."[40]

After leaving Los Alamos, Condon worked from August 1943 to February 1945 as a part-time consultant on uranium isotope separation at the University of California at Berkeley. In October 1945 President Truman nominated him for the directorship of the National Bureau of Standards on the advice of Henry Wallace. Condon was subsequently

Edward U. Condon (*right*) lectures Senators Brien McMahon (*left*) and Bourke B. Hickenlooper (*center*) about uranium, November 8, 1945. (National Archives; reproduced by permission of Corbis-Bettmann)

confirmed by the Senate without dissent and took the helm at the Bureau of Standards in November 1945.

Politically, Condon described himself as a liberal. An independent thinker, he subscribed to no set ideology beyond a faith in the ability of human beings to act generously and rationally and an impatience for opinions he found small minded and self-serving. He was confident in his own opinions and unafraid of controversy. Unimpressed by authority, he felt no reluctance in confronting or criticizing individuals in powerful positions if he felt them misguided or wrong.

While he held a generally liberal outlook, Condon, like most politically active scientists, took up causes more narrowly focused around science. Condon was particularly vocal on the subjects of atomic energy, secrecy, and international cooperation in science, and he was a prominent figure in the postwar atomic scientists' movement. At the heart of his political interests stood a firm commitment to internationalism. As discussed in chapter 1, Condon believed that secrecy would not only stifle scientific research, but it would also create a suspicious mind-set damaging to international relations. In Condon's view, open cooperation in science between nations provided the only sensible alternative, and he promoted such cooperation through his public writings and his membership in the American-Soviet Science Society. Cooperation, he insisted, would create good will as well as spread the benefits of science to the entire world. In 1946 Condon predicted that if an internationalist course was followed, the result would be "world friendship and cooperation and not atomic war and the destruction of civilization."[41] In his later writings, Condon continued to promote an internationalist outlook, and he lamented the rise of anticommunist hysteria and its attendant obsession with atomic secrets.[42]

Condon's views were not very radical, but his internationalism, opposition to secrecy in science, and positions on atomic energy, combined with his prominence in government and a partisan political atmosphere, were sufficient to earn him long-term and intense scrutiny from J. Parnell Thomas and the House Committee on Un-American Activities. Thomas was Condon's political opposite. As we have seen, in general, he opposed anything that smacked of liberal politics. According to Lewis H. Carlson, "Besides the Communists, [Thomas's] particular aversions were organized labor, New Deal 'bureaucrats,' and 'fuzzy-minded' liberals, all of whom, he considered, in one way or another, to be a threat to his America."[43] Anti–New Deal, antilabor, and anticommunist attitudes characterized his congressional career from the very beginning. Campaigning in 1936 for the House seat in New Jersey's seventh district, he attacked the New Deal as the product of "socialist dreamers," and as a freshman congressman in 1937, he accused the CIO of being dominated and controlled by communists.[44] In his first term, Thomas joined the newly formed Dies Committee, which later became HUAC. Thereafter, most of his twelve-year congressional career centered on containing the supposed communist threat.

Thomas's views on atomic energy were typical of the postwar belief in the "atomic secret." He favored military control of atomic energy and saw exchanges of scientific information with other nations as the surrender of American technology and military advantage. In an article about patents, he argued that the legal acquisition of descriptions of U.S. patents by the Soviet Union constituted "legal espionage." While scientists contended there were no real secrets in science, J. Parnell Thomas warned that knowledge of patents might help the Soviets discover "the great secret of the atomic bomb."[45] Thomas's fears reflected a mind-set that the atomic scientists opposed as a total misunderstanding of how science worked.

Although Thomas opposed all liberals, he was particularly suspicious of scientists. He wrote in *Liberty*, "Our scientists, it seems, are well schooled in their specialties but not in the history of Communist tactics and designs. They have a weakness for attending meetings, signing petitions, sponsoring committees, and joining organizations labeled 'liberal' or 'progressive' but which are actually often Communist fronts."[46] He cited the efforts of individual scientists and scientists' organizations to oppose secrecy and military control of atomic energy as evidence of the naive and unreliable character of scientists, and he insisted that many of the atomic scientists had fallen prey to communist influence. Swayed by his faith in the military, Thomas concluded that the only way to safeguard atomic energy was to insist on strict military security.

Thomas's views shaped the HUAC subcommittee's March 1 report on Condon. The report listed several reasons for HUAC's suspicions of Condon but provided little support for its claim that he was a security risk. Most of the committee's criticisms concerned either Condon's political views or his associations with persons HUAC deemed of dubious loyalty. The report attempted to portray Condon as untrustworthy and suspect because of the pending status of his security clearance with the Atomic Energy Commission, his association with "an individual alleged, by a self-confessed Soviet espionage agent, to have engaged in espionage activities," his general associations with foreigners, and his connection to the American-Soviet Science Society, allegedly a communist front. The subcommittee concluded the report with a recommendation that Condon either be removed from his position or that the secretary of commerce prepare a statement defending the continued presence of Condon in the Bureau of Standards despite the evidence against him.

The shortcomings of the report have been discussed in great detail by Robert K. Carr, so I will not consider all of the committee's accusations point by point here.[47] The HUAC subcommittee wrote the report in such a way that its allegations sounded grave, but in actuality they carried very little weight. For example, the charge that as of November 1947 Condon only had "pending" security clearance status and had not received permission from the AEC to work on atomic projects reflected no more than a routine matter of bureaucratic backlog. The Atomic Energy Act of 1946 allowed employees cleared under Manhattan Project regulations to retain their security clearance while the newly created AEC carried out its own investigations. "Pending status" simply meant that the AEC had not yet completed a new investigation of Condon, not that he had been denied security clearance.[48] As for Condon's "associations," one commentator noted that HUAC's allegations were "like saying that Condon is alleged to have associated with a man who is alleged to beat his wife."[49] "Associations," as discussed earlier, served as a tactic to cast Condon in a subversive light without having to accuse him of any actual impropriety or involvement in intrigue.

Possibly the most serious charge in the report concerned Condon's association with alleged communist-front organizations. After noting there was no evidence that Condon belonged to the Communist Party, the subcommittee went on to denounce his membership on the executive committee of the American-Soviet Science Society. The House Committee on Un-American Activities contended that the American-Soviet Science Society was affiliated with the National Council of American-Soviet Friendship, Inc., which had been placed on the attorney general's November 1947 list of eighty-two organizations designated totalitarian, fascist, communist, or subversive. This claim was potentially problematic for Condon, not because any evidence indicated that the National Council of American-Soviet Friendship was actually dangerous or subversive but because under President Truman's executive order that defined the government's loyalty program, membership in any organization on the attorney general's list was sufficient to justify dismissal from a government position.[50] When later confronted with the claim that the American-Soviet Science Society had become independent in 1946 and therefore had no ties to the National Council of American-Soviet Friendship, HUAC contended that any claimed separation constituted mere window dressing, and the fact that the two groups shared the same address and telephone number in the Manhattan telephone directory clearly demonstrated their continued affiliation.[51]

The available evidence indicates that the separation was actual and unequivocal. The May 28, 1946, minutes of the Executive Committee of the American-Soviet Science Society stated, "It was formally agreed that since the Society has been in practise [*sic*] an independent organization, the society express its gratitude to the Council for having furnished most of the aid for the organization to grow and henceforth the Society be considered disaffiliated and independent." The confusion over the address and telephone number seems due primarily to a clerical error. After the separation, the American-Soviet Science Society moved from 114 East 32d Street to 58 Park Avenue, but since it was in financial disarray, it did not order a new phone number and neglected to notify the phone company to change the old listing.[52]

The HUAC subcommittee, however, preferred sounding the alert to the menace of internationalism to formulating a carefully crafted, technical analysis of Condon's membership in the American-Soviet Science Society. The subcommittee claimed that Condon's participation in the American-Soviet Science Society indicated "the dangerous extremes to which Dr. Condon has gone in an effort to cooperate with Communist forces in the United States."[53] But in specifying what those extremes were, the subcommittee quoted Condon's remarks in an address he gave on March 5, 1946 (incidentally the same day on which Winston Churchill gave his Iron Curtain speech), in which he emphasized the need for scientific cooperation between the United States and the Soviet Union.[54] To most scientists, who believed scientific knowledge transcended national borders and viewed international cooperation as an important way of promoting research, Condon's words were only reasonable. But to J. Parnell Thomas, with his low regard for scientists, Condon's speech must have seemed naive at best and subversive at worst. For Thomas, the American-Soviet Science Society, with its purpose of promoting contact and cooperation between Soviet and American scientists, posed a threat regardless of its relationship to the National Council for American-Soviet Friendship. Given Thomas's faith in secrecy and the subcommittee's focus on Condon's membership in the American-Soviet Science Society and his remarks in favor of scientific cooperation, it is reasonable to conclude that Condon's internationalist beliefs provided the real focus of HUAC's attack. To those who thought that "atomic secrets" had to be preserved at all costs, U.S.-Soviet reciprocity in science could only mean subversion.

Condon responded to HUAC's allegations immediately. He said, "I have nothing to report. If it is true that I am one of the weakest links in atomic security that is very gratifying and the country can feel absolutely safe for I am completely reliable, loyal, conscientious and devoted to the interests of my country, as my whole career and life clearly reveal."[55] A few days later he addressed the more general issue of the effects of loyalty investigations on science. In an open letter to Senator Bourke B. Hickenlooper (R.-Iowa), chairman of the Joint Committee on Atomic Energy, he warned that there had been "a mounting tension of threat of purges, spy-ring exposures, publicity attacks and sudden dismissals without hearings," all of which made scientists "increasingly reluctant to work for the Government."[56]

The Commerce Department expressed its confidence in the Bureau of Standards director when it announced on March 3 that its loyalty board had already cleared Condon on February 24.[57] Over the next several months Condon received public support from the American Civil Liberties Union, Henry Wallace, and Representatives Helen Gahagan Douglas (D.-Calif.) and Chet Holifield. Scientists and scientists' groups rallied in defense of Condon. Albert Einstein and Harold C. Urey, speaking for the Emergency Committee of Atomic Scientists, pronounced HUAC's accusations "a disservice to the interests of the United States."[58] Robert E. Marshak, chairman of the Federation of American Scientists, charged the committee with laying a "deliberate smear" against Condon "contrary to the American instinct for fair play and to the democratic ideal expressed in the Bill of Rights."[59] The Federation of American Scientists also sent out a letter to members urging them to write to Senator Hickenlooper, Representative Thomas, and Secretary Harriman.[60] President Truman received scores of letters and telegrams expressing confidence in the scientist. Among them were letters from Hans T. Clarke, president of the American Society of Biological Chemists, the entire Physics Department faculty of Harvard University, the Physical Society of Pittsburgh, the Association of Pittsburgh Scientists, and a resolution from the Association of New York Scientists with the signatures of 250 members.[61] The American Physical Society issued a statement in support of Condon on March 5. The American Association for the Advancement of Science followed suit in late April, and the National Academy of Sciences in mid-May. Both the AAAS and NAS criticized HUAC's failure to provide a hearing for Condon and warned of the dire consequences the case might have for science-government relations. The AAAS statement declared, "The continuation of American scientific achievement for the purposes of both peace and war depends upon the freedom and peace of mind of our scientists. They have no right to ask for special privileges, but they should have the rights accorded every citizen and should be protected against treatment of the sort accorded to Dr. Condon." The National Academy of Sciences, for its part, cautioned that HUAC's treatment of Condon "may have the effect of deterring scientists from entering government employ and may diminish the respect with which citizens regard opportunities for service to their government."[62] On April 12, the Emergency Committee of Atomic Scientists held a dinner to demonstrate its faith in Condon. The dinner was sponsored by some 150 scientists, including nine Nobel Prize winners and 70 members of the National Academy of Sciences. The list of sponsors included Hans Bethe, Karl T. Compton, Albert Einstein, Linus Pauling, Leo Szilard, Edward Teller, and Harold C. Urey, among many others.

The Condon case provided a major rallying point for scientists' attempts to curb government-sponsored restrictions on scientists' civil liberties. But HUAC's attack on Condon had ramifications beyond science-government relations. The committee did not target Condon merely out of a desire to persecute a well-known physicist for his political beliefs. Many scientists held similar views; Condon's internationalism did not provide sufficient incentive for HUAC to single him out. Larger political concerns also motivated the release of HUAC's March 1 report. Committee appropriations, opposition to President Truman, and election-year politics provided the primary incentives for fingering Condon as a security risk. Embedded in all of these immediate issues, however, were differing conceptions of the meaning of national security in the nuclear age.

The most clearly visible reason for HUAC's attack was that it formed part of the committee's strategy to support its fiscal year 1949 appropriations request. The committee had asked for a record $200,000 appropriation, double the previous year's budget. Representative Arthur G. Klein (D.-N.Y.) noted on March 8, "A sure sign of preparations for making a new request for funds is the release of sensational charges against some public figure."[63] The next day, debate over HUAC's high bid for appropriations drew bitter comments about the committee and statements in defense of Condon from several members of Congress. Among them Representative Leo Isacson (American Labor-N.Y.) declared, "It is no coincidence that this attack on Dr. Condon—an attack abhorred and shamed in all responsible opinion of press and science comes at this moment when the House Committee on Un-American Activities seeks the most swollen appropriation it has ever ventured to ask of a Congress."[64]

The rising Cold War tension between concern for security and protection of democratic freedoms shaped the congressional debate over the Condon case and its relationship to HUAC's budget request. Representative John A. Blatnik (D.-Minn.) accused HUAC of attempting to stifle political dissent by denouncing scientists in order to prevent them from speaking to the American public about the perils of the nuclear age. "Scientists," he said, "who know better than anyone else the destructive forces which might be let loose by another war, sometimes feel it their duty to speak their minds. The Committee on Un-American Activities has taken it upon itself to guard the American people from such uncensored statements from men whose opinions might command respect."[65] On the other side, Representative Rankin stood by his committee's handling of the Condon case and announced, "There are enemies in this country today plotting to get their hands on the atomic bomb," enemies who hated HUAC almost as much as some of his fellow members of Congress. He concluded, "This committee is rendering one of the greatest services ever rendered by any committee of the Congress of the United States. We are not out to persecute anybody, but we are trying to protect this country."[66] The conceptual gulf between Blatnik and Rankin reflected the growing conflict between the atomic scientists, who insisted on internationalism as an utter necessity to avoid nuclear destruction, and antiradical nativists, who argued that national security required vigilant guarding of atomic secrets that could otherwise be quickly lost to espionage.

In the end, the vote was not even close. The House voted 337 to 37 to grant the $200,000 budget to HUAC. Only the two American Labor Party members, thirty-four Democrats, and Republican representative Jacob K. Javits of New York cast their votes against the appropriation.

In addition to assisting HUAC's quest for funds, the Condon attack represented a way of discrediting Truman in a presidential election year. For HUAC, it was enough that Condon was a Truman appointee. But when the president decided to withhold the May 15, 1947, Hoover-Harriman letter from Congress, he played right into the committee's hands.

The March 1 report on Condon included an excerpt from a confidential 1947 letter from FBI director J. Edgar Hoover to Secretary of Commerce W. Averell Harriman. The letter concerned Condon's associations. The most serious allegation appeared in the opening paragraph: "The files of the Bureau reflect that Dr. Edward U. Condon has been in contact as late as 1947 with an individual alleged, by a self-confessed Soviet espionage agent, to have engaged in espionage activities with the Russians in Washington, D.C. from 1941 to 1944."[67] The HUAC subcommittee omitted the following crucial sentence in the letter: "There is no evidence to show that contacts between this individual and Dr. Condon were related to this individual's espionage activities." Exactly how HUAC obtained the material from the Hoover-Harriman letter in the first place is unclear. The committee claimed that in December 1947 it sent an investigator to the Department of Commerce who was able to copy part of the letter before being asked to stop.[68] It is just as likely that the excerpt came directly from the FBI, a source the committee could not reveal without jeopardizing its relationship with the bureau. In either case, HUAC had to find another way to get a copy of the letter in its entirety. Immediately after issuing the March 1 report, HUAC attempted to obtain a full copy of the Hoover-Harriman letter by gaining access to the records of the Commerce Department board that had cleared Condon. In response, Truman issued an executive order on March 14 forbidding the release of departmental loyalty files to Congress on the grounds that permitting access to confidential files would violate the civil rights of federal employees and compromise the FBI's investigatory capabilities. Ten days later, Secretary Harriman refused to hand over the Department of Commerce records on the grounds of executive privilege. On April 22, the House passed House Resolution 522 by a vote of 300–29 and ordered Truman to release the loyalty files on Condon. Truman refused. On April 27, Truman ordered all copies of the Hoover-Harriman letter transferred to him from the Commerce Department and the Justice Department.[69]

The resolution never came to a vote in the Senate, so the president had no legal obligation to release the letter. But H.R. 522 quickly created a political uproar. Truman's actions became both a major source of criticism for the president and a defense for HUAC. The committee attempted to shift the blame to the Truman administration for its failure to provide Condon a hearing. Committee members contended that if Truman would only hand over HUAC's main source of evidence against Condon, the committee would be only too happy to meet the embattled Bureau of Standards director in an open forum.[70]

Other statements threatened more serious political fallout. In early March, Representative Rankin made a dramatic appeal calling for the impeachment of "top-flight bureaucrats who are protecting people on the Federal payroll whose loyalty is questioned."[71] On May 1 Republican national chairman Carroll Reece, sensing an election-year opportunity, echoed Rankin's sentiment and listed impeachment as one of several steps that Congress could take to force the issue over the Hoover-Harriman letter.[72] Condon perceived the damage being done by the situation and sent a note to Undersecretary of Commerce William C. Foster on May 5 asking that Truman make the letter public. He wrote, "The situation resulting from the President's action is somewhat embarrassing to me, because Congressman Thomas is using this to create the impression I am trying to conceal something." He then expressed confidence that, although he had not seen the letter, it contained no damaging information about him.[73]

Truman did not release the letter, but on June 1 he drafted a speech to Congress detailing his reasons for refusing to obey H.R. 522. The draft was consistent with previous statements that release of such confidential letters infringed upon the rights of government employees and inter-

J. Parnell Thomas, here brandishing a copy of H.R. 522, April 22, 1948. (New York World Telegram and Sun Collection, Library of Congress; reproduced by permission of Corbis-Bettmann)

fered with the FBI's investigative duties. By this time, however, the controversy surrounding the letter had abated, and Truman never delivered the speech.[74] No long-term political damage to the Truman administration resulted from the incident, but it did serve to lend an air of legitimacy to HUAC's position.

More than a few scientists and legislators suggested that HUAC's denunciation of Condon constituted the first step in the renewal of attacks on the McMahon Act and civilian control of atomic energy. As discussed in chapter 2, in 1946, Thomas had been a staunch opponent of the McMahon bill. He had condemned the bill as "the creature of impractical idealists," and he had fingered Condon as an example of the unreliable character of supporters of civilian control.[75] To the FAS and Representatives Chet Holifield and Helen Gahagan Douglas, the correspondence between Thomas's opposition to civilian control and Condon's role in the formulation of the McMahon bill seemed to indicate that civilian control of atomic energy was also about to come under fire.[76] Condon himself categorized HUAC's accusations as part of "an undercover attempt to smear civilian control of atomic energy by smearing the scientists who assisted the development of the Atomic Energy Act of 1946."[77]

For reasons that will be discussed in chapter 5, the imagined battle over civilian control of atomic energy never materialized. Richard L. Meier, the FAS's executive secretary, admitted that the belief among reporters, most scientists, and Condon himself that the Condon case foreshadowed renewed opposition to civilian control was mistaken. In a March 10 memorandum he informed FAS members, "Since the first blast [at Condon], however, there has been no follow-up and the diligence of a dozen or more top grade reporters has not uncovered any real connection between the two issues as yet. By now it seems probable that no coordinated program exists in that quarter."[78] Although Condon's positions on atomic energy provided Thomas with the initial impetus to target him publicly, Thomas does not appear to have planned a larger assault on the AEC. But while the McMahon Act emerged unscathed, some scientists believed the attack on Condon suggested the existence of a larger plot to persecute scientists for their political views. The Washington Association of Scientists noted that the growing importance of scientific research to military advancements had been accompanied by "a tendency for distrust of scientists to arise," and it worried that scientists, because of "their deep belief in freedom of intellectual activity and free exchange of information," were "regarded with suspicion by some who do not understand how science operates."[79] By this time, the Condon case was only the most visible of an increasing number of instances in which scientists, especially those engaged in atomic energy research, had faced difficulty in pursuing their work because of their political views. Pleas for scientific freedom were becoming confronted more and more often with the doctrine of national security and the countervailing claim that atomic secrets had to be protected. Scientists had accepted security restrictions as a short-term necessity during World War II. But as the Cold War became firmly embedded in American political life, scientists began to find the security requirements of the war virtually impossible to reverse in peacetime.

Meanwhile, the Atomic Energy Commission had conducted its own review of Condon's security clearance. On March 29, after consultations with the Commerce Department about its evaluation of Condon, the AEC directed its Personnel Security Review Board to make a recommendation. On June 7, the board advised the AEC to defer the matter "pending reasonably prompt action by the House Committee on Un-American Activities with respect to its investigation and hearing as to Edward U. Condon."[80] Lilienthal and the other AEC commissioners rejected the board's proposal out of fairness to Condon, concern that a "prolonged delay will tend to discredit our whole personnel security review system," and a belief, reinforced by Truman's March 14 executive order, that the AEC, as an executive branch agency, had to take primary responsibility for making a decision.[81] The AEC quite properly refused to give any jurisdictional ground to HUAC. On July 15 the AEC cleared Condon for access to restricted information required for his duties as director of the

National Bureau of Standards. In a three-page memorandum concerning the decision, the AEC stated, "After examining the extensive files in this case, the commission has no question whatever concerning Dr. Condon's loyalty to the United States."[82] The *Bulletin of the Atomic Scientists* applauded the AEC's decision and sharply criticized the House Committee on Un-American Activities. The *Bulletin* declared, "This House Committee has the practice of never stating clearly that a man they have 'investigated' has proved his innocence; but drops the investigation if it becomes 'unprofitable,' leaving lingering suspicions."[83]

With the AEC report, the Condon case was temporarily laid to rest. But those "lingering suspicions" remained to be rekindled in later years. Throughout the early 1950s, Condon had to defend himself repeatedly from the accusations of HUAC and submit to cumbersome reevaluations by the federal government's loyalty and security apparatus. Representative Richard B. Vail revived HUAC's charges in April 1951. Frustrated by the prospect of another ordeal and attracted by a high-paying position at Corning Glass Works, Condon left the Bureau of Standards in August. Even so, he could not escape HUAC's attention. The following year, not long before the midterm congressional election, the committee finally subpoenaed him for a hearing. Although Vail promised that shocking material would be revealed, Condon answered all questions candidly and vehemently denied ever having violated security regulations. No evidence of wrongdoing on Condon's part emerged. Nevertheless, his troubles continued. In 1954, Condon successfully applied for security clearance for a classified project Corning was conducting for the navy. But in October, when the news that he had been cleared became public, the secretary of the navy suddenly revoked his clearance. Vice President Richard M. Nixon, then campaigning in Montana, publicly took credit for the withdrawal of Condon's security clearance. Tired of having to fight a seemingly endless battle, Condon left Corning for university life. He had some difficulty finding a permanent position, but in 1956 he was hired by Washington University. Several years later, he went to the University of Colorado, where he completed his scientific career.[84]

Unlike the encounters of Eugene Rabinowitch, John and Hildred Blewett, and Robert H. Vought with the army and AEC security systems, well-known scientists such as Shapley and Condon had certain advantages in their confrontations with HUAC. The source of their troubles was clear. Shapley and Condon did not have to improvise their way through a vague and confusing bureaucracy. Instead, they could respond directly and publicly to their critics. Both were self-confident and individualistic and had few fears about condemning HUAC's attitudes and methods. Nor did they have to face the problem of isolation. As elder statesmen of science, they could count on backing from the scientific community. Condon in particular enjoyed a great deal of support from other scientists and scientists' professional organizations. In a poll on the Condon case taken by the Atomic Scientists of Chicago, one scientist pointed out the value of such support: "At least Condon had well known men to stand up for him. What if the same charges were placed against some obscure scientist? His whole career would be jeopardized."[85] Unlike many of the scientists who needed to obtain security clearances, Shapley and Condon were well-established scientists who could afford to confront domestic anticommunism with little risk to their personal security.

Professional prominence did, however, leave Shapley and Condon open to repeated denunciation in public forums. Because they were outspoken scientists in conspicuous political positions, attacks on them served political ends. Anticommunists used Shapley's name as a symbol of the communist threat throughout the late 1940s and early 1950s. Conservative anticommunists also tried to use the Condon case as an election-year tool in 1948, 1952, and 1954. Condon finally had to leave government service, and then an industry position that required a security clearance, in order to break free from HUAC's harassment.

Nor did fame guarantee immunity from the psychic toll of loyalty attacks. Shapley and Condon both remained resolute and fearless in their willingness to confront their accusers, but their outward, public resistance was not necessarily free of inward costs. Regarding Condon, physicist Frederick Seitz later observed, "The ordeal took a heavy toll on him, both physically and emotionally. He was never again quite the same cheerful, carefree man I had known in earlier years."[86] A later, much better known security clearance case left J. Robert Oppenheimer angered and bewildered. After the loss of his security clearance in 1954, Oppenheimer still held a comfortable position as director of the Institute for Advanced Study to provide him with financial support and intellectual challenge. But his ordeal with the AEC aged him prematurely and burdened him with a private pain that he carried for the rest of his life.[87] Professionally established scientists did not have to worry about material consequences, such as long-term unemployment, but they still had to deal with the frustration, pain, and anger that were an inherent part of having one's loyalty questioned.

The controversy over the Condon case, as well as reports of individuals being denied security clearances by the AEC, produced concern and fear throughout the scientific profession. In the late 1940s, the major scientists' organizations attempted to find a way to halt or mitigate the abuses of the security system. To a certain extent, they succeeded. But they did so only by adopting constrained political strategies and pursuing limited goals. The early success of the FAS in the debate over civilian control of atomic energy had depended on a highly visible, public approach. As the federation deliberated over how to respond to security clearance cases within the AEC, however, its political style would undergo a marked transformation.

CHAPTER 5

Negotiating Security

The FAS, the AEC, and the JCAE, 1947–1948

• • •

By mid-1947, the Federation of American Scientists viewed security clearance as a pressing problem. Reports of clearance denials at AEC installations, combined with signs that J. Parnell Thomas intended to launch investigations against both E. U. Condon and scientific personnel at Oak Ridge, portended difficult times for scientists with any record of liberal-left political activity. Although the FAS's leaders had implemented direct measures to diminish the influence of the progressive left within the federation and were themselves uneasy about political radicals, they also opposed actions by the state that threatened to confine dissent to overly narrow boundaries. Members of the FAS were appalled that advocacy of civilian and international control of atomic energy, support of labor protest and civil rights, exploration of left-wing political philosophies, and other noncentrist political views increasingly led congressional committees and government agencies to impugn the loyalty of individual scientists and refuse them security clearances.

Personnel security became an issue that defined the relationships between scientists, the Atomic Energy Commission, and the Joint Committee on Atomic Energy. It also helped mold the character of scientists' interactions with the Cold War state. Throughout 1947 and 1948, the FAS engaged in quiet diplomacy with the AEC to intervene on behalf of particular individuals and to formulate general security clearance policy.[1] The federation also made tentative attempts at garnering publicity to sway public sentiment in the scientists' favor. The AEC, with liberal New Dealer David E. Lilienthal at the helm, provided a receptive forum for the atomic scientists' concerns. Over time, the FAS grew to rely on quiet diplomacy over publicity, out of both the success of its dealings with the commission in implementing procedural reforms and concern not to endanger the public image of the AEC.

Faced with HUAC's antipathy toward civilian control of atomic energy, the atomic scientists, the AEC, and the Joint Committee on Atomic Energy (JCAE) initially had common cause in the protection of the commission. But as time progressed, the relationship between the AEC and the JCAE became increasingly strained. Although the joint committee opposed HUAC's attempts to assert jurisdiction over atomic energy, it also differed sharply with the AEC's approach to security clearance. Bourke B. Hickenlooper, JCAE chairman, gradually became more and more frustrated with Lilienthal's civil libertarian stance on clearance and what he perceived as the undue influence of scientists over the AEC. As a result, relations between Hickenlooper and Lilienthal deteriorated throughout 1948. By the end of the year, scientists had developed a cooperative relationship with the AEC over personnel security, but it came at the expense of growing tension betweeen the AEC and the joint committee.

Oak Ridge and AEC Security, Part I

When the AEC officially took over the operation of the Manhattan Engineering District on January 1, 1947, the new agency faced enormous organizational and administrative tasks. The AEC inherited from the army a massive enterprise that employed some two thousand military personnel, four thousand civilian government employees, and thirty-eight thousand contractors scattered in production facilities, laboratories, and offices throughout thirteen states. In addition to managing existing sites, programs, and personnel, the commission also had to oversee the expansion of old installations and the creation of new ones. Scientists were especially interested in research facilities, and they anxiously awaited the establishment of Brookhaven and the reorganization of Clinton Laboratories and Argonne as part of the AEC's system of national laboratories.[2]

Of all the AEC's initial concerns, however, few absorbed as much of the commission's attention as personnel security. According to Walter Gellhorn, in the AEC's first two years, the commissioners devoted about a third of their meeting time to personnel security and related matters.[3] The Atomic Energy Act of 1946 required the FBI to reinvestigate all former Manhattan Project employees transferred to the AEC, as well as investigate applicants for new positions. Inquiries by the FBI generally took four to six weeks. After the FBI completed a field investigation, the AEC had to evaluate the results, which took even longer. Limited resources meant that former Manhattan Project personnel (among them, E. U. Condon) could not be reinvestigated before undertaking duties for the AEC without causing massive delays in the start-up of the commission's operations. To prevent such holdups, the AEC usually granted emergency clearances to former Manhattan Project employees, thus reserving the resources of the security division for expediting the cases of new applicants.[4] Reinvestigations could then wait until the initial disarray involved in launching the new agency receded, operations were routinized, and security clearance policy was codified.

The threat of action by HUAC, however, meant that the AEC could not defer consideration of security clearance policy for long. As discussed in chapter 2, during the 1946 debate over the McMahon bill, HUAC had sent investigators to Oak Ridge and claimed to find evidence of subversion there. A year later, Thomas resumed his attacks on scientists and civilian control of atomic energy in the print media. The House Committee on Un-American Activities appeared to be gearing up for a major new initiative on atomic energy and national security.

In June 1947, as Thomas sent up his first trial balloons against E. U. Condon, he also published an article, "Reds in Our Atom-Bomb Plants," in *Liberty* magazine. Thomas opened his piece with a grim assessment of AEC security. "The Atomic Energy Commission," he declared, "must come to grips shortly with pro-Soviet infiltration of its own organization." He then contended that facilities at Oak Ridge, most notably Clinton Laboratories, were "heavily infested" with communists. Scientists, wrote Thomas, were especially to blame for the lax security situation and the inability of security officers to dismiss suspicious employees expeditiously. He reported, "The commanding officer [at Oak Ridge] assured me that the matter was very delicate, for if certain of the suspected physicists were discharged scores of other scientists had threatened to walk out."[5] Scientists' objections to unfair security measures, Thomas indicated, constituted a security problem.

The HUAC chairman went further and pinpointed the scientists themselves as a threat to American security. He claimed that according to security files at Oak Ridge, a large number of scientists belonged to communist-front groups, so defined by HUAC. How Thomas gained access to the files was unclear. Lilienthal suspected General Groves, but the FBI constituted a more likely source. In March, Thomas had given the names of five Oak Ridge scientists to Louis B. Nichols, head of FBI public relations and third in command at the bureau. Nichols reported to Clyde Tolson, Hoover's right-hand man, that he felt inclined to provide Thomas with information from the scientists' FBI files on "a personal and confidential basis."[6]

Whatever the source of his information, given his general enmity toward scientists, Thomas reached predictable conclusions. Scientists, he insisted, were "gullible" and "politically naive." Thomas equated the advocacy of information exchange by the Association of Oak Ridge Engineers and Scientists with espionage by Alan Nunn May and warned ominously, "Each of hundreds of scientists has knowledge that could shorten an enemy's time in perfecting the bomb. If five key men could be subverted, years could be saved."[7] As he had done during the debate over the McMahon bill, Thomas continued to insist on the reality of the "atomic secret," the unreliability of scientists, and the ease with which the "atomic secret" could be lost.

Thomas wrote at length of the danger posed by the atomic scientists' movement. The Federation of American Scientists, the Emergency Committee of the Atomic Scientists, and the National Committee on Atomic Information, Thomas contended, while not actual communist fronts, were highly susceptible to communist influence. Thomas insinuated that Albert Einstein, chairman of the Emergency Committee of the Atomic Scientists, was a potential subversive who had "lent his name to a score of organizations classified as fronts by the Committee on Un-American Activities."[8] The NCAI was suspect because of the supposed communist connections of present and past members and employees. The Association of Oak Ridge Engineers and Scientists, the Association of Oak Ridge Scientists, and the Association of Los Alamos Scientists indicated the dangerously radical nature of the scientists' movement through their opposition to military control of atomic energy. Thomas portrayed the "science lobby" as a conspiracy that had not only opposed military control during the McMahon bill debate but had also supported Lilienthal's appointment as AEC chairman in order to place a person inexperienced with security in control of the AEC. To Thomas, the solution to protecting the atomic bomb from espionage appeared obvious. The only way to ensure the safety of the "atomic secret" was to place the military in charge of atomic energy. He concluded, "I believe that in the present chaotic world situation our only solution is to repeal the act and return [the] Manhattan district to the Army, which can best administer security."[9]

Both the JCAE and AEC considered Thomas's diatribe a serious threat to the McMahon Act. Hickenlooper met with AEC commissioners Lilienthal and Strauss the afternoon of June 5 to discuss immediate responses. Hickenlooper decided to dispatch JCAE personnel to Oak Ridge, and Lilienthal directed members of the AEC staff to discover how HUAC had gained access to the Oak Ridge installations and Oak Ridge security files.[10] Lilienthal had already spent a difficult morning discussing clearance cases, including that of the Blewetts, and in his diary he wearily described the day as "one of those bad days when I wonder why in the name of hell and good sense I am willing to have anything to do with so ugly and insane an enterprise, much less accept chief responsibility for it."[11]

The need to resolve clearance cases such as that of the Blewetts, combined with the threat posed by HUAC, forced the AEC to accelerate the codification of its security clearance procedures. Throughout June 1947 the commissioners discussed procedural reforms, including the need for specific criteria and review policies, and the possibility of creating a personnel security review board to make decisions about particularly controversial cases. Meanwhile, JCAE staffers David S. Teeple and William Sheehy reported back to the joint committee concerning Oak Ridge and recommended the removal of fifteen laboratory employees on security grounds.[12]

In July, a new series of security-related crises broke out. Beginning on July 9, widespread press reports announced that important documents relating to the atomic bomb were missing from Oak Ridge. Most of the lost documents had vanished into a morass of accounting errors left over from the Manhattan Project; in a few instances, army personnel had taken the records as souvenirs at the end of the war. Press coverage, however, created the impression that the AEC had allowed a major breach of security. Shortly thereafter, on July 17, the *Washington Times-Herald* reported HUAC's plan to subpoena E. U. Condon. The committee, it appeared, had taken another step toward initiating a major assault on the AEC and civilian control of atomic energy.[13]

On July 14 and 17, members of the JCAE conferred over the news stories about the supposed missing documents, physical security at AEC sites, and the status of personnel security procedures. Representatives from the AEC joined their discussions on July 22. In early August the AEC responded to the pressure to tighten up personnel security by withdrawing security clearances from several employees, including two chemists at Oak Ridge, whom the commisssion suspended.[14] Both chemists were interviewed by AEC officials within two weeks of having their clearance removed, but, as in Vought's and the Blewetts' cases, neither could discover the exact reasons for the commission's actions. It appeared that the AEC had withdrawn the two chemists' clearances because of their political attitudes. According to a description of the AEC's interviews by the Association of Oak Ridge Engineers and Scientists (AORES), one of the chemists, identified only as Mr. G, was questioned about his political beliefs regarding the Soviet Union, communism, atomic energy, and security, as well as his marriage and family life. In the interview, Admiral John E. Gingrich, director of AEC security, reportedly made dismissive remarks about scientists' "attitude toward security and the fuss being raised over the Negro question." In a comment apparently aimed at discouraging political activity, Gingrich also "personally advised Mr. G. against the unfortunate practice of scientists acting outside their own field." The second scientist, a Mr. W, was asked mainly about his associations, especially if one particular friend was a communist and if Mr. W had ever discussed Marxist philosophy with that person.[15]

A copy of the Oak Ridge association's account of the cases caused some anxiety on the part of G. O. Robinson of the AEC's Office of Information at Oak Ridge. If the news media became aware of the AORES's report and published stories based on it, Robinson feared the creation of "a public relations problem for the Commission."[16] The AORES, however, did not spread publicity about the two cases beyond its membership. Instead, scientists at Oak Ridge relied on the quiet application of pressure through internal channels rather than appeals to the public. The Oak Ridge association recommended procedural changes to the AEC, scientists from Clinton Laboratories met with AEC officials, and the AORES urged its members to write in protest to the AEC.

On August 15, a week after the second chemist lost his clearance (and about the same time that the FAS and the Mohawk Association of Scientists and Engineers appealed to the AEC on behalf of the Blewetts), John H. Bull, chairman of the AORES, wrote to the AEC to urge the adoption of fair clearance procedures to protect employees' rights. The Oak Ridge association recommended that employees denied clearance be entitled to a clear statement of charges, the right to a hearing before an impartial board, time to prepare a defense, assurance of a decision within a reasonable period of time, and the right to appeal adverse decisions directly to the AEC commissioners. Fair procedures were necessary not only to protect the rights of individuals but to ensure the success of the AEC's atomic energy program. Recent cases of "undesirable practices" threatened the morale of AEC employees and the work of the commission. Furthermore, clearance difficulties might drive scientists away from atomic energy research. Bull warned, "If the matter of clearance is handled in a way which seems arbitrary or terroristic, we greatly fear that the Project will suffer loss of personnel."[17]

On August 20, three division directors from Clinton Laboratories met with AEC officials in Washington, D.C., to discuss the clearance situation at Oak Ridge. The Clinton scientists stressed the need for suspended employees to be supplied with specific information and to be allowed a hearing. The AEC officials informed the scientists that the commission was in the process of establishing a review board to develop criteria for handling security clearances and to consider individual cases, and they promised to take the scientists' concerns into account. General manager Carroll Wilson, however, also made reference to concurrent international tensions and Americans' widespread belief in the need to preserve atomic secrets. The AORES's account of the meeting did not have to mention the primary source of these international tensions by name. With Truman's announcement back in March of what later became known as the Truman Doctrine, the Cold War entered a new stage of heightened confrontation. To anyone who still hoped that U.S.-Soviet relations might reach accommodation, the Truman Doctrine speech was a disappointing blow that signaled America's commitment to a state of cold war and an international order based on bipolar power politics. External political tensions, the AORES concluded, meant that the AEC was under heavy pressure "to employ only lily-white individuals."[18]

Alvin M. Weinberg, acting director of the physics division at Clinton Laboratories, was unable to attend the meeting in Washington, but from what he heard of it, he felt "neither the scientists nor the commission attempted to avoid the delicate and difficult questions connected with AEC personnel security policies." He thought the threat to morale at Oak Ridge needed to be considered more thoroughly, however. He wrote to Fletcher Waller, head of the AEC personnel office, that confidence at Oak Ridge had "deteriorated so badly" that, barring the creation of conditions under which scientists could work without fear of suspension, "I seriously doubt that the Physics Division, and indeed the AEC laboratories generally, can avoid disintegration." Weinberg already knew of three section chiefs in the physics division who were prepared to accept academic positions unless "an immediate improvement in the personnel security situation" occurred, and he feared that the "widespread defection among the senior staff" might spread to those in junior positions, or at least adversely affect the quality of their work.[19]

Theoretical physicist Eugene Wigner, who had just resigned as director of Clinton Laboratories several weeks earlier to return to Princeton University, wrote to James B. Fisk, director of research at the AEC, with similar concerns. Wigner felt that morale was high and found the AEC's attitude sympathetic, but he also observed that the security clearance revocations at Oak Ridge had created an atmosphere of fear and uncertainty among scientists there, even those who were longtime employees of the atomic energy project. Wigner wrote, "It is almost fantastic to hear people, who have been with the project since its inception and whose work is part of its foundation, get to wondering lest they be considered untrustworthy in the future." Wigner acknowledged the AEC's tenuous political status, but he urged the commission to make a strong statement to assure scientists that political affiliations were irrelevant to clearance decisions. He concluded, "Everyone knows that the Commission is in a difficult position and that its actions on these questions are closely scrutinized. However, to yield easily to pressure from outside may jeopardize the Commission's standing inside."[20]

The Oak Ridge scientists' discussions with AEC officials paid off, and the two chemists resumed their duties. Alvin Weinberg was heartened by their reinstatement and the AEC's efforts to codify its clearance procedures. He assured Fletcher Waller, "This turn of events has already had a marked effect in improving the morale of the scientists in the Physics Division of Clinton Laboratories and in strengthening the confidence which they hold in the Atomic Energy Commission."[21]

The Oak Ridge cases demonstrated that the application of pressure through internal channels could assist individuals and possibly influence policy as well. The two Oak Ridge chemists benefited from the nearby presence of an active FAS chapter and the efforts of top-level scientists at Clinton Laboratories. Oak Ridge scientists, working individually and through the AORES, successfully negotiated and cooperated with the AEC to reach a result agreeable to all parties. In doing so, they established a precedent for the FAS. Quiet diplomacy appeared capable of rapidly producing positive results, and it looked like a viable strategy for future action.

Evaluating the Situation: The FAS's Observations, 1947–1948

As FAS members contemplated their response to security policy and related problems, they took great care to gauge the position of the AEC. In the debate over the McMahon bill, the atomic scientists had invested both emotional attachments and political capital in the commission. Given the difficulty of establishing a civilian atomic energy agency in the first place, they feared that if they criticized AEC security procedures too strongly, they might rattle further the new agency's already shaky political ground. Throughout 1947 and 1948, the atomic scientists worried about rumors of renewed congressional efforts to discredit the AEC and place atomic energy under direct military authority. These concerns pulled the FAS toward a middle ground between outright compliance with existing procedures and strong overt protest against the injustices of the security system. Members of the FAS were profoundly disturbed by cases in which scientists were denied security clearance on political grounds, but they did not wish to risk actions that they believed might jeopardize the AEC's existence.

By March 1947, the FAS council had already pinpointed overly restrictive secrecy and security regulations as urgent problems.[22] In June, FAS secretary Joseph H. Rush wrote to the AEC to express the FAS's uneasiness about security clearance policy. Members of the FAS, Rush stated, were greatly concerned about clearance procedures, especially "the failure in some quarters to discriminate between the legitimate requirements of the situation and an hysterical fear of any kind of unorthodox thinking." The FAS council, however, was ready with a solution: "No scientist should suffer the loss of his position, or failure to get a position for which he is otherwise eligible, or the damage to his feelings and to his professional standing that comes from being rejected by security clearance authorities without having (1) the right to know on what basis he was rejected, and (2) the right to a hearing at which the evidence against him must be presented and he be given the opportunity to present evidence in refutation." Only if due process were guaranteed, the federation stressed, could the AEC create an appropriate balance between security considerations and individual rights.[23]

The suspensions of the two Oak Ridge chemists in August 1947 spurred further action. By October several local FAS chapters were working independently, with some loose coordination by the FAS leadership, to address security clearance problems. At Oak Ridge, Katharine Way and Gerald H. Goertzel continued to direct the AORES's efforts throughout 1947 and 1948. At Cornell University, FAS members formed the Cornell Committee on Secrecy and Clearance, which sought to collect information on individual cases and study security clearance policy in different government agencies. In Chicago, the Atomic Scientists of Chicago prepared a memorandum with recommendations for procedural reforms in AEC clearance policy.[24] Informal coordination of activities by the AORES and the Atomic Scientists of Chicago was made possible through a close friendship between Katharine Way at Oak Ridge and Beth Olds in Chicago. Richard L. Meier and William Higinbotham also attempted to keep the different FAS chapters apprised of each other's activities and to make sure the national organization did what local organizations could not. As Higinbotham informed Robert S. Rochlin at Cornell, the main FAS office in Washington had the ability to "ride herd on these [government] agencies personally."[25]

At the same time, the FAS made certain it adopted a decidedly non-confrontational attitude toward the AEC. The federation was determined not to do anything that might have some kind of negative effect on the commission. It could have been otherwise. In March 1947, Philip Morrison, Melba Phillips, and Arthur Roberts drafted a statement on "military secrecy and security" for consideration by the FAS council. Writing in a deliberately polemical manner that reflected a progressive left political perspective, Morrison, Phillips, and Roberts sharply condemned the AEC's personnel security operations. Security was, they conceded, a legitimate issue, but in practice the AEC's actions were "extra-legal, arbitrary, and often subversive of every right of the individual in a democracy." Calling attention to the repression of scientists and other intellectuals in Argentina, Greece, and Portugal, they expressed a hope that "the situation here will never be comparable." "But," they concluded, "when a scientist is refused clearance, as in one case at hand, on the single basis of his belonging to an inter-racial committee working for fair employment practices for Negroes [here they were possibly thinking of Robert H. Vought], or when the existence of long-standing professional relations with foreign scientists of openly Marxist sympathies may be used to block an appointment, it is time to re-examine the working standards of loyalty to the United States."[26]

Oscar K. Rice, a member of the AORES, liked the draft and hoped the FAS would release it "as soon as possible." Higinbotham and Victor F. Weisskopf, however, objected strongly to the statement's strident tone.[27] The federation delayed action, partly because the confirmation of the AEC commissioners was pending and the FAS did not want to risk upsetting the politically delicate proceedings. In addition, the federation expected to be able to work with the commissioners once they were confirmed. Rush told Rice, "As it appears now that the Commission will be confirmed today, we shall then be free to go ahead an[d] negotiate with the Commission and take other action as soon as necessary to try to develop reasonable security policy."[28] It was best, the FAS's leaders had concluded, to proceed with quiet diplomacy before considering stronger measures.

Acting out of similar concerns, Higinbotham quashed the efforts of the Cornell Committee on Secrecy and Clearance to gather information by distributing a questionnaire to the AEC and other government agencies on loyalty and security evaluations. Not only did he feel the questionnaire of little use, but he feared it would have harmful repercussions. Speaking for the FAS administrative committee, Higinbotham wrote to Rochlin, "I am in fact afraid that the questionnaire may actually do some damage of this sort: copies will undoubtedly get into the hands of the Un-American Affairs Committee [HUAC], the Joint Congressional Committee [JCAE], etc. These important groups are not sympathetic to the personal interests of scientists but are pathological on the subject of security. They are constantly riding the AEC and other government agencies in regard to clearance." He added that government agencies tended to have a cooperative attitude toward the FAS on loyalty and security issues, and both scientists and administrators would be served best if the Cornell committee did not distribute the questionnaire. He continued, "At present, the agencies I know about are trying to work out reasonable procedures. They are glad to talk to representatives off the record. This [the questionnaire] tends to put them on the spot, perhaps at a time when it may make their problem more difficult."[29] Higinbotham believed that the AEC and other government agencies stood at the mercy of congressional committees, especially HUAC and the JCAE. His perceptions were particularly apt in the case of the AEC, for the Atomic Energy Act of 1946 granted the joint committee a statutory oversight role that allowed Congress to exercise an unusual level of control over the commission beyond power over appropriations. Higinbotham felt that the best strategy was to work with the executive agencies and avoid any actions that might place them at a disadvantage with respect to congressional committees obsessed with security.

Other FAS members shared comparable opinions. Richard L. Meier had explicit information, presumably from AEC sources, of tensions between the AEC and the JCAE. In February 1948, he wrote to Rochlin about reports that Senator John W. Bricker (R.-Ohio) had been "savagely attacking" the AEC in JCAE meetings. Personnel security was a matter of particular concern. He commented, "The Joint Committee has been pressing particularly hard on the subject of personnel with 'derogatory' items in their dossiers. They are tired and impatient from hearing the AEC plaint that 'Scientists are different. Scientists are scarce. You can't treat them like ordinary government employees.' "[30] Katharine Way also blamed the JCAE for its intransigence. In May 1948, when more Oak Ridge scientists were suspended, she wrote to Beth Olds, "It seems that this struggle is really one with the Cong. Comm. My guess is that the

Commission is probably ashamed of the whole thing but were pushed into it by the Joint Comm." Unlike Higinbotham, however, Way felt publicity provided the best response. She added, "Our idea is to make the Joint Comm. relax its pressure by making things publicly ridiculous. New stories about ridiculousness should help a lot."[31] Olds replied that in Chicago there were fears, especially from members of the University of Chicago Law School who provided legal advice for the scientists' movement, that sharp criticisms of AEC clearance procedures would make the situation worse. She wrote, "[Edward H.] Levi is especially worried about driving the charges 'underground.' He feels that too much criticism now will force the Commission to make the charges even more vague and less defensible for the accused." Olds herself did not agree and felt that lawyers had "a tendency to be too obscure and cautious."[32]

A delicate political balance had to be maintained. Although FAS members were critical of the JCAE, they also recognized that the joint committee served as a valuable buffer against HUAC. In a February 1948 memo, Meier blamed the JCAE for the AEC's preoccupation with security and the attitude that "almost anyone is a potential Alan Nunn May." He also noted, however, that "if there were no Joint Committee it could easily be even worse."[33] As discussed in the previous chapter, HUAC's March 1948 attack on E. U. Condon led scientists to fear that J. Parnell Thomas was planning a new assault on civilian control of atomic energy. After several weeks of waiting, however, Meier concluded such an attack was unlikely. He observed in a March 24, 1948, memo, "J. Parnell Thomas is pretty much hindered in the atomic energy field because he is intruding on the jurisdiction of another committee, and one that guards its domain quite jealously."[34] The JCAE had severely flawed attitudes toward security, but at least the joint committee's dominance over atomic energy within Congress provided protection for the AEC from HUAC and other opponents of civilian control of atomic energy.

JCAE-AEC Relations and Tensions

Members of the FAS had an astute understanding of the relationship between the Joint Committee on Atomic Energy and the AEC. Atomic scientists, AEC officials, and JCAE members were all united in their commitment to build a vigorous, politically healthy program in atomic energy research and development. Each group, however, had different interests and different conceptions of what such a commitment meant.[35] With the exception of Lewis L. Strauss, the AEC commissioners tended to share the atomic scientists' liberal notions about personnel security and information exchange. Several influential members of the JCAE, on the other hand, especially Bourke B. Hickenlooper, had a more conservative and restrictive conception of security. Throughout 1948 and 1949, disagreements over security led to growing discord as JCAE members, particularly Hickenlooper, became increasingly dissatisfied with the attitudes of the AEC commissioners and atomic scientists toward security clearance. Although HUAC's sensationalist and self-aggrandizing methods set it apart, HUAC was far from alone in its fears of losing the "atomic secret." Congressional conservatives in general tended to stress security via concealment as the best way to protect information about the atomic bomb from espionage, an issue that they perceived as the primary threat to the American atomic monopoly and the nation's security. Scientists and AEC officials, on the other hand, preferred to emphasize the idea of security through achievement. They argued that unduly strict security measures would stifle progress in research and allow American science to lose its lead over Soviet science.[36] Over time, these basic attitudinal differences created a widening gap between members of Congress, and scientists and the AEC. As time passed, the divide grew more and more difficult to bridge with mutual interests.

When David E. Lilienthal took the chairmanship of the AEC, he initially doubted whether or not the commission could form a good working relationship with the JCAE, especially on matters related to security. Unlike Hickenlooper and the other security-minded members of the JCAE, Lilienthal was extremely troubled by the personal and professional damage wreaked on individuals when the AEC denied them clearances. His own confirmation hearings provided a powerful memory of the tormented feelings produced by having one's loyalty questioned. Much of the hearings had revolved around Lilienthal's liberal record as head of the Tennessee Valley Authority (TVA) and senators' charges that he had stood too far to the political left and neglected to keep his agency alert to the danger of communist influence. In his diaries, Lilienthal vividly chronicled his reactions to the hearings. His graphic description of Senator Kenneth D. McKellar (D.-Tenn.) as "hateful" and "creating suspicion against me" would have approached paranoia, except McKellar's bitter animosity toward Lilienthal, dating back to the TVA, was well known. Lilienthal recorded physical reactions: back pain, exhaustion, he "slept badly," and he was in a nervous state of "near-panic inside."[37] When McKellar pressed him on his inability to remember the exact location of his parents' birthplace in Austria-Hungary, Lilienthal recalled feeling "trembly inside" and disgusted by the senator's barely disguised "meanness." McKellar, he recalled, had "the generous complacent look when the victim is about to be taken in hand: some taunt about being leftist—Communist."[38]

The confirmation process took more than two months, during which Lilienthal oscillated between hope and despair. On March 25, Senator Bricker introduced a motion to have the AEC nominees undergo FBI investigations. The proposal went down in a Senate vote on April 3. Lilienthal greeted the news of Bricker's defeat with little more than exhausted relief. He wrote, "The last ten days have been the worst I can remember. Very tired, low, difficult to maintain perspective, agonized by the prospect of weeks of 'FBI' investigations and butchery and wanting nothing so much as release, and yet seeing that there was no way out except to go through with it. And now this, which is a considerable victory."[39] When he was finally confirmed six days later, Lilienthal felt it was an "anticlimax though very good to taste."[40]

Mindful of the suffering security inquiries could impose, Lilienthal became extremely uncomfortable whenever the commission had to discuss security clearance cases. He once described a typical meeting about cases as "punishing as hell." The use of FBI reports to make personnel security decisions led to "wear and tear on the soul." Lilienthal had little faith in the accuracy of what was technically described as "derogatory information" in FBI reports. Such information, Lilienthal wrote, usually consisted of "cases of people who have a mother or a brother or wife who is, or is reputed to be, a Communist or the equivalent, and the 'evidence' to confirm these conclusions is only some FBI agent's rendition of what someone has said, or a conclusion from some very flimsy, thin stuff indeed." He continued, "And so we sweat and agonize about the injustice to these people by such a travesty as our examination of these files must be."[41] In walking the fine line between protecting individual rights and preserving national security, Lilienthal was deeply concerned that AEC policy maintain respect for the former. At the same time, however, he feared the negative consequences for the AEC should the commission clear someone who later committed a major breach of security.

Given his solicitude for individual rights, Lilienthal had considerable reservations about his ability to work with the JCAE. As several historians have noted, JCAE chairman Bourke B. Hickenlooper shared the common assumptions of other congressional conservatives about the need to protect atomic secrets from espionage. He had a single-minded preoccupation with security, particularly the reliability of AEC personnel.[42] By the summer of 1947, however, Lilienthal had grown more confident about the AEC-JCAE relationship. After a July 21 meeting with the joint committee to discuss the physical security of AEC installations, Lilienthal wrote in his diary, "This was the first really good day we have had for a long time. My anxiety over the Joint Committee hearing was dispelled. . . . I believe we are developing some confidence in this committee, and that is very important—most important." Lilienthal felt especially pleased about Hickenlooper's performance. He observed, "Hickenlooper seems to be growing, working hard, and trying to be helpful, trying to make our job possible."[43]

The relationship was not completely smooth. Disagreements over the granting of security clearances continued to be a source of tension. On November 29, Lilienthal recorded his reactions to the previous day's meeting with the JCAE in his diary. He described the meeting as a "difficult afternoon" because "the Joint Committee was meeting in Washington to raise hell about some people we had 'cleared.'" His diary entry, however, implied that the joint committee was more concerned about HUAC's potential response than any deep-seated objections of its own to AEC decisions in specific instances. Deputy general counsel Joseph Volpe Jr., Lilienthal wrote, reported that "the concern of the committee as they went through the cases was to acquaint themselves with the cases, because they expect J. Parnell Thomas to blow off about it, and to warn us against taking on a burden, a public relations burden, of cases that could be exploited by our enemies."[44] Lilienthal received further encouragement in a conversation with Hickenlooper himself on November 30. Lilienthal noted in his diary, "He [Hickenlooper] said the Friday meeting on 'clearances' had gone well; everybody laid things right on the line, got things off their chest, but 'no one got mad.' He suggested that perhaps the committee should later next week issue a statement on the whole subject to 'take the edge off J. Parnell Thomas if he blows up' and to reassure the public that the Commission was on top of these things, and that the record was a good one." The JCAE chairman appeared worried about HUAC, but he seemed to have full confidence in the AEC. Lilienthal again felt optimistic about Hickenlooper. He wrote, "This is a remarkable attitude and bodes good for the future."[45]

Although Lilienthal felt the commission and the joint committee were learning to work well together, records of the two groups provide a very different picture of AEC-JCAE relations. Perhaps misled by the JCAE's determination to protect its jurisdiction from HUAC, Lilienthal failed to realize the full extent of Hickenlooper's and other conservatives' own apprehensions about national security. Sharp disagreements over security policy, particularly the AEC's decisions to grant clearances in a number of controversial cases, rankled joint committee members, especially Hickenlooper. Although the AEC-JCAE relationship remained cordial and cooperative throughout 1947 and early 1948, AEC officials and joint committee members had fundamentally different beliefs about the nature of security and the weight to be accorded to individual rights. These differences were initially submerged under common interests, but they gradually became more pronounced until they were irreconcilable during the last year of Lilienthal's tenure as AEC chairman. As will be detailed in chapter 7, a dispute between the AEC and the JCAE over whether or not to require scientists granted AEC fellowships to obtain security clearances erupted into public controversy in 1949 and eventually led to Lilienthal's resignation.

From late 1947 through the first months of 1948, Hickenlooper constantly, sometimes several times in a single day, barraged Lilienthal and other AEC officials with requests for information about the granting of emergency clearances, the establishment of security clearance criteria, delays in the disposition of current cases, and other aspects of the personnel security system.[46] The AEC replied slowly to the inquiries, when it replied at all. William W. Waymack, one of the commissioners, later criticized the JCAE's inquiries as no more than an onerous bureaucratic task: "The greater part of communication [between the AEC and the JCAE] became 'paper work.' It centered increasingly on security matters. It tended to take the form of time-consuming replies to letters apparently designed to make a written record."[47] Brian Balogh, in his study of the AEC and nuclear power, maintains that security provided a basis for JCAE authority to compensate for the committee's lack of expertise in atomic energy, but this view of the JCAE's motives seems overly functionalist and reductionist.[48] Fears of encroachments by HUAC, as well as Hickenlooper's and other committee members' belief in the primacy of national security, were equally important, if not more so.

The growing division between the AEC and the JCAE became clearer in the spring of 1948, as both organizations struggled to respond to the Condon case and the possibility of an investigation of Oak Ridge by HUAC. In March and April, several executive session meetings of the JCAE focused on the Condon case and other personnel security matters. In its March 8 meeting, the committee was especially concerned with how to handle the threat HUAC posed to the authority of the joint committee and the AEC. Hickenlooper related a discussion he had had two days earlier with fellow JCAE member Sterling Cole (R.-N.Y.), HUAC committee members Richard M. Nixon (R.-Calif.) and Karl Mundt (R.-S.D.), and HUAC chief investigator Robert E. Stripling on the Condon case and the question of proper congressional jurisdiction over atomic energy. At that meeting, they had discussed the possibility of future coordination and cooperation between the two committees and reached a tentative agreement that HUAC would inform the JCAE before taking any action on matters related to atomic energy. Even so, Hickenlooper still feared that Thomas might "cut loose" on whatever cases at Oak Ridge he had information about, just as he had launched his crusade against Condon. Senator McMahon also stressed that incidents such as the attack on Condon had to be avoided in the future for the sake of atomic energy research, and he alluded to fears of what HUAC might do: "I understand he [Thomas] has derogatory information on seven or eight. We can't stand repetition of the kind of headlines we had in the last week or two. I don't think the project can stand it—we'll be in a fine state of demoralization." The meeting concluded with a cautious decision that the JCAE would consult further with Thomas to define more carefully HUAC's jurisdiction on investigations related to atomic energy.[49]

On March 31 and April 1, the joint committee held two more executive session meetings to discuss security clearance procedures. At these meetings, committee members paid special attention to the twenty-odd security clearance cases (including those of Vought, the Blewetts, and the Oak Ridge scientists) that the AEC had identified earlier as controversial. Their discussion revealed the nature of the growing rift between the joint committee and the AEC over security. Hickenlooper was not merely concerned with possible action by HUAC; he was impatient with what he viewed as the undue influence of scientists in the AEC. He also advocated an absolutist approach to security, arguing that individuals should not be granted clearance unless they were above all possible reproach. Nor did he express any sympathy for the rights of the individuals involved in disputes.

At the outset of the March 31 meeting, Hickenlooper emphasized the need for the JCAE to preempt HUAC on the twenty borderline cases. As he discussed the AEC's handling of the cases since the previous summer, he noted, "We were anticipating, and still are, that the Thomas Committee might come out with a blast of some kind." The AEC, Hickenlooper felt, was not taking the matter seriously. He complained that most of his written inquiries had gone unanswered, and the rare replies he had received were "evasive and nebulous."[50]

The committee then went on to discuss specific individuals. Both the Vought and Blewett cases irritated Hickenlooper. Since Vought was a new applicant, rather than a former Manhattan Project employee reapplying for clearance, Hickenlooper thought the AEC should stick to its original decision and not bother with an elaborate reconsideration of Vought's case. Hickenlooper commented, "He [Vought] is an applicant for a job with [the] AEC—he is a poor security risk and should not be employed by the AEC or its contractors—that happens to be the recommendation of the Security Board to the AEC. Now, mind you, he has never been employed by the Commission—just applying for a job. Yet, the AEC goes through the whole process. Why consider him further? Why go through all this business?" Senator William F. Knowland (R.-Calif.) added, "He certainly hasn't any God-given right to a job with the Commission."[51] The Federation of American Scientists insisted that applicants and employees deserved the same treatment, but Hickenlooper and Knowland saw no reason to grant job applicants the benefit of the doubt or to examine their situations more closely. Government employment, in their view, constituted a privilege, not a right, and they believed the AEC ought to be able to refuse to hire an individual on any grounds whatsoever. Political freedom, or even basic fairness, was not a relevant consideration.

Hickenlooper blamed the AEC's reluctance to take a tougher position on pressure from scientists. He told the committee, "There is a group of scientists that have [*sic*] put Vought and the Blewetts in and backed them for employment. That group of scientists is putting tremendous pressure on the Commission to employ these people."[52] When asked to identify a specific group, Hickenlooper declined, but he noted that the pressure came from both within and outside the AEC. Hickenlooper named no names, but he could not have been thinking of anything other than the FAS and its member organizations, and the efforts of AEC scientist-administrators such as Philip Morse on behalf of the Blewetts or Alvin Weinberg on behalf of the Oak Ridge chemists.

The joint committee continued the discussion the next morning in an executive session meeting that also included members of the recently formed AEC Personnel Security Review Board, as well as AEC commissioners Sumner T. Pike, Lewis Strauss, William Waymack, and Robert Bacher and general manager Carroll Wilson. Lilienthal, exhausted and on vacation in Florida, did not attend. The primary purpose of the April 1 meeting was for the JCAE to acquaint itself with the functions of the Personnel Security Review Board and to discuss the AEC's handling of controversial clearance cases. At the meeting, Hickenlooper and Bricker expressed their unhappiness with the slowness of the AEC's responses to the joint committee's requests for information, but Hickenlooper also assured the AEC officials that the committee had only the best interests of the AEC at heart. Hickenlooper and Bricker also complained openly about pressure from scientists to clear Vought and the Blewetts, and they, along with McMahon, expressed their fears of action by HUAC and its possible effects on public opinion.[53]

Throughout the discussion, references were made to the clearance of specific scientists. Former Supreme Court justice Owen J. Roberts, chairman of the Personnel Security Review Board, reported that the board had reviewed thirty-one controversial security clearance cases. Among them were those of Waldo Cohn, James G. Stangby, Herbert Pomerance, Oscar K. Rice, M. M. Shapiro, Ernest Everett Minett, Cuthbert Daniel, and H. H. Goldsmith, all of whom were identified as "security risks" according to an April 1947 document quoted in the meeting.[54] Just how many of these scientists were actually aware of their questionable security status is unclear. Goldsmith certainly knew; his security clearance had been revoked in August 1947, although it was restored a month later.[55] Readily available evidence does not indicate if the others endured similar treatment.

Exactly why these Oak Ridge scientists were considered "security risks" is also unknown. Federal Bureau of Investigation files shed some light on the cases of Rice, Cohn, and Daniel. According to the FBI's Knoxville office, Rice had "a violent distrust of the military control of atomic information." He also "advocated any advancement of the negro [*sic*], no matter what social consequences might result," and he was a member of the Southern Conference for Human Welfare and the Independent Citizens Committee of the Arts, Sciences, and Professions in addition to being active in the FAS. Cohn, according to an FBI informant, was "intolerant of security regulations," and he expended a great deal of effort informing co-workers at Oak Ridge about the atomic scientists' movement, as well as the persistent racial discrimination in Tennessee. Another informant, according to the FBI's summary, thought Cohn was "a devout follower of the Communist Party line" because he opposed Truman's increasingly firm stance toward the Soviet Union during 1945 and 1946. As for Daniel, as noted in chapter 2, the FBI believed he was a radical and onetime member of the Socialist Party, although it also acknowledged his strident opposition to the Soviet Union and Stalinism. Unfortunately, it is not clear which of the FBI's concerns about Rice,

Cohn, and Daniel were shared by the AEC and JCAE. Nor has specific information emerged about the other controversial cases at Oak Ridge.[56]

It is worth mentioning, however, that Cohn, Stangby, Pomerance, Rice, Shapiro, Minett, Daniel, and Goldsmith were all involved in the atomic scientists' movement. Shapiro had been chairman of the Association of Los Alamos Scientists during the debate over the McMahon bill and was currently active in the AORES; Goldsmith was coeditor and cofounder of the *Bulletin of the Atomic Scientists*; Cohn, Stangby, Pomerance, Rice, and Minett were all members of the AORES, and Daniel had been a member until he left Oak Ridge in 1947. Whether their FAS activities were relevant to their being defined as "security risks" is not known. At the very least, evidence suggests a positive correlation between being a member of the FAS and being considered a "security risk," although the connection may have resulted from the tendency of a significant proportion of FAS members to support other liberal-left causes. For example, in addition to his membership in the Independent Citizens Committee of the Arts, Sciences, and Professions and the Southern Conference for Human Welfare, Oscar K. Rice was also active in the American Association of Scientific Workers and served as AASW vice president from 1942 to 1945. In some military security clearance cases, the causal link between FAS membership and denial of clearance proved direct. In at least one army case, military intelligence included FAS membership as one of the reasons for denying an individual security clearance.[57] As discussed in chapter 3, Eugene Rabinowitch also felt the army refused to grant him clearance because of his activities in the FAS.

Two weeks after the April 1 meeting, the AEC implemented its Interim Procedure to formalize the handling of personnel security. It seemed the AEC had taken an important step toward answering the JCAE's demands to clarify and streamline its operations. Only days later, however, AEC-JCAE relations hit a new low. Lilienthal's relationship with Hickenlooper reached the breaking point. In an April 23 meeting with Senator Arthur H. Vandenberg (R.-Mich.) to discuss the approaching debate over his reappointment to another term on the AEC, Lilienthal was shocked to learn that Hickenlooper had complained to Vandenberg about his and the AEC's performance, especially in regard to the borderline security clearance cases. Lilienthal grew livid. Later he wrote in his diary at length about his reactions upon learning of Hickenlooper's complaints: "Then I really opened up. I said [to Vandenberg] that I would have thought a great deal more of Hickenlooper if he had not poured over me all this guff to make me feel that he was for me and for our work; it would have improved my opinion of him as a man if he had told me the things he thought were wrong, so we could respond to such criticism, rather than pretend." Lilienthal was particularly incensed by the joint committee's anxieties over personnel security, and its willingness to sacrifice individual rights out of fears of adverse publicity from HUAC. He recorded in his diary, "I said that as to the personnel clearances, Hickenlooper was terrified of Parnell Thomas and his criticism of the Joint Committee, and was prepared to outdo them if necessary; that for my part I knew enough about the institutions of freedom and how they are undermined that I would be God damned if I would start lynching these poor devils just because Hickenlooper or anyone else didn't have the backbone to insist on decency in these things. . . . I really blazed."[58] From then on, Lilienthal's references to Hickenlooper in his diary typically consisted of complaints or disparaging remarks, such as describing him as "full of nonsense" and a "lightweight statesman," and commenting on his long-windedness.[59] Apparently, Hickenlooper held a similarly low opinion of Lilienthal. Thirteen months later, he would lead the charge that eventually drove Lilienthal out of the AEC.

Oak Ridge and AEC Security, Part II

Late in March 1948, after the first suspensions at Oak Ridge, reports of security clearance cases such as those of Vought and the Blewetts, and HUAC's attack on E. U. Condon, frightening rumors spread among FAS members about what the future held for scientists. In addition to worrying about reports that HUAC would go after the FAS, the federation's leaders feared that sometime in the following month the government would summarily dismiss all politically suspect scientists. According to scuttlebutt, when this happened there would be no charges, no hearings, and no appeals.[60]

What actually followed did not prove quite so dire, but it was disturbing enough. In April, scientists learned of five new AEC security clearance cases at Oak Ridge. By this time, in response to the August 1947 suspensions at Oak Ridge and similar incidents, the AEC had implemented the Interim Procedure, which had been designed with due process rights in mind. Under the newly enacted guidelines, two Oak Ridge scientists were suspended with pay, two were informed that there were concerns about their security status but not yet suspended, and one was simply questioned. In the first four cases, the commission supplied the scientists with a statement of the specific charges against them, asked them to submit a written response to the charges, and informed them of their right to request a hearing before a local AEC board. The board would then make a recommendation to the AEC general manager. If the board advised against granting clearance, the employee in question could appeal to the AEC's new Personnel Security Review Board. The AEC commissioners themselves would make the final decision.[61]

The new security policy seemed adequate in its respect for procedural rights. The problem was the nature of the evidence that caused the scientists' security status to be questioned in the first place. Vague and arbitrary charges abounded in these cases. Most of the accusations, based on FBI interviews, dealt with individuals' associations with alleged members of communist or subversive organizations or reports of the individuals' own political views. Typical allegations included such statements as, "A former landlord of yours has reported that in 1943, after you moved . . . certain magazines and pamphlets which may have been left on the premises by you may have included a copy of the magazine 'New Masses,'" and "A neighbor has stated that she believes a close relative by marriage is a Communist." For one scientist, the sole charge against him alleged, "A person with whom you associated closely in the years 1943–47 said you were very enthusiastic about Russia and seemed to be pro-Russian in your point of view."[62] In general, the charges were indefinite, indecisively worded, and based solely on undistilled statements from FBI interviews without further corroborative efforts; they smacked more of rumor and gossip than serious allegations.

The Federation of American Scientists began to contemplate its response to the new cases. For several weeks, members of the Association of Oak Ridge Engineers and Scientists discussed whether or not to seek publicity for the Oak Ridge scientists. Unfortunately it is not known if any of the five Oak Ridge scientists facing revocation of their security clearances were members of the AORES, but if they were, it must have added extra urgency to the deliberations. Finally, the AORES contacted the *New York Herald Tribune*, which dispatched science writer Stephen White to cover the story.[63]

In late May, White published four articles about the situation at Oak Ridge. He based his findings on interviews with scientists under investigation and AEC officials and on transcripts of hearings before the local review board.[64] He observed that at the procedural level the investigations were generally reasonable. Employees under investigation were not given the identities of FBI informants or allowed access to FBI dossiers, but the hearings themselves were "eminently fair."[65] In that respect, at least, the AEC provided a vast improvement over the media events staged by HUAC.

The main problem, according to White's account, consisted of the nature of the charges, which were often trivial, grounded in factual errors, or based on unsubstantiated hearsay. Although the AEC wanted to respect individual rights, its standards for questioning employees' loyalty seemed to differ little from those of HUAC. White discovered that even members of the local review board found the charges unconvincing. One of the scientists under investigation commented sarcastically, "Any self-respecting radical would be ashamed of charges like those."[66] Charges made by anonymous informants were the worst, for without the right to cross-examine accusers, they were virtually impossible to refute. The scientist cited for being "enthusiastic about Russia," for example, could do no more than deny the charge and hope to be believed. S. Frank Fowler, a member of the local review board, commented, "I have to admit that I squirmed a little listening to him try to defend himself against a charge of that sort. He said, 'Who is this man that says I am a Communist? Who am I defending myself against? He has no name, no face, no social security number.' I must say that I sympathize with him."[67]

The security situation, White reported, was creating a tense atmosphere at Oak Ridge. In addition to the five scientists under investigation, rumors indicated that scientists could expect up to thirty more cases in the near future. As a result, an atmosphere of fear, a precipitous decline in laboratory morale, and concerns about the destruction of Oak Ridge as a center of scientific research spread among Oak Ridge scientists. Scientific personnel were already tense over the possibility of a strike by unionized workers at Oak Ridge and, on top of that, the loss of the nuclear power program to Chicago. White wrote that the security investigations constituted "a new and staggering blow to morale that was already scraping the ground," and he predicted a mass exodus of scientific personnel from Oak Ridge.[68] A third of the senior physicists and chemists had already resigned, and others were surely in search of alternative positions in industry or academia.[69]

As White's stories began to appear in the *Herald Tribune*, the AORES released a statement warning that scientists at Oak Ridge were "seriously demoralized" by the security cases and the prospect of further investigations. Although scientists conceded the need for security restrictions, they felt that the Oak Ridge charges were frivolous, false in some instances, and generally "insignificant statements that could be made about almost any one who has the varied contacts normal in college and

Cartoonist F. O. Alexander's rendition of events at Oak Ridge, May 1948. (Association of Oak Ridge Engineers and Scientists Papers, University of Chicago Library)

work." Without changes, the Association of Oak Ridge Engineers and Scientists predicted, "these present procedures will make it increasingly difficult to enlist the services of scientists on government projects."[70]

The suspensions at Oak Ridge, while not front-page national news, received limited coverage in other papers. The *New York Times* paid only minimal attention, but the *Washington Post* kept track of the story.[71] In general, leaders of the AORES were pleased about the press coverage.

Katharine Way told Beth Olds, "It now seems to us that this kind of publicity is the best weapon."[72] The Washington office of the FAS was less pleased, because lack of adequate coordination from the AORES had botched the possibility of much greater media attention. In a lengthy memorandum, R. L. Meier berated the AORES executive committee sharply for having missed an important opening:

> This time I have a gripe. At the time you decided to publicize the hearings and the cases you should have warned me what was coming. I could have been prepared then for the barrage of telephone calls from U.P., A.P., NEWSWEEK, TIME, SCIENCE SERVICE, etc. The FAS office has many defects but its press connections have been maintained, at the expense of other functions.
>
> The last memo to the associations dealt with just this subject and hinted at the techniques necessary when using publicity for winning a battle. We need follow-through, some at Oak Ridge, some in Washington, some elsewhere, if we are going to do any good at all. We have to make official stupidities too hot to handle, and, once these things are out in the open, *continuous* publicity is the only means. For instance, if we had been warned, we could have used Condon's speech here in Washington on "Science and Security" to put in a good lick for you; it would have been front page stuff, but the opportunity was missed.
>
> Then when I wired Shapiro trying to get a running account of what happened and what is likely to happen so as to try to get integrated, nothing arrives but a mimeographed sheet!
>
> If you are able to supply enough news on secrecy and clearance to hit the HERALD TRIBUNE once or twice a week till next July then the situation is adequately taken care of. But if you are not, then Washington, Cornell, Chicago, and other areas will have to pitch in and help out. Otherwise we are sure to muff this opportunity.[73]

Meier's reproof indicated both a willingness to court public opinion and considerable media savvy. The Federation of American Scientists not only had the resources to generate press coverage but also considered publicity a powerful means of persuading policy. Even with the decline of the NCAI, the FAS did not think exclusively in terms of elite-level politics. Quiet diplomacy was important, but it was not the only option. With the Condon case recently in the news, the timing for a publicity campaign seemed especially opportune. James S. Stewart, executive committee chairman of the AORES, could only reply to Meier's scolding, somewhat sheepishly, "We agree that you have a gripe coming. We can only use the cadet's answer 'No excuse, sir.'" Stewart explained that the AORES committee on security had thought of publicity in limited terms, "essentially a factual statement," and "no one had thought of a planned campaign of publicity." The press inquiries after the appearance of the first article had caught the AORES by surprise. Stewart promised Meier, "We're going to coordinate better here in Oak Ridge in the future."[74]

But publicity did not rule out quiet diplomacy. In addition to working on press coverage, the AORES contacted the AEC directly. On May 20, Stewart sent a letter and a copy of the AORES's recommendations on future security policy to the AEC commissioners and AEC general manager Carroll Wilson. Seeking to clarify the rationale behind security proceedings, the Oak Ridge scientists outlined the need for procedural rights beyond those granted under the Interim Procedure, as well as a higher burden of proof on the part of the AEC. They requested that specific charges be accompanied by an explanation "as to why it is expected that these [charges] may lead to future disloyalty." All too often, the AEC presented charges as if they required no explanation, even though the connection between the allegations and the likelihood of future violations of security restrictions was not at all clear. The Oak Ridge scientists also recommended that individuals undergoing a hearing be guaranteed due process, including the right to cross-examine witnesses and the right to appeal adverse decisions in federal court. They reiterated the need for a high standard of proof: "The AEC must not only prove that the charges against the accused are true, but also that the explanation offered as to why these might lead to a future disloyal act is logically consistent." The Association of Oak Ridge Engineers and Scientists perceived the hearings as essentially equivalent to criminal legal proceedings and concluded that the punitive nature of denial of clearance therefore required the same burden of proof and the same procedural protections found in the courts. The association also insisted that new applicants for AEC positions have the same rights as former Manhattan Project employees seeking AEC clearance.[75]

Wilson replied at length on June 16. His letter was respectful but not encouraging. Regarding the need to show the relationship between charges against an individual and future disloyalty, Wilson missed the central point about standards of evidence and proof. Instead, he merely stated that security decisions did not just involve judgments of loyalty but had to take into account character and associations. The AEC often explained this distinction by pointing out that a person who was habitually drunk was not necessarily disloyal, but such a person's behavior constituted a defect of character that might compromise the individual's ability to protect confidential information. Wilson commented that most of the AEC security clearance cases involved character or associations, but he did not respond to the thrust of the Oak Ridge scientists' argument that there existed a fundamental need to make explicit the relationship between charges and maintenance of security. As for the AORES's other recommendations, Wilson noted that the Interim Procedure granted a wide range of procedural rights, but the commission could not go so far as to guarantee that decisions would be based solely on sworn testimony. Local AEC boards would evaluate FBI reports in as impartial a manner as possible, but the AEC would not discard important information if it came from confidential sources. Judicial review by the federal courts was a matter beyond the AEC's jurisdiction. With respect to applicants, the AEC had to consider the matter further but at present felt inclined not to grant applicants denied clearance the right to a hearing. Wilson's reply was cordial and detailed, but it seemed to indicate that the AEC had little desire to liberalize the Interim Procedure. Wilson did, however, invite the AORES executive committee to discuss security clearance policy the next time he visited Oak Ridge.[76]

Before Wilson's reply arrived, Katharine Way and Henri Levy wrote to Lilienthal and Personnel Security Review Board chairman Roberts. In their letter to Roberts, they were especially critical of the charges against the Oak Ridge scientists. The accusations were generally not "well-founded or significant." Of the eighteen charges in the five Oak Ridge cases, they acknowledged eleven as true but irrelevant to security concerns. In one case, for example, the affiliation in question involved an organization that was not only noncommunist but went so far as to bar Communist Party members from leadership positions. Of the other charges, two were false, two were partly true and partly false but in any case irrelevant, two were no more than the "anonymous appraisals of the thoughts and opinions of the accused by his acquaintances," and one was the "now famous [charge] that the accused 'may' have possessed a copy of the New Masses in 1943." The vast majority of the allegations, Way and Levy contended, could have been eliminated easily by more careful consideration. Evaluations of individuals' thoughts and opinions, they felt, should be discounted unless they were based on sworn testimony and the informants were willing to be cross-examined, while only "common sense" would remove the *New Masses* charge. Way and Levy recommended that hearings be held only after charges had been fully investigated, and only if informants were willing to make them in sworn testimony subject to cross-examination. Such reforms, they suggested, would greatly improve morale and slow employee turnover at Oak Ridge.[77] On July 17, Roberts replied that the Personnel Security Review Board would give their advice "due consideration when, as and if it is requested by the Commission to give its opinion in this connection." The AORES responded with follow-up letters to both Roberts and Lilienthal and suggested that other FAS chapters do likewise.[78]

Meanwhile, the Federation of American Scientists continued to seek press coverage, with mixed results. Two magazines requested articles from Harold C. Urey on security clearance, and Beth Olds explored the possibility of finding ghost writers for Urey.[79] In the end, the articles never materialized. However, the *New Republic* and the *Bulletin of the Atomic Scientists* did prepare articles on the situation at Oak Ridge that were published in July.[80] When the limited press coverage proved insufficient to mobilize the scientific community, Meier advised Henri Levy on drafting a letter to the major scientific journals. Meier noted, "Due to poor A.P. and U.P. reporting we will not have the scientists with us until they [the charges] are carefully documented in a scientific journal and they can see for themselves how silly the charges are. We can't get any useful pressure on the Joint Congressional Committee until scientists have a relatively solid front."[81] Levy wrote a letter, dated June 16, that detailed the charges against the Oak Ridge scientists and warned of the "grave implications concerning civil rights of scientists, and the relation of scientists with the government" embodied in the events at Oak Ridge. He sent it to half a dozen major scientific journals, both professional and popular. Only *Chemical and Engineering News* saw fit to publish his comments.[82] The Oak Ridge story prompted some press reports, but nowhere near as many as the Condon case. As a news item, it was difficult for the unnamed Oak Ridge scientists and the relatively unknown AORES members to attract as much coverage as the beleaguered National Bureau of Standards director and the House Committee on Un-American Activities.

There matters lay until September 6, when eight well-known scientists, Harrison Brown, Karl T. Compton, Thorfin R. Hogness, Charles C. Lauritsen, Philip M. Morse, George B. Pegram, Harold C. Urey, and John C. Warner, sent a telegram to President Truman detailing their concerns about political attacks on scientists. As the FAS had done earlier, these scientists laid the blame squarely on HUAC rather than the AEC. They accused HUAC of "creating an atmosphere that makes men shun government work" and making it "increasingly difficult for scientists and engineers to function." The AEC, on the other hand, was exonerated: "We wish to stress that in our opinion the Atomic Energy Commission is in no way to blame for the unfortunate situation that now exists." Rather, HUAC had turned atomic energy into a "political football," thereby placing the AEC on the defensive.[83]

The message from the eight scientists received widespread press coverage and attention from the highest government officials. President Truman soon commented publicly. At the centennial meeting of the American Association for the Advancement of Science on September 13, four days before embarking on a coast-to-coast whistle-stop tour for his reelection campaign, Truman greeted Condon on stage and delivered a vigorous condemnation of attacks on the loyalty of scientists. In the address, broadcast nationwide on radio, Truman backed heartily the views of the eight scientists. He also blasted smears on scientists as "unfounded rumors, gossip and vilification" that created an atmosphere that "is un-American, the most un-American thing we have to contend with today."[84] Lilienthal echoed Truman's sentiments four days later in his own speech to the AAAS and warned that important scientists were leaving government laboratories to prevent the possibility of "public humiliation and smears on their character and patriotism."[85] A HUAC subcommittee consisting of J. Parnell Thomas, John McDowell (R.-Penn.), and Richard B. Vail countered that "in spite of" Truman, Lilienthal, and "a few misguided scientists," the subcommittee would reveal within a few days "a shocking chapter in Communist espionage in the atomic field."[86] Although HUAC issued a *Report on Soviet Espionage Activities in Connection with the Atomic Bomb* on September 28, the report contained no new information, and the promised shock was never delivered.

Truman's speech revitalized the FAS's efforts. On September 14 the AORES released a statement warning that the dangerous situation described by the president was "already being realized at the Oak Ridge National Laboratory." The statement continued, "During the last few months about forty percent of the senior physicists and chemists have resigned. One of the chief reasons is the feeling that at any time atomic scientists may be attacked on flimsy or unsubstantiated grounds."[87] R. L. Meier followed up the AORES press release with a letter to the editor in the *New York Times* on September 23. Now Meier explicitly named the JCAE, rather than HUAC, as a source of scientists' difficulties. Meier stressed that although the precise effect of security investigations on government employment of scientists was difficult to quantify, "harassment stemming from forces within the Joint Congressional Committee on Atomic Energy and the Security Division of the A.E.C." was a definite factor in the decisions of senior scientists to leave Oak Ridge.[88]

Scientists also wrote to the JCAE and AEC directly. L. W. Nordheim, professor of physics at Duke University and former director of Clinton Laboratories, informed Senator Hickenlooper, "Perhaps the most important single factor that deters scientists from working in atomic energy and other government laboratories is the security situation."[89] Ray W. Stoughton, now chairman of the AORES, wrote Senator Hickenlooper on behalf of the Oak Ridge association and explained that although official resignation letters from scientists tended to specify alternative job offers as the reason for leaving, the AORES knew from personal contacts that "the harassing security environment played an important role in the making of their decisions."[90] Stoughton also sent a memorandum to the scientists' associations, urging them to write to Hickenlooper. He predicted cautiously, "It really seems that some progress in the security business may be made through such efforts."[91] Alvin M. Weinberg, now director of the physics division at Oak Ridge National Laboratory, wrote Lilienthal about a theoretical physicist who felt interested in working at Oak Ridge but worried that the quality of personnel at the laboratory was on the decline because of the security clearance situation and the threat of investigators from Congress.[92] Stoughton wrote to Lilienthal as well, but, unlike Weinberg, he held the AEC directly responsible for the poor working conditions at Oak Ridge created by overzealous concerns about security. Speaking for the AORES, Stoughton complimented Lilienthal on his statements before the AAAS but then declared, "We would also at this time like to say emphatically that the security procedures of the Atomic Energy Commission have been a much more important factor in the demoralization of project scientists than have the actions of the Thomas Committee. . . . The Atomic Energy Commission procedures are far from sufficient to avoid harassment on unsubstantial and unsubstantiated charges."[93] With the exception of Meier's *New York Times* editorial, in public, scientists held HUAC in reproach for unfairly questioning scientists' loyalty. In their private communications with the AEC and joint committee, however, they were beginning to hold AEC security procedures equally responsible. The ideology of anticommunism, whether displayed in a demagogic or more temperate manner, tended to have the same effects of equating left-of-center politics with disloyalty and thereby unjustly endangering scientists' careers.

Hickenlooper took note of the eight scientists' telegram to Truman and wanted to know more. On September 14 he requested from Lilienthal "a specific analysis" of points raised in the telegram, including whether or not the AEC's supply of scientific personnel was "dangerously low" and the extent to which the political atmosphere surrounding atomic energy impeded AEC operations. He also wanted to learn if the AEC had any direct evidence that scientists had resigned or refused government job offers because of either the activities of HUAC or the security procedures of the AEC.[94]

On the last point Hickenlooper was unlikely to be sympathetic, and the scientists' subsequent comments further irritated and alienated him. Meier met with Hickenlooper and a JCAE staff member on September 27 and learned of the joint committee's displeasure with his letter in the *New York Times*. He wrote to Ray Stoughton, "They wanted to rake me over the coals for saying they were 'harassing the AEC.' . . . Technically, I admit, they have a good case. They only press for decisions. . . . Nobody gets fired or clubbed with publicity." The joint committee, Meier had to concede, was not as bad as HUAC. "Nevertheless," he observed, "the pressure on the AEC is considerable." Meier also informed Stoughton, "They completely disbelieve your story about the 40% figure," and he asked how the figure had been determined.[95] The breach between the scientists and the JCAE had widened.

In actuality, although scientists objected to burdensome security regulations, it was not all that clear that problems with personnel security caused the exodus of scientists from Oak Ridge. On September 30, in an initial reply to Hickenlooper's letter of September 14, Lilienthal told the senator that it would be difficult to determine scientists' reasons for leaving the AEC from their formal statements of resignation. Unofficial sources, however, indicated that personnel security presented a major deterrent to government employment.[96] Yet, as Lilienthal told Hickenlooper in a lengthier reply two weeks later, personnel security constituted only one of several factors that persuaded scientists to leave Oak Ridge. Others included concerns about the nature and quality of the AEC research program at Oak Ridge National Laboratory and unhappiness with the quality of life in the town of Oak Ridge. More generally, scientists traditionally preferred university or industrial jobs over government positions; red tape and low pay made federal jobs unattractive. On the one hand, Lilienthal felt that recent reforms in security clearance procedures meant scientists "now have greater confidence in the fairness of these procedures than at the outset." On the other hand, Lilienthal also believed that security clearances and classified research, which carried with them the accoutrements of fences, guards, badges, procedures for the handling of documents, and other cumbersome rituals, inevitably carried "an aspect of personal wear and tear which makes such jobs relatively unattractive to many scientists and engineers."[97]

Equivocation characterized other analyses as well. The Association of Oak Ridge Engineers and Scientists' figures concerning the percentage of senior scientists who had resigned from Oak Ridge were accurate, but the AORES's explanation of the reasons behind those resignations was problematic. Although security policy contributed to scientists' desire to leave, they departed mainly because the transfer of the reactor program to Chicago made prospects for research at Oak Ridge far less interesting. Stoughton explained to Meier, "As to the reasons, some people feel that the 'security' problem was not a terribly important factor; others feel that it was *the* main reason (I don't agree with the latter). Everyone who read our release before it was sent out agreed that the 'security' problem was *one of the* chief reasons. Most of us agree that a still more important chief reason was the Dec. 27th (1947) decision of the AEC to move the reactor program to Chicago & to give the laboratory to Carbide & Carbon Chemicals Corp. It was not the purpose of our release, however, to discuss other reasons."[98]

Since the list of senior scientists who resigned is available in the FAS papers, it is possible to trace their subsequent career paths. By October 1948, thirty-five of seventy-four senior chemists and physicists planned to leave Oak Ridge. Using the 1949 and 1955 editions of *American Men of Science*, thirty-one of the thirty-five could be traced. Of this group, four ultimately stayed at Oak Ridge, six went to work at other AEC installations, three went to navy laboratories, and one went to the RAND Corporation, which had recently been founded by the air force. Thus, fourteen of these scientists chose new positions that required remaining within the federal loyalty-security system. Of the remaining scientists, ten went to universities, and seven went into industry or consulting. Among these seventeen, one of the university scientists later spent a year at Livermore Laboratory, and one of the industrial scientists later went to work at MIT's Lincoln Laboratory; thus, these two scientists apparently had no strong objections to classified research.[99] Only fifteen of the thirty-one, just under half the sample, could conceivably have left Oak Ridge to escape the security restrictions, but it is equally likely that they simply found better career opportunities in academia and industry.

In general, although many critics insisted throughout the 1940s and 1950s that adverse security regulations would lead scientists to vacate government laboratories, few scientists went so far as to vote with their feet and leave positions out of opposition to the security clearance system. A study of scientists who resigned from government posts in 1948 found that the majority quit because of low pay and frustration with bureaucracy. Although some scientists cited loyalty and security investigations as a source of dissatisfaction, the burdens of the loyalty-security system were not a primary reason scientists left government employment.[100] General employment data for the 1950s indicate that the percentage of scientists and engineers employed by the federal government, as compared with the universities and industry, actually rose slightly between 1950 and 1952 and declined only slowly after that.[101] Although some scientists may have avoided government jobs out of fear of persecution or unhappiness with security regulations, clearance requirements did not provoke a mass exodus from government laboratories.

As of the fall of 1948, the first round of negotiations over security clearance between the FAS, the AEC, and the JCAE had come to an uneasy close. The threat from HUAC, at least, had been neutralized. The House Committee on Un-American Activities did not need to tangle with the powerful JCAE over turf when it could go after Hollywood, labor unions, the universities, and other tempting targets without competition. J. Parnell Thomas's legal problems also may have changed HUAC's priorities. Although reelected in November 1948, Thomas was indicted on charges of payroll padding shortly thereafter. His indictment made it difficult for him to continue to play a very active role on the committee. A year later, he pleaded no contest. He resigned from the House in disgrace in January 1950 and, to E. U. Condon's eternal glee, went to prison for nine months. With Thomas sidelined, no other member of HUAC stepped forward to carry on the fight against the atomic scientists and the McMahon Act. It was easier to leave atomic energy alone. In the following years, HUAC did little more than revive old, discarded allegations of wartime atomic espionage. Although Condon continued to suffer on-again, off-again attention from the committee throughout the early 1950s, HUAC never dared undertake a concerted, organized campaign to expose supposed postwar threats to America's atomic security.

But HUAC's exit from the scene did not remove the main source of scientists' difficulties. The implementation of President Truman's federal loyalty program did more to strengthen domestic anticommunism than any of HUAC's actions. The loyalty program constituted Truman's partisan response to Republicans' allegations that the Democrats were soft on communism. Truman hoped to control the domestic security issue by balancing anticommunist policies with respect for civil liberties. Ultimately, he failed. Rather than confining antiradical nativism to the fringes of American politics, the loyalty program made domestic anticommunism mainstream. Truman and other Cold War liberals never fully appreciated the way their own anticommunist rhetoric validated the more extreme actions of HUAC and, later on, Senator McCarthy.[102]

For scientists, the security clearance situation remained troubling, but not as bad as it had been. By September 1948, three of the Oak Ridge scientists had received clearance, but four cases were still pending.[103] Katharine Way received a scare in August, when she was initially denied employment at the National Bureau of Standards despite Condon's desire to have her work there. Paul C. Tompkins of the AORES noted, "No explanation was given but we naturally suspect security or loyalty questions are involved."[104] Tompkins's suppositions were significant—scientists had begun to assume automatically that their political activities were responsible for delays or denials of employment, whether or not they had explicit information. Whatever the problem in Way's case, it was cleared up, and she went to work at the Bureau of Standards in 1949.

Among the chapters of the FAS, the Association of Oak Ridge Engineers and Scientists had shown the greatest willingness to respond vociferously and publicly to the security clearance situation. By the end of 1948, however, several of the AORES members who were most active on security problems, including Way and Gerald Goertzel, had departed from Oak Ridge. Oscar K. Rice, Maurice M. Shapiro, and other key members of the AORES also left.[105] Although Ray W. Stoughton, Henri A. Levy, and others remained, the AORES would do little on loyalty and security issues in the future. With Condon cleared and the Oak Ridge cases slowly being resolved, however, the security clearance system seemed to have improved from the confused situation a year earlier. The executive committee of the Atomic Scientists of Chicago informed its membership on December 9, 1948, "The relatively enlightened treatment of employees of the Atomic Energy Commission, in comparison to those of other government agencies, is doubtless due largely to the efforts of our organization and of other members of the Federation of American Scientists."[106] Surface reforms in the processing of security clearance cases, however, did nothing to ameliorate deeper political rifts. Indeed, the AEC's willingness to address scientists' concerns only heightened the conflict between the commission and the JCAE. The Federation of American Scientists successfully influenced the AEC to make procedural changes and expedite individual cases, but at the price of alienating the joint committee.

The first year and a half of AEC operations constituted a transitional phase, in which the political relationships between scientists, the AEC, and Congress were in flux. During this period, the FAS tried to address scientists' security clearance difficulties with both quiet diplomacy and publicity, all the while gauging the relative positions of the AEC and the JCAE. At the same time, America's commitment to domestic anticommunism was also relatively unsettled, although Cold War conflict between the United States and Soviet Union escalated quickly after the Truman Doctrine speech. In 1948, the freedom to challenge the ideological foundations of Cold War anticommunism still existed in American politics, although it was disappearing rapidly. For scientific organizations, 1948 was a critical time, a last opportunity to contest vigorously the underlying assumptions of the security system and, by extension, the fundamental postulates of domestic anticommunism. In the end, they would fail to seize the day.

CHAPTER 6

Responses

Scientists' Associations and Civil Liberties, 1948–1949

• • •

The Federation of American Scientists took the early lead in addressing the effects of loyalty and security investigations on scientists, but other scientific organizations soon followed. Concerned about the repercussions of the federal loyalty program, the American Association for the Advancement of Science formed a committee on civil liberties at the end of 1947. The Condon case provoked the National Academy of Sciences into forming its own civil liberties committee late in 1948. In the meantime the FAS formalized its existing efforts to protect scientists from attacks on their loyalty.

The responses of all three organizations to the impact of Cold War anticommunism on American scientists were curiously limited. Caught up in disagreements over the proper boundaries of political conduct by scientists and disputes about realistic expectations and strategies given the state of Cold War politics, scientists' groups did little more than study the personnel security problem and recommend procedural reforms through the appropriate official channels. Even the FAS drew inward and abandoned the public approach that had served the atomic scientists well in the early debate over civilian control of atomic energy.

Where government agencies were receptive, pressure for limited internal reform achieved modest success, especially in the AEC. Quiet diplomacy also provided a relatively safe strategy that minimized the possibility of a red-baiting backlash against the scientists' organizations themselves. At the same time, however, the narrow political role scientists built for themselves undercut their ability to critique the fundamental assumptions on which the loyalty-security system was based. In the long run, limited procedural reforms could not shield American scientists from the nuclear fear and anticommunist determination that grew to dominate Cold War politics and culture before the end of the 1940s.

The National Academy of Sciences, the Condon Case, and Civil Liberties

When HUAC attacked Edward U. Condon, the National Academy of Sciences had an opportunity to rush to the aid of an academy member, lead by example, and guide the scientific profession in a vigorous defense of scientists' political rights. The academy had, potentially, more political influence than any other nongovernmental scientific organization in the United States. At least in theory, it admitted only the most illustrious scientists for membership, and it was the preeminent scientific body in the nation. It functioned as both an honorary association and an agent of science policy. Empowered by its 1863 Act of Incorporation to provide expert advice to the federal government, the NAS actively assisted in the mobilization of scientific research in two world wars. After World War II, Frank B. Jewett and Detlev Bronk boldly predicted the academy would be a powerful force in science-government relations. But the NAS faltered in its handling of loyalty and security issues, becoming mired in deliberations over its proper function and legitimate sphere of action. Unwilling to expand its role and risk controversy, the academy ultimately adopted only a weak stance in the area of civil liberties and scientists.

The academy's long, drawn-out debate over how to address threats to the political freedom of scientists began with the Condon case. On March 10, 1948, nine days after J. Parnell Thomas released his subcommittee's report on Condon, NAS president A. Newton Richards distributed a draft statement on the Condon case to members of the NAS council and to Frank B. Jewett, Vannevar Bush, and James B. Conant. The statement, drawn up by Richards with the assistance of several other persons, including Owen J. Roberts, former Supreme Court justice and chairman of the AEC's Personnel Security Review Board, delivered a refined, yet biting, critique of the rhetorical tactics deployed by HUAC in the March 1 report on Condon: "The report contains a recitation of associations, recounted in such a way as to imply improper disclosures of secret information; of incidents, so described as to imply guilt. In it are many insinuations but no direct accusation; while it contains no avowed evidence of wrong-doing, it calls the record which it describes 'derogatory.'" The statement condemned HUAC for using methods that prejudged individuals as guilty without presenting them the opportunity for a hearing, and it concluded that episodes such as the attack on Condon, if repeated, would make scientists reluctant to accept government service, leading to "incalculable harm" to the nation.[1]

Richards informed the NAS council, as well as Jewett, Bush, and Conant, that he believed the statement should be published, but first he wanted their advice.[2] Most who replied expressed varying degrees of approval for the statement and its release. Bush, however, objected strenuously. He spoke to Richards on March 11 and warned against emphasizing any possibility that scientists would "run out on Government needs because they don't like the way Government behaves." He added that he had consulted with JCAE chairman Hickenlooper, who also, Bush reported, "doubts very much whether a statement should be made."[3] Richards met with Hickenlooper the next day, however, and found the senator without major complaints. According to Richards, Hickenlooper believed the statement "sound" and "thought its publication might do good, particularly if it were made to contain a sentence favoring full and factual examination of legitimate charges."[4]

Bush's opposition was not unexpected. Bush and Condon disagreed strongly when it came to civilian control of atomic energy and the National Science Foundation, and Condon's outspoken liberalism did not mesh well with Bush's innate conservatism. Regarding a conversation with Condon several months earlier, Jewett, then NAS president, had noted that Condon "is of the impression that Bush might not be too friendly with him because in different matters they have been on opposite sides of opinion."[5]

Taking Bush's and Hickenlooper's comments into account, Richards revised the statement and circulated the new version to the same group that had seen the original. Among other changes, the revised draft still warned that incidents such as HUAC's release of the report on Condon would discourage scientists from federal employment, but it also provided assurance that "the time will never come when the help of American scientists . . . will not be freely available to our Government."[6] Jewett thought the changes in wording would satisfy Bush, but Bush objected even more forcefully to the new version. Any intimation that scientists might leave government laboratories, Bush said, citing recent comments by AEC chairman Lilienthal, was bound to "backfire both against the Commission and against scientists generally." Bush then argued that supporting Condon was a losing proposition and public opinion would ultimately be on the side of HUAC. He explained, "Certainly Lilienthal, when all the facts are out, is going to have difficulty convincing people that in this case he, and Condon, have been rigorously careful. The people will see carelessness at least, assume much more is present and uncovered, believe that it would have remained uncovered if it had not been for the Committee's investigation, and regard Lilienthal's protest as an attempt to stop a move that was necessary for the public security."

Although he did not specify the reasons, Bush clearly distrusted Condon and felt that he had been careless in some way. He wrote, "I fear that Condon has been very foolish, and has brought this on himself, and that those who go to his support will be caught in the undertow, no matter how fine their motives." The academy would do better, Bush argued, to "wait until someone is attacked where it is not only unfair but where it is clearly spite on their part."[7]

Condon's earlier sense of Bush's less than flattering opinion of him was on the mark. In general, Bush was leery of those whose politics flirted too closely with the progressive left, and relations between the two men were already antagonistic. In 1947 the Physics Department of Stanford University had approached Condon about becoming the department's next chairman. Stanford's president, Donald Tressider, scotched the appointment, however, after Bush advised Tressider that Condon lacked the appropriate personal and political temperament to guide the department's vigorous expansion through the aggressive acquisition of postwar military funding.[8] Now, a year later, Bush thought Condon was also the wrong type of person for the academy to back in a public confrontation with HUAC. The academy faced too dangerous a political situation, Bush insisted, to throw its weight and prestige behind Condon, a scientist who was all too far from being above suspicion.

Despite Bush's opposition, Richards evidently felt there was enough enthusiasm for the statement to canvass the rest of the academy. On March 31 he sent the revised draft to the entire academy membership and requested the members' opinions on whether or not to publish it. Unless the statement encountered "overwhelming disapproval," Richards thought he would disseminate it "as widely as possible."[9]

Richards's pronouncement received solid backing from NAS members. Out of 168 replies received after a week, only 15 opposed its release. Yet despite this unmistakable show of approval, Richards reversed his previous position and decided that publication of the statement fell outside the academy's mandate. He explained to the council that in delivering unsolicited advice, the academy might "be accused of attempting to incite public opinion against a legally constituted committee of Government and that, I have come to believe, is something we have no business to do. The most unfortunate outcome would be to jeopardize our relations with Government."[10] Richards was by nature a cautious person. Already unnerved by Bush's vehement objections, he also gave extra weight to the views of the fifteen other academy members who opposed releasing the statement. Now Richards worried that the

A. N. Richards, 1947.
(National Academy of Sciences)

publication of unsolicited opinion violated the academy's prerogatives as defined by its Act of Incorporation and might create a breach in science-government relations. He decided it would be more proper to meet with J. Parnell Thomas in person to express the academy's concerns rather than circulate the NAS's criticisms publicly.

Richards met with the HUAC chairman on April 14. He later reported that Thomas read the NAS statement, "delivered an oration on the dangers and prevalence of communistic sympathies among government employees," and assured him that Condon would receive a fair hearing, although cross-examination of witnesses would not be allowed. The meeting convinced Richards that Thomas was "a zealot with a mission" and that publicizing the academy's criticisms would do nothing to encourage HUAC to reform its procedures. Richards had originally wanted to speak out, but he now felt "it would be completely unwise to publish the statement."[11]

The NAS president reached this conclusion despite nearly unanimous support for releasing the statement. By April 20, Richards had received opinions from 310 of the 401 members of the academy. There were 275 approvals and only 35 disapprovals. Geneticist George W. Beadle felt the pronouncement was "a fine statement of the situation & my feeling in the matter," biologist E. G. Conklin favored it "heartily," and psychologist Edwin G. Boring enthused, "The Academy ought to exercise its responsibilities in just this manner!" Several members accepted the revised draft but felt it could have gone further. Immunochemist Michael Heidelberger thought it was "unnecessarily weak in its condemnation of this outrage to decency and its damage to the reputation of one of our honored members." Physicist Merle A. Tuve wrote, "I would prefer to see it *stronger*," and added that HUAC ruined "a) men's reputations, b) chances for Gov't. to get good men, c) Any *real* proof of subversive action."[12] Geneticist L. C. Dunn objected to the "timidity and half-hearted tone of the support given to Dr. Condon" and expressed his hope that the final version of the statement would "give voice to the indignation which so many of us feel."[13]

Most of the disapprovals that contained explanatory comments expressed faith in Condon or criticism of HUAC but registered wariness about the wisdom of the academy's involvement. A few, like Bush, had doubts about Condon. According to a summary of members' comments, C. Guy Suits, who had earlier been so unhelpful in Robert H. Vought's efforts to obtain clearance at GE, feared that "should derogatory and reliable information be revealed, the Academy would be put in a very poor light." Based on his personal acquaintance with Condon, Suits did not "doubt [Condon's] loyalty, but is not so sure about his discretion & judgment and feels they may suffer in forthcoming hearing."[14] Some disapprovals, on the other hand, objected to the statement's caution. Harlow Shapley condemned the statement "because it declines 'to pass judgment' on the quite patent injustices . . . That is, it is too too kind."[15]

Despite the overwhelming sentiment in favor of publication, the council voted against release of the statement. In light of Richards's meeting with Representative Thomas, council members, like Richards, felt publication without prior approval from HUAC would be at best useless and possibly counterproductive. Richards tried to reassure members of the academy: "I think that the interview was more effective than publication would have been and feel sure that by refraining from publication possible danger to relations between the Academy and Government has been avoided."[16]

Many academy members objected to the council's course of action.[17] I. I. Rabi, a member of the council, had earlier written to Richards that "silence at this time would be interpreted as consent and establish intimidation and unjudicial procedure as a tolerable and perhaps normal condition."[18] Mathematician Marshall H. Stone angrily denounced the council's decision as "pussyfooting."[19] Medical researcher John P. Peters, who a few years later would sue the Public Health Services over the termination of his consultantship on loyalty grounds, pronounced himself "extremely disappointed" by the council's inaction. The near unanimous support expressed by the academy membership, Peters argued, obligated the academy to publish the statement. Furthermore, Peters felt the academy had a responsibility to inform the public of its distaste for HUAC's methods. Unlike Richards, Peters perceived the academy as "an organization which could and should rightly advise the people in matters in which it is expert, even if this advise [*sic*] requires criticism of governmental agencies." He added that given Richards's belief that the statement alone would not sway Thomas, the academy had an even greater duty to appeal directly to the public. He wrote, "If he is, as you suggest, impervious to such criticisms as those which have been published by various groups of scientists, it is the more incumbent upon the scientists to place their cause before the public to which Mr. Thomas ultimately can not be impervious."[20]

A tense and heated discussion of the Condon statement took place at the academy's April 27 spring meeting. Richards summarized the actions the council had taken up over the past several weeks to the approximately one hundred academy members in attendance. Then he opened up general discussion with the announcement, "And now my breast is open for your spearpoints." Speaking for the physics section, E. C. Kemble wanted to know more about the council deliberations that had led to "a rather disappointing result," namely the decision to withhold the statement from publication. At this point, E. U. Condon left the room. As he exited, he proclaimed, with cheerful, yet acerbic, sarcasm, "Mr. President, I have been a student of this question for some time and have some knowledge concerning it, but I simply want to call attention to the fact that I am withdrawing from the room so that this discussion can proceed without my being present, and that is in accordance, I believe, with the policy of the Academy in not giving advice unless it is asked for. On the other hand, my specialized knowledge is available to the Academy if it has any questions it would like to ask of me."[21] The remarks represented a none too subtle jab at Richards's narrow interpretation of the role of the academy. The audience greeted Condon's declaration with laughter followed by applause.

Mathematician Oswald Veblen spoke next and voiced his support for publication. He viewed dissemination of the statement as an urgent matter, less for the sake of Condon himself than for the future of American science. Referring to the experiences colleagues had endured in Nazi Germany, he commented, "I know that such men, in general, feel that they made a mistake in not coming out at an early stage of the progress toward fascism." He warned that the situation in the United States was not too dissimilar: "We are now living to a very large extent under a police state. Nearly every man in this room has a dossier in the FBI to which he has no access, and which presumably includes all kinds of gossip which he probably knows nothing about."[22] After Veblen spoke, geologist Charles S. Piggot added that the academy needed to send a signal to the scientific community. American scientists, he felt, were "looking to the National Academy of Sciences for an expression on behalf of the scientists of this country and if we have not got the guts to give that expression, I think we are losing our right to the confidence of the scientists of this country."[23]

Next, L. C. Dunn introduced a motion for the academy to approve Richards's actions to date as an expression of general support for the NAS president. The motion passed. Then Veblen moved for the members of the academy to direct the president or the council to publish the statement. Vannevar Bush asked if publication would "involve any lack of good faith" between Richards and Representative Thomas, given Richards's earlier indication that he did not intend the statement's release. Richards answered that if instructed by the academy to publish, he would not feel bound by any prior commitment to Thomas. At this juncture, medical biologist George W. Corner suggested that at the very least the academy should issue a statement to the press that indicated the academy's consultations with Thomas and called for fair treatment of Condon.[24]

The discussion turned away from Veblen's motion. Nobel Prize–winning biochemist Wendell M. Stanley protested the idea of the academy becoming "all tangled up" in public criticism of HUAC. As an alternative, he urged the academy to study the more general question of civil liberties and aim for "constructive criticism" and "constructive recommendations" rather than engage in a futile critique of HUAC's actions. J. Robert Oppenheimer agreed with Stanley. In mid-March, he had discussed the Condon statement with Stanley and suggested to Richards that the academy "should perhaps address itself to the basic issues" of Cold War relations between scientists and the federal government.[25] At the business session, he affirmed his support for Stanley's position and suggested that the academy "engage in a broad study of the problems involved of this special type." He noted, however, that academy sponsorship of such a study and release of the Condon statement were not mutually exclusive, and he suggested that the academy members vote on a proposal for a study after considering Veblen's motion.[26]

The debate continued. Some members favored publication, but others remained opposed. Rabi suggested releasing the statement along with the names of those who endorsed it, but then physical chemist William D. Harkins objected that HUAC might take advantage of the fact that there were members of the academy who did not support it. Aeronautical engineer and National Advisory Committee for Aeronautics chairman Jerome C. Hunsaker announced that he had previously favored publication, but given what he had learned at the business session, he could not consent to an action that would "have our President break his word." Richards again indicated his willingness to accede to the academy members' wishes and reiterated that he did not regard his letter to Thomas as "a promise not to publish." Chemical engineer and petroleum industry executive Robert E. Wilson, however, backed Hunsaker's view and added, "The one constructive thing we can do, and probably get a unanimous vote on, is along the line of Dr. Corner's suggestion that we ask for a prompt hearing and not let a man sit under charges." At this point, Bush jumped back into the discussion. He announced, "I think that Mr. Corner has made an excellent suggestion." Corner's proposal, Bush explained, would allow the academy to voice its concerns about HUAC publicly without placing Richards in a compromising position.[27]

Veblen's motion came closer to a vote. When Richards asked to have the content of Veblen's proposition clarified, Bush commented, "I think his motion is simply an instruction to the President to publish a statement." Veblen caught Bush's implication that another, weaker pronouncement could be substituted for the one under consideration and objected, "To publish *this* statement." Bush then offered an amendment to the motion "to the effect that the publication take the form of a statement to the press by the President, approved by the Council, embodying the ideas expressed by Mr. Corner." Several persons seconded Bush's proposal, and surprisingly the transcript contains no indication of further objections from Veblen. The afternoon was growing late, and perhaps Veblen sensed that the momentum of the debate had shifted. The amended motion passed, and the meeting adjourned.

Bush's compromise resulted in a press release issued May 3. The release was much weaker than the statement originally distributed to academy members. It quickly summarized the original as an expression of "grave concern" that HUAC's procedures might discourage scientists from accepting government employment, but it also took care to report that "Mr. Thomas authorized [Richards] to assure the members of the

Academy that Dr. Condon would be treated with complete fairness."[28] The condemnatory tone of the original statement had been completely excised. The press release simply provided a neutral account of the academy's dealings with Thomas without any specific criticisms of the procedures followed by HUAC. The press release, a product of weeks of protracted deliberations, was so lackluster that few, if any, news outlets even took notice of it.[29] Condon lamented, "I doubt very much that Prof. Richards will do anything effective. . . . It is amazing to me how some of the older scientists seem to be so completely lacking in perception of what is going on."[30]

Political differences help explain the split between scientists who opposed vigorous action in the Condon case and scientists who advocated it. Those who preferred a low-key effort were usually, if not conservative in the strict political sense, from backgrounds at the center of power in the United States. They tended to be part of the governmental advisory elite, leaders in industry, or both. Bush and Conant fit this description, as did Hunsaker, Suits, and Robert E. Wilson. By comparison, many of the scientists who spoke out most strongly in defense of Condon were either liberal or further to the left. Dunn, Peters, Shapley, and Veblen were all active in the progressive left, and Condon himself was an outspoken liberal. Stone's politics are less clear, but he was an adamant civil libertarian. Influence within the academy lay with the more conservative group, which populated the council and NAS committees. In 1948, not one member of the council was associated with the progressive left or known for being particularly liberal. From a political perspective, then, it is not surprising that the NAS council consistently acted to water down the academy's response to the Condon case.[31]

There were exceptions, of course. I. I. Rabi was an elite scientist-administrator—he had been associate director of the Radiation Laboratory at MIT during the war, and he was a member of the AEC's General Advisory Committee. But Rabi never cultivated the calculated politesse of so many policymakers. He was fairly conservative, but he also distrusted anticommunism as an ideology. He consistently opposed abusive investigatory practices, and he was dismayed by the academy's failure to take a firm stand in the Condon case.[32] Political beliefs and institutional interests only defined general tendencies; they did not rigidly dictate individual scientists' political positions.

Scientists' general discomfort with and disdain for loyalty-security investigations crossed political boundaries. Investigations threatened to undermine the autonomy and intellectual freedom of all scientists, and conservatism as an ideology could just as easily accommodate antistatist, libertarian attitudes toward personal freedom as anticommunist-mandated subordination to the state. Frank B. Jewett, for example, was just such a libertarian conservative. Although he felt the scientific community should not back Condon unreservedly until all the facts were known, he did favor a strong condemnation of HUAC's methods. When HUAC first attacked Condon in 1947, Jewett even approached friends in New Jersey Republican Party circles to see if anything could be done to turn J. Parnell Thomas out of office.[33]

Nor did scientists who opposed publicly criticizing HUAC necessarily refrain from doing so in private. As Bush told Jewett, "I feel as strongly as anyone else that the methods of the Un-American Activities Committee are entirely improper, and I have said so forcibly under circumstances where I think it will have a good deal of effect."[34] Some scientists also differentiated between what they could do as individuals and what steps they felt the NAS could take. Wendell M. Stanley, for example, thought the academy should avoid a messy public confrontation with HUAC but felt no personal compunctions about tackling HUAC on his own. In a public statement issued shortly after HUAC's March 1 report, Stanley sharply criticized the committee for "actually destroying the very thing it professes to defend." He minced no words and declared HUAC "should be abolished or its methods modified drastically."[35] Conservative, liberal, and progressive left scientists, then, could agree on the basic need to protect scientists' civil liberties. Their political differences, however, led to more subtle distinctions in the types of efforts they were willing to embrace. When it came to organizational responses, more conservative scientists consistently advocated a temperate political manner and low-risk actions, and they rejected the methods of overt protest.

The National Academy of Sciences' pallid efforts did not end with the press release on the Condon case. Stanley's comments at the business session led the NAS to appoint a committee on civil liberties "which would investigate, study and report to the Council regarding civil rights, including clearance of scientific personnel, secrecy, and loyalty investigations."[36] In November, Richards appointed James B. Conant chairman of the Committee on Civil Liberties, with J. Robert Oppenheimer and Oliver E. Buckley, president of Bell Telephone Laboratories, as the other committee members.

Neither Conant, Oppenheimer, nor Buckley could be described as ardent civil libertarians or vocal liberals. Richards appointed the three because, "in addition to being wise," their experience with loyalty and security issues dated from the early days of World War II.[37] The committee's conclusions about civil liberties, however, reflected little in the way of wisdom gained through experience. The committee's sole accomplishment was a short, two-and-a-half-page statement of recommended changes in personnel security policies. Conant, Oppenheimer, and Buckley urged the government to make a stronger distinction between employees whose positions had national security implications and ordinary federal workers, and they advocated the transfer without prejudice of employees who might pose security risks to positions that did not involve access to secret information or otherwise affect national security. These changes, they suggested, would better protect employees' rights without compromising the nation's safety.[38]

Like the academy's press release on Condon, the statement on civil liberties was extremely cautious. It expressed concern but offered no specific criticisms of existing personnel security procedures. Generally vague, the statement showed no evidence that the committee had made a serious attempt to investigate and study problems related to personnel security. It was so nebulous that the exact government policy being addressed—personnel security, the loyalty program, congressional investigations of scientists, or all three—remained unclear. Not only was the statement poorly conceived, but the NAS council decided to limit its distribution to the White House rather than release it publicly. Ultimately, the effort yielded nothing more than a cursory acknowledgment from President Truman.[39]

Oppenheimer admitted to Richards that the council considered the statement "not very hot stuff and that many members of the Council would have wished to say more and deeper things."[40] Physicist P. W. Bridgman found the decision not to publicize the academy's position "regrettable" and lamented the academy's increasingly constrained view of its own role. Bridgman wrote to Richards, "It seems to me that during the last few years a philosophy has been growing up with regard to the functions of the Academy and what it may appropriately do which I regard as seriously restrictive." While he recognized that the academy had originated as a quasi-governmental agency, it had developed into "the highest representative of science in this country" and "should not be bound by the fetters of its origin." The academy's sense of self-restraint, Bridgman felt, was "seriously limiting its opportunities for usefulness."[41]

So constricted an interpretation of the academy's prerogatives was not inevitable. Less than two years earlier the NAS leadership, under the presidency of Frank B. Jewett, had advanced a much more expansive vision of the academy's purpose. During World War II, the NAS had played a major role in administering scientific research, and at the end of the war, the academy had energetically sought to ensure continued science-military cooperation and expand its influence over the postwar organization of scientific research. By 1947, the NAS seemed more than willing to act without a prior request from the government. In the 1946–47 annual report of the academy's National Research Council, Chairman Detlev Bronk promised that in the future the NRC would be "more than a waiting agency through which government and private organizations seek assistance from the scientists of the country." Instead, the NRC would be "adventurous in seeking opportunities for leadership and useful action in all fields."[42] Similarly, in November 1947, then president Jewett delivered a forceful address to the academy on the changed relationship between the NAS and the federal government. According to Jewett, the NAS charter granted sweeping powers to the academy. The Civil War–era charter provided a broad constitution that recognized that "times and conditions change" and gave "not the slightest hint of any attempt to shackle the Academy to the problems or philosophy of 1863." Under the charter's mandate, the academy could advise the government at its own discretion; the charter did not "dictate how the advice should be obtained or given." The prerogatives delegated to the academy under its Act of Incorporation, Jewett declared, made the academy "potentially the most powerful organization in the field of science in the United States, and probably in the world." He concluded with ebullient confidence, "There is practically no limit to the distance the Academy and Research Council can go in influencing the broad trends of scientific development in America."[43] In his history of the NAS, Rexmond C. Cochrane observed that Jewett's address "banished the long-held notion that the Academy could act for the government only when called upon and had no power of initiative or privilege of providing advice."[44]

Despite this commitment to enlarging the NAS's postwar role, the academy did little to ensure that the civil liberties of scientists were protected. In part, the leadership style of Jewett's successor narrowed the academy's options. A. N. Richards consistently held to a strict construction of the academy's powers and responsibilities. With respect to the Condon case, it was his initial inclination to take vigorous action that was uncharacteristic, not his ultimate decision to draw back. Under Richards's leadership, the NAS tended to be reluctant to stretch the boundaries of its charter.

Moreover, the institutional configuration of postwar science constrained the academy's actions. Many of the elite leaders of American science held powerful positions on government boards in which they helped define and build the Cold War linkages between science and the national security state. Jewett and Bush, for example, both participated in postwar discussions with the military about how best to ensure that scientific research would continue to serve the nation's security needs. Bush eventually became the first chairman of the Research and Development Board, established under the National Security Act of 1947 to provide advice on scientific research and development to the secretary of defense. Detlev Bronk served on the Naval Research Advisory Committee of the Office of Naval Research, established in 1947. Oppenheimer served as the first chairman of the AEC's prestigious General Advisory Committee (GAC). Conant was also a founding member of the GAC, and Oliver E. Buckley joined the committee in 1948. These are only a few of the high-level positions held by these men and other prominent academy members.[45] When Bronk and Jewett spoke of the expanding influence of the academy, they were referring more to the growing status of scientist-administrators within the national security state than the ability of the NAS to bring public pressure to bear on repressive government policies.

This is not to say that these scientists acted simply as agents of the national security state or readily accepted the values of Cold War anticommunism. Their reactions were often more complex. For example, as discussed in chapter 1, both Bush and Conant had serious reservations about the strong postwar relationship between science and defense.[46] Nevertheless, despite their doubts, the elite scientific leaders took actions that helped foster the postwar dependence of science on military patronage. With the academy tied to a set of institutions in which influence depended on commitment to the goals of the national security state, the NAS's efforts to address Cold War anticommunism foundered less on fear of or sympathy with domestic anticommunism than on a more basic reluctance to speak out and criticize Cold War politics.

The AAAS Special Committee on Civil Liberties

The American Association for the Advancement of Science, like the National Academy of Sciences, also had the potential to influence loyalty and security policy, although for somewhat different reasons. The AAAS was the largest scientific organization in the United States. In 1948, it had approximately forty-three thousand members representing a diverse range of disciplines in the natural sciences and the social sciences. The association concentrated primarily on professional development and the spread of scientific knowledge, and it did not have a strong record of political activism.[47] But its large and heterogeneous membership meant it had the potential to mobilize a substantial portion of American scientists. Size and diversity, however, proved its undoing in its tentative efforts to formulate a response to the Condon case and the threat domestic anticommunism posed to the rights of scientists. Lacking a robust political tradition, the AAAS was ultimately unable to withstand protests from a small minority of its membership when it tried to address the dangers of the security system and the federal loyalty program.

At the December 28, 1947, meeting of the AAAS council, University of Minnesota School of Medicine physiologist Maurice B. Visscher presented a resolution on behalf of the American Association of Scientific Workers that called for the council to establish a committee to examine the impact of loyalty-security investigations on the scientific community. The proposed committee would prepare "a factual report" on the effects of the federal loyalty program and the activities of HUAC on scientists and make policy recommendations to the AAAS council.[48] The resolution passed, and Harlow Shapley, then serving as president of the AAAS, appointed Visscher chairman of the Committee on Civil Liberties for Scientists, with Cornell legal scholar Robert E. Cushman, chemical engineer and university administrator Frederick Hovde, James R. Newman, and Richard L. Meier as the other members. Apparently Hovde chose not to serve and was replaced by Johns Hopkins School of Medicine physiology professor Philip Bard, and Columbia University legal scholar Walter Gellhorn joined the new committee as a consultant.

The precise impetus for the creation of the AAAS committee is unclear. Visscher's resolution did not mention any specific cases prompting the proposal, but by late 1947 growing numbers of scientists had tangled with executive branch loyalty-security probes, and both Shapley and Condon had come under the scrutiny of HUAC. It is surprising, however, that the resolution did not explicitly call attention to security clearances. Perhaps as a medical researcher, Visscher was less attuned to problems related to security clearances, or he may have erroneously assumed security clearance policy was subsumed under the loyalty program. Nevertheless, given the committee's composition, especially the presence of Meier, the executive secretary of the FAS, security clearance would necessarily absorb a major portion of the committee's attentions.

Unlike the NAS council and the academy's Committee on Civil Liberties, the AAAS committee had a liberal-left bent. Visscher was a leading member of the American Association of Scientific Workers and sympathetic to the progressive left. Both Cushman and Gellhorn were prominent civil libertarian legal scholars, and Gellhorn was active in the National Lawyers Guild, the progressive left's alternative to the American Bar Association. Newman was a New Deal liberal, and Meier was an FAS moderate deeply concerned about preserving scientists' individual rights. Given its composition, the AAAS Committee on Civil Liberties for Scientists promised to take loyalty and security issues seriously.

The new committee held its first meeting in Washington, D.C., on March 12, 1948. In addition to formulating its long-term plans, the committee also recommended that the AAAS respond publicly to the Condon case. Visscher told Shapley that such an action would constitute "a great service to the cause of science." In sharp contrast to Richards and the NAS council, Visscher concluded that "for the AAAS to be silent in this matter is unwise, perhaps dangerous." Shapley apparently agreed. Much of the language of the AAAS's subsequent April 25 statement on the Condon case was drawn directly from Visscher's letter.[49]

After months of work, the Special Committee on the Civil Liberties of Scientists submitted its report to the executive committee of the AAAS on December 18, 1948. The report delved much deeper than the statement of the NAS Committee on Civil Liberties. It was seventy-seven pages long, with detailed discussions of secrecy, personnel security, and the federal loyalty program, and its analyses reflected both scientists' concerns as well as the legal expertise of Cushman and Gellhorn.

Most of the report dealt with the problems posed by security clearance and the loyalty program. The committee conceded that inquiries into scientists' character and attitudes were permissible where matters of national security were concerned, a position that Cushman and Gellhorn shared with other civil libertarian legal scholars in the postwar period. Existing security clearance procedures required reforms, however. Both the AEC and the military, the committee contended, lacked formal procedures that respected due process rights. In the AEC, problems remained despite the implementation of the Interim Procedure. No formalized standards existed for rendering decisions, charges were still vague, there continued to be an overreliance on secret testimony in FBI reports, and AEC security officers had no obligation to provide a statement of their reasoning in the event of negative decisions. Applicants for positions, if denied employment, were especially vulnerable because they had no right to know if the reasons they were not hired

Maurice Visscher, 1947.
(American Physiological Society)

involved security. The situation in the armed forces was similar. The Industrial Employment Review Board, responsible for clearance decisions involving military contractors, imposed the additional burden of classifying hearings and forbidding employees from publicizing their situations. (The AEC allowed employees to make their cases public if they so wished.) Finally, the AAAS committee found a growing tendency, within the AEC and elsewhere, to require security clearance for positions that did not involve access to restricted data. To address this disturbing trend, the committee advised that "the stringent application of personnel security clearance should be limited to smaller numbers of scientists rather than extended to ever larger groups."[50]

Whereas personnel security investigations served a necessary, though regrettable, purpose, the committee found no legitimate grounding for the loyalty program. On this point, the AAAS took a stronger position than either the academy or, as we shall see, the FAS. The committee found the loyalty program "basically objectionable because it seeks to determine an employee's loyalty by inquiring into his supposed thoughts and attitudes," and argued that the loyalty order ought to emphasize overt behavior rather than beliefs. Ordinary dereliction of duty, however, could easily be dealt with through administrative supervision, and exisiting laws already covered criminal behavior sufficiently. The legitimate functions of the loyalty program, then, merely duplicated practices already in force. The committee did not openly call for the outright abolition of the loyalty program, although that position was implicit in its critique. Instead, the report advocated the revision of the loyalty program to ensure the protection of individual rights.

The committee made a number of concrete recommendations. Regarding AEC clearance procedures, the committee proposed measures similar to those supported by the FAS, such as publication of the standards used to decide cases, release of precise statements of charges to employees, and written explanations of the reasoning behind decisions to withhold clearance, as well as the extension of the procedural rights of employees to applicants. The committee also urged that security clearances be required only for individuals who were to have access to classified information. The report suggested that the military take similar measures, plus the added stipulation that it not classify hearings. The committee also advised comparable procedural safeguards for the loyalty program, assuming its retention, and suggested that organizations on the attorney general's list be given the right to contest their listing.[51]

The committee recommended that the AAAS formally transmit the report to President Truman, publish it "in an effort to stimulate broad consideration of its contents by scientists and their organizations as well as by newspapers, columnists, journals of opinion, radio commentators, and other informers of public opinion," and send it to the relevant government agencies and officials.[52] The report was not published, however. The first delay in its distribution occurred in January 1949 when Visscher received an inquiry from the loyalty board of the Federal Security Agency in connection with his consultantship to the cardiovascular study section of the National Institutes of Health. Visscher felt the inquiry was probably a routine matter, but he feared that if he was "declared ineligible to serve as an occasional consultant . . . on a 'guilt by association' count," the report would be compromised. The alternatives were either to postpone publication or for Visscher to withdraw from the committee. Visscher worried that if he became involved in an appeals process there might be a long delay, so he felt inclined to withdraw. He told the other committee members, "It is self-evident that to have the name of one not cleared in a loyalty case appear on the report would seriously impair its effect. Therefore as long as I am not cleared my name cannot be used effectively. Since appeals may take a long time I am inclined to the view that we should settle on a policy now. . . . The only really totally safe course, as far as the report is concerned, is for me to withdraw from it completely."[53]

The other members of the committee stood up for Visscher and argued against his resignation. Cushman wrote Visscher, "I feel strongly, and the others agree, that we should lie low and do nothing at all about this matter for the time being. There is every probability that there is no real issue in your case at all, and that you will be cleared as promptly as red tape permits. It is in expectation of this result that I should urge you to stay on the committee, even if it means holding up the report." Cushman reassured Visscher that based on his consultation with Eleanor Bontecou, an expert on the federal loyalty program, the inquiry that Visscher had received was indeed a purely routine matter.[54] In a brief message, Meier wrote Visscher and the other committee members that based on "a careful independent inquiry" he had reached the same conclusion as Cushman. He noted, "I am sorry the report must be held up, but it seems to be the only way of handling the situation."[55]

Visscher's concerns turned out to be unfounded, and he was cleared on February 24.[56] But his response to the Federal Security Agency inquiry illustrates very clearly the intimidating effect loyalty and security investigations had on individuals even under ordinary circumstances. Until he received clearance, Visscher felt he could not participate fully as a member of the committee lest an adverse decision undermine the report's legitimacy. Visscher's reaction to a routine loyalty investigation provides a telling example of how the anxiety produced by the normal operation of the loyalty program impaired people's freedom of action and promoted political conformity. The negative effects of the loyalty program reached far beyond the numbers of individuals dismissed or not hired in the first place. The act of investigation itself instilled fear, and it discouraged deviations from Cold War political orthodoxy. After his encounter with the loyalty program, Visscher, like Eugene Rabinowitch, decided to preserve his political autonomy by resigning from the study section and refusing ever again to accept a position that required loyalty or security clearance.[57]

Even after Visscher received loyalty clearance, the report on civil liberties went unpublished. In March 1949, the report was sent to members of the AAAS council, along with a questionnaire detailing possible options for its publication and distribution. Approximately half the council members replied, voting more than four to one that the report be adopted as the official statement of the AAAS, formally sent to President Truman, published, and the recommendations distributed to the appropriate government bodies.[58] Despite this unambiguous show of support for the widest possible distribution, the executive committee of the AAAS, meeting on April 23 and 24, voted to return the report to Visscher's committee for condensation. The civil liberties committee was unable to abridge the report in a satisfactory manner, and in July Visscher suggested that *Science* simply publish the report's recommendations. His suggestion was accepted. The conclusions and recommendations of the report appeared in the August 19, 1949, issue of *Science*, and mimeographed copies of the full report were available upon request.[59] But the report never received the wide distribution beyond the scientific community advocated by the AAAS council.

Why did the AAAS report on civil liberties fail to reach a larger audience? Cost may have been a factor, but the failure even to send copies to President Truman or the agencies that administrated the loyalty-security system suggests a more deeply rooted opposition. According to an FBI report, Harlow Shapley blamed "reactionary conservatives" on the executive committee for bottling up the report and refusing to publish it. Years later, Visscher also cited the presence of "some timid souls" and "a few apparent supporters of Joseph McCarthy" on the AAAS council and executive board. Neither Shapley nor Visscher provided more details, however.[60]

Unfortunately, there are few documents in the AAAS archives or Visscher's personal papers that shed further light on the nature of this controversy. In addition to anticommunist sentiment among individual committee members, the executive committee probably also feared opposition from the AAAS's member organizations. The mere publication of the conclusions and recommendations did result in fierce opposition from at least one member organization, and several other AAAS affiliates also considered voicing objections. After the Visscher committee's recommendations appeared in *Science*, the American Association of Petroleum Geologists (AAPG) "took vigorous exception to the Visscher report and was even urged by one of its representatives on the AAAS Council to disaffiliate."[61] At its April 1950 annual meeting, the AAPG publicly disassociated itself from the report and issued the following resolution:

> The American Association of Petroleum Geologists recognizes that, as Secretary Acheson has said, the United States is "engaged in a struggle that is crucial from the point of view of the continued existence of our way of life."
>
> The AAPG further recognizes the urgent necessity, under these circumstances, of preventing information of scientific or other nature, vital to our safety, from falling into the hands of a possible enemy.

> Totalitarianism has proved itself far more destructive to the liberties of scientists and restrictive of scientific progress than any laws or regulations imposed by the United States in the interest of security or national defense.
>
> The fact that restricted data can be readily and secretly transmitted to unauthorized persons and that the national security can be endangered by even a single disloyal scientist has been amply demonstrated. The Fuchs case offers an example.
>
> The AAPG believes that all loyal citizens of the nation, whatever their category, must be united without reservation in support of the measures deemed necessary by the Federal authorities for the security and defense of the nation. We believe that no loyal citizen, whether scientist or not, should object to investigation of his loyalty. Therefore, although secrecy may for a time impede our scientific progress, we shall abide by such security requirements. We take pride in our readiness, cheerfully and wholeheartedly, to prove our loyalty and patriotism in case of inquiry.[62]

The resolution, adopted unanimously by the fifty members of the AAPG business committee present at the meeting, explicitly embraced the logic of Cold War anticommunism. It accepted the notion of the vital secret and expressed the belief that the state had the right to take whatever action it deemed necessary to protect the national security. Furthermore, it rejected the Visscher committee's argument that loyalty investigations were invalid when no national security interest was at stake. The AAPG resolution made no mention of individual rights; it simply proclaimed that citizens had the duty to submit themselves to loyalty inquiries. These attitudes were incompatible with the tenets of the Visscher report.

The AAAS leadership took the petroleum geologists' protest seriously. Discussions between the AAPG and AAAS resulted in the publication of the former's resolution in the June 9, 1950, issue of *Science*. Then AAAS administrative secretary Howard Meyerhoff traveled to Ardmore, Oklahoma, to visit the retiring president of the AAPG and heal the rift. The AAPG president returned the visit in September "to assure the Administrative Secretary that, by formal action of their Executive Committee at Banff, the matter had been dropped and that resumption of amicable relations is desired."[63] Diplomacy had resolved the crisis.

How much bearing the AAPG's vehement objections had on the failure of the executive committee to distribute the Visscher report cannot be ascertained from AAAS records. Perhaps the AAPG merely confirmed most executive committee members' preexisting attitudes, or possibly the AAPG's reaction led the executive committee to fear an even stronger outcry from other AAAS member associations if it took more energetic efforts to influence loyalty-security policy. Although the AAAS participated in the postwar debate over the formation of the National Science Foundation, it was an organization without much experience in politics, and its leadership may have been unwilling to act forcefully without unanimous support from its member organizations, an unrealistic standard given the size and diversity of the association.[64] Instead, the Visscher report was quietly tucked away. Despite a promising start, the AAAS failed to mount an effective challenge to the loyalty-security system.

The FAS: The Scientists' Committee on Loyalty Problems

As discussed in the previous chapter, by 1947 several local FAS affiliates, under loose direction from the FAS Washington office, had already taken measures to respond to the problems scientists had encountered with the federal loyalty-security system. In May 1948, the FAS council began to consider mounting a more centralized effort. In June, R. L. Meier and J. M. Lowenstein, Walter Gellhorn's assistant, wrote a memo to notify the FAS chapters of the federation's plans. The personnel security situation, they explained, had serious political ramifications with respect to the growing militarization of scientific research and other restrictions on scientists, and the federation had to act if it was to preserve its own political viability. Meier and Lowenstein observed, "For the first time in our experience, practically everyone in the scientific fraternity is affected [by security restrictions] — the scholars are finding academic freedom and independence dissipated by military contracts and external social pressures, while industrial and applied scientists meet with new forms of arbitrary dicta. . . . The long term solution we believe must be international security through international collaboration, and the FAS continues to work toward those ends, but one of the pressing problems that has now come up is that of maintaining our effectiveness and freedom of action over the next few years." Meier and Lowenstein then observed that although scientists were especially susceptible to loyalty inquiries, they were also in a position to take action. They noted, "Scientists are particularly vulnerable through being in the most sensitive security areas, atomic energy and national defense. But for this very reason, because of the current demand for high grade scientific talent, they are in a good position to demand fair clearance and loyalty procedures."[65] In light of these circumstances, the FAS decided to form a new committee to address loyalty, security, and related issues.

Meier and Lowenstein's observations are important for understanding the place of scientists within the politics of anticommunism. Scientists were not simply passive victims of unjust Cold War political repression. They were also agents of the national security state, who offered crucial expertise for the execution of Cold War policies. As such, they had extra room to maneuver, even as the rise of domestic anticommunism made political dissent increasingly difficult. In contrast to the NAS's more submissive attitude, the FAS believed scientists could take their fate into their own hands. But in 1948, as Meier and Lowenstein realized, the scientists' situation was approaching a critical juncture. If scientists did not insist on their full political autonomy in the face of anticommunist attacks, they could expect a dim future for nuclear arms control, freedom in science, and other goals of the atomic scientists' movement.

The federation started with high aspirations for the new committee on civil liberties. The FAS council hoped such a committee would not act merely in favor of scientists but help "improve the situation for all citizens." Meier and Lowenstein conceived of the new group as an "action committee," and they stressed the need for aggressive measures, including publicity and direct intervention in individual cases. In September, as the newly formed Scientists' Committee on Loyalty Problems (SCLP) was just about to start operating, Higinbotham and other FAS officers reiterated the importance of keeping the abuses of the loyalty-security system before the public eye. Calling attention to the 1948 election, they declared, "Candidates, especially, will be alert to public opinion. It is not enough that the local chapter has sent strong telegrams or issued a statement. We ARE news today, but we won't be news tomorrow. We must all continue to point out that the importance of atomic secrets is overemphasized, that secrecy conflicts with scientific progress, that no professional class can work effectively in an atmosphere such as that which exists for us today." At a meeting called by Harlow Shapley to discuss responses to recent hearings by HUAC on wartime atomic espionage, representatives from the FAS, the American Association of Scientific Workers, and the National Council of the Arts, Sciences, and Professions decided that the SCLP would "assume the lead in getting out publicity." After the November elections, Melba Phillips hoped the new committee would take scientists' concerns about security clearance straight to Congress and the Truman administration.[66] With the SCLP, the federation appeared to be preparing for a major public initiative to shore up individual political rights in the face of domestic anticommunism.

Formed during the summer of 1948 and centered at Princeton University, the Scientists' Committee on Loyalty Problems was composed of scientists in the greater Princeton–New York City area, including Princeton astrophysicist Lyman Spitzer Jr. as chairman and William Higinbotham as associate chairman. The committee held its first meeting on September 25.[67] At about the same time, other FAS chapters gradually shut down their activities on security issues. The Cornell Committee on Secrecy and Clearance transferred its files to the Princeton group, and at the Association of Oak Ridge Engineers and Scientists, operations declined as active members left Tennessee in the wake of the transfer of the nuclear reactor program to Argonne.[68] For the next three years, the SCLP directed the bulk of the FAS's civil liberties efforts.

Despite the initial grand conceptions behind the Scientists' Committee on Loyalty Problems, the FAS's civil liberties efforts contracted under the direction of the new committee. The SCLP followed a highly constrained, unobtrusive strategy. It emphasized procedural reforms, limited its activities primarily to policies directed specifically at scientists, and preferred the use of internal contacts within government agencies over publicity as a way to expedite cases. This carefully circumscribed, low-risk approach meant that the committee never directly contested the ideology of Cold War anticommunism or the evolving meaning of security under the postwar national security state.

The restrained nature of the SCLP was apparent from the beginning. The SCLP stated it would provide information and criticize procedural problems, but it would not directly defend individuals. A stand on an individual's loyalty would interfere with the committee's purposes. As noted in the minutes of the SCLP's first meeting, "[The committee] could take no stand on whether any individual was 'subversive,' or [a] 'security risk,' it did not propose to defend any individual's loyalty, character, or discretion. There was a general feeling among the Committee that this attitude toward the individual would not and should not interfere with the committee's duty to use the most effective and vigorous methods to help an individual get a fair hearing, nor should it interfere with the committee's responsibility to criticize what it regards as bad procedure or faulty criteria."[69] Under this policy, aid to individuals was usually limited to information about federal loyalty-security procedures, advice, and a legal referral. In some instances, the SCLP also used internal contacts to expedite the processing of cases. A March 18, 1949, report on SCLP activities noted that the committee had provided thirty scientists with information and, where necessary, legal referrals, and in an additional ten cases, the use of "personal contacts" led to "the result that either the processing was speeded up or the status was clarified."[70] The available documentation, however, indicates that even when the SCLP used quiet diplomacy, it rarely reached the level of direct intervention that had earlier benefited the Blewetts and the Oak Ridge scientists. As the FAS formalized and routinized its operations, and as AEC clearance procedures simultaneously became more streamlined, individuals received less assistance from the FAS, not more.

Individuals' politics may have helped determine the level of aid they received. Samuel A. Goudsmit, a member of the SCLP, had special access to AEC intelligence, and he used his connections to obtain inside information about scientists' cases. But he was also suspicious of scientists he considered too far to the left, and, unknown to his scientific colleagues, he regularly provided the AEC and the Central Intelligence Agency with his assessment of various scientists' political inclinations. Given his access to information and propensity for political labeling, he probably had a great deal of influence over whether or not scientists received help from the SCLP beyond advice and a referral for legal services.[71]

In addition to its limits on aid to individuals, a restrained manner also characterized the SCLP's policy recommendations. In mid-November 1948, the committee began to consider a policy statement on the federal loyalty program and government security clearance procedures. The original statement articulated positions similar to those taken in the AAAS's report on civil liberties. It conceded that government employment was not a right and that in areas in which national security was at stake the government had a legitimate purpose in conducting security investigations. At the same time, however, the proposed statement contended that investigations of all federal employees were unnecessary and costly. Such investigations, the SCLP warned, "may interfere with the civil rights of large numbers of people, may create a secret police atmosphere, and may have a damaging effect on the morale of government workers. Investigations of employees working in classified areas may be similarly dangerous if whole professions become involved."[72]

Based on these observations, the SCLP made several concrete proposals. The committee recommended that the federal loyalty program be "modified or rescinded," that the application of security clearance requirements be "minimized in government, industry, and research institutions," that security clearance cases be founded on "thorough investigations," and that individuals be guaranteed "all possible procedural safeguards so that a minimum of unjustified dismissals will take place, damage to [their] reputation and ability to find employment elsewhere will be avoided, and good morale will be maintained." To carry out this program, the SCLP had a number of specific suggestions. In clearance cases, the committee proposed that prior to suspension or discharge of an employee or a decision not to hire an applicant based on security grounds, the individual be given a statement of the charges against him or her, granted a hearing within thirty days, and have the right to appeal an adverse decision. In addition, the administrative process needed to run more smoothly so as to avoid unreasonable delays.[73]

Over the next few months, the SCLP debated the proposed statement. The committee also solicited outside opinions. Katharine Way submitted a detailed critique. The SCLP also received comments from Eugene Rabinowitch, John P. Peters, the Atomic Scientists of Chicago, the Association of New York Scientists, and other respondents, but their replies unfortunately do not appear to have survived in archival records.

Way identified a central shortcoming in the SCLP's basic approach to loyalty and security issues. She pointed out that the primary problem scientists faced was not procedural but the lack of clear standards and the relationship between the loyalty-security program and the Cold War. She wrote, "All the present personnel security programs are based on the idea that possible 'disloyalty' is connected with leanings toward Communism. No one is very sure just how the two are connected. However, the fear of Communism is quite obviously the basis for all loyalty checks and can be expected to continue as such as long as the state of tension between ourselves and Russia continues."[74] The crucial problem, Way contended, was the absence of criteria regarding what constituted a threat to American security. She continued:

> The greatest progress in making sense out of a program based on this central fear should come from some clear thinking and more general agreement on just what "Communistic tendencies" constitute a real threat to the country. Many people, including AEC officials, apparently consider interest in Negro problems, consumer co-operatives, or civil liberties as possible danger signals. To most such interest would indicate only a strong desire to preserve democracy and to make it work. However, real connections with foreign powers could certainly be dangerous to the national welfare. It seems important to try to say just what connections could be dangerous and why.[75]

In other words, a deeper analysis of what it meant to be a "security risk" would go far in ending the abuses of the loyalty-security program. Way observed, "Clarification of ideas as to where real dangers lie would serve to eliminate harassment of many employees who have nothing more than a well developed sense of social responsibility."[76] Way's comments identified the central failing of the FAS's and other leading scientific organizations' policies on personnel security: a reluctance to confront directly the ideology behind the Cold War.

Although Way laid greater emphasis on the need for standards, she also had some comments about procedural issues, especially the use of unsworn testimony in security investigations. In one of the cases at Oak Ridge, she recalled, six witnesses who had supplied the FBI with derogatory information in unsworn interviews agreed to give sworn testimony in a hearing. Five of the six recanted their earlier stories, and the sixth was "completely confused as to times and places." The incident suggested to Way that "sworn testimony, confrontation, and cross examination cannot safely be dispensed with." The FBI could still use interviews to obtain leads, but formal charges required either sworn testimony or independent corroboration to protect the rights of the individual in question.[77]

Finally, as she had done earlier in responding to the Oak Ridge cases, Way stressed the need for publicity. The lack of clear standards and fair procedures was not a new problem. She noted that a number of scientists "have been making statements and writing letters to this effect for over a year now with no noticeable results." Her solution: "The best chance for improvement now seems to lie in the maximum amount of publicity." She suggested that "a healthy public pressure is sure to be built up" if the AEC's policy of questioning the security status of individuals who already had long records of reliable government service became widely publicized. She explained, "The AEC apparently believes that it has the right to inquire into the political and social philosophies of its employees and that it can call them to account at any time before a 'security officer' for their ideas or social actions. A public which so heartily condemns the inquisition methods of a dictatorship will surely insist on their elimination in a democracy."[78] Like Meier, Higinbotham, and others who hoped for a high-visibility program from the SCLP, Way viewed publicity as central to the committee's purpose.

Lyman Spitzer Jr., the SCLP chairman, had very different thoughts about the proposed policy statement. He advocated a limited approach that would avoid political controversy, and he utterly opposed criticism of the loyalty program. He wrote SCLP secretary Arthur S. Wightman that the comment on the loyalty program in the statement was "highly controversial," and he worried that "if we adopted it, the effectiveness of our attempts to improve clearance procedures in security areas might be considerably impaired." Instead, Spitzer felt it would be better and safer for the FAS to direct its activities at security clearance alone.[79]

Spitzer feared a red-baiting backlash against the FAS if the committee took a position on the loyalty program. Although he personally did not object to the employment of communists in government positions that did not involve national security, he felt there was little point in trying to persuade the public of such a position. If the committee did so, he predicted, it would "immediately be strongly supported by all communists and fellow travellers." He continued, "While this might not be the same kiss of death for us as it was for Henry Wallace, it would not be particularly helpful. Army and Navy men would immediately conclude from our position that we were communist dominated. . . . If the repeal of the present loyalty requirement were our first objective, we should have to take these risks of support by the communists and attack by the military." Rather than have the SCLP tackle the all too risky problem of the loyalty program, Spitzer advocated that the committee limit its concerns to reforming security clearance procedures and restricting the application of the security system to as narrow a range of situations as possible. He stated, "This is a program where I believe we shall not get involved in any political fights, and where there is much chance of success."[80]

Others took a strong stand against the loyalty program and provided their reasons at the December 13, 1948, meeting of the SCLP. Spitzer missed the meeting, but those present discussed his letter to Wightman. Walter Gellhorn, attending as an invited guest, pointed out that the Hatch Act, combined with other existing laws, made most of the loyalty order superfluous. The almost completed report on civil liberties by the AAAS, for which Gellhorn was a consultant, shared this reasoning. Higinbotham proposed that discussion of the loyalty program be separated from discussion of possible action by the SCLP, and he argued vigorously for the program's abolishment. Truman's loyalty program, he contended, wasted government resources, discouraged people from government service, diverted resources from more important security investigations, and violated individual rights. Wightman's minutes of the meeting noted that after discussion, "it seemed to be the general opinion of those present that, if anything, Executive Order 9835 weakens, rather than strengthens, the security of the country."[81]

Most members of the SCLP shared a general sense of disapproval for the loyalty program, but they disagreed over whether or not the committee should take action. According to Wightman's minutes, some committee members agreed with Spitzer that speaking out against the loyalty program would "handicap [the committee's] attempts at securing better security procedures," but others felt "that a forthright statement on the loyalty program might be an aid rather than a handicap." In the end, Wightman noted, the discussion was "rather inconclusive."[82]

At the next SCLP meeting on February 12, 1949, the committee agreed to confine its actions to sending a letter to President Truman detailing the scientists' concerns about the loyalty program. The committee also explicitly rejected a high-intensity publicity campaign as part of its future activities. By doing so, it turned away permanently from the expansive hopes the FAS and other scientists' organizations had originally held for the SCLP. Lack of finances was part of the problem, but Wightman's minutes of the meeting also indicate that the committee believed gaining influence over government officials and highly placed scientists was of greater necessity than "education of the man in the street."[83] For its remaining lifespan, the committee would concentrate on a more limited administrative strategy rather than a large-scale publicity effort.

Thus, as a result of the SCLP's decisions, by early 1949, the FAS's policy on security clearance had narrowed in a matter of months. The federation abandoned widespread use of publicity as a possible strategy, provided only limited assistance to individual scientists, and restricted its concerns to government policies that primarily affected scientists and not Americans in general. The constricted boundaries of the SCLP's willingness to act became strikingly apparent in the committee's response to SCLP member David Bohm's testimony before HUAC on May 25, 1949.

The last major foray of the House Committee on Un-American Activities into atomic energy occurred in 1948 and 1949 when it held a series of hearings and issued reports on alleged wartime espionage among physicists at the Berkeley Radiation Laboratory. The investigations simply rehashed information that the FBI and Justice Department had known of for years and uncovered no evidence of espionage. In the context of the Cold War, however, the hearings offered HUAC the opportunity for another public spectacle.

In the spring of 1949, Giovanni Rossi Lomanitz and David Bohm, both former students of J. Robert Oppenheimer, and Oppenheimer's brother Frank were subpoenaed to testify before HUAC about their wartime connections to the Communist Party. Lomanitz and Bohm invoked the Fifth Amendment when asked if they were members of the Communist Party or acquaintances of party member Steve Nelson. Frank Oppenheimer testified that he had been a member of the Communist Party from 1937 to 1941 but refused to answer questions about the political affiliations of any of his acquaintances. The hearings led all three to lose their university positions. The University of Minnesota fired Oppenheimer from his assistant professorship only an hour after his testimony, while Lomanitz had a day's grace after his HUAC appearance before losing his assistant professorship at Fisk University in Tennessee. Bohm was not fired immediately, but after he was indicted for contempt of Congress, Princeton refused to renew his contract, despite his distinguished record of publications. Both Bohm and Lomanitz were indicted for contempt; both were ultimately acquitted.[84]

After his appearance before HUAC, Bohm immediately offered to resign from the Scientists' Committee on Loyalty Problems, and the committee debated its response. Aside from the contempt citation, Bohm had not been charged with any illegal act. Like many liberal-left academics in the San Francisco Bay area during the 1930s and 1940s, he had been attracted to the intellectual ferment and local political program offered by the Communist Party. He joined the party briefly from 1942 to 1943 but left because the "meetings were interminable."[85] Nothing about his political history was particularly covert, but he refused to testify under the Fifth Amendment because he did not want to involve others in similar predicaments before HUAC. The House Committee on Un-American Activities, for its part, had no new evidence to present and no truly legitimate investigatory purpose in the Radiation Laboratory cases. The committee's intimations of wartime atomic espionage merely provided an excuse for subjecting individuals with past (or present) Communist Party connections to invasive inquiries into their political beliefs. If witnesses took the Fifth, so much the better—the committee could imply they were dangerous subversives who were hiding their wrongdoing.

Given the circumstances, members of the SCLP might have entered the fray to defend a fellow colleague. Instead, the committee preferred to steer clear of controversy. As Samuel A. Goudsmit told Spitzer, "I believe that any incident which might hamper the effectiveness of the functioning of our committee should be avoided." If the SCLP could not expect the publicity surrounding Bohm to pass quickly, Goudsmit argued, the committee should let Bohm step down. Although Goudsmit declined to express his opinion on the likelihood of the Bohm case receding from public view, he leaned toward taking Bohm's voluntary resignation. He wrote, "I wish to point out that the acceptance of Bohm's resignation does not mean at all that we believe that there is anything against him. I'm merely afraid that his presence will obstruct further work."[86] Goudsmit feared that some government agencies already considered the FAS too radical, and he felt the committee could not risk supporting Bohm. He wrote:

> I base my fears on some of the experiences I have had in Washington, in very close dealings with certain government agencies. Occasionally the FAS is mentioned and they definitely do not like this organization because, as they say, some key members are too far to the "left." They give instructions to their employees not to have any direct dealings with the FAS. These men know that I am a member and admit that the organization is not subversive but that one has to be very careful with which individual one is dealing. . . . I believe that our committee cannot afford to have this ever so slight veil of suspicion interfere with its activities.[87]

Goudsmit did not name the government agencies he had consulted; nor did he mention that he sometimes shared their views about left-wing scientists.

The SCLP held a special meeting on June 4 to discuss the Bohm case. Unfortunately, minutes of the meeting do not appear to have survived, but apparently the committee reached a unanimous decision to resign en masse. The mass resignation served as a purely symbolic gesture that avoided an appearance of condemning Bohm while not lending him direct support either. The resignation took place without any interruption in the SCLP's functions. Spitzer informed FAS chairman Hugh C. Wolfe that the committee expected to reorganize as soon as possible, and in the interim the current committee members would continue to serve.[88]

The FAS administrative committee accepted the SCLP's resignation on June 17, and the committee reassembled itself in late July. Spitzer took the mass resignation as an opportunity to restructure the committee and restrict its membership to those who had time to serve actively. He did so with a particular political conception in mind. He noted to Robert Marshak, when considering the appointment of biologist Clifford Grobstein to the committee, "As you emphasized at the outset, our Committee can be most effective if it can be composed of responsible conservatives who support liberal viewpoints."[89]

Spitzer's offhand remark to Marshak is particularly striking in the context of postwar liberalism. As Athan Theoharis has observed, intellectuals who defined themselves as liberal anticommunists "identified with 'responsible' conservatives, not 'sentimental' liberals."[90] Spitzer similarly preferred conservatives who respected civil liberties to popular front liberals. Even so, the FAS cannot be characterized purely as an exemplar of ADA-style liberal anticommunism. For example, the federation did not argue, as the ADA often did, that anticommunism had to take priority over civil liberties. Nevertheless, the SCLP was not inclined to be activist under Spitzer's leadership, especially when more vigorous initiatives carried the risk of being red-baited. Higinbotham displayed greater desire for the SCLP to take on a larger role, but given his characteristic fear whenever the FAS faced the threat of red-baiting, he likely assented to Spitzer's cautious program. Reluctant to act more forcefully, the SCLP failed to provide a stalwart voice against the abuses of the loyalty-security system.

That is not to say the SCLP accomplished nothing. For scientists in trouble, the committee provided a constrained, but valuable, resource. The committee also played a part in scientists' protests against the spread of loyalty and security investigations to unclassified research. Throughout 1949 and 1950, the SCLP continued to press the AEC and JCAE to implement reforms in the security system, especially the extension of procedural rights to applicants. On September 19, 1950, the AEC announced that job applicants would have the same rights as employees when applying for security clearances. Federation of American Scientists executive committee members Clifford Grobstein and Alan H. Shapley (physicist and one of Harlow Shapley's children) viewed the policy change as a sign that "the patient work" of the SCLP had "borne fruit."[91] Such procedural reforms, however, still failed to address the security system's use of liberal political activity as an indicator of disloyalty.

Although scientists claimed that loyalty and security investigations constituted a major source of friction between scientists and the federal government and posed issues of utmost concern to the scientific profession, the scientists' organizational challenge to the loyalty-security system was relatively weak. Neither the prestigious National Academy of Sciences, nor the large and diverse American Association for the Advancement of Science, nor the smaller and more specialized Federation of American Scientists mounted a vigorous campaign against the abuses of the security system. All three groups sought to restrict investigations of scientists' political beliefs and activities to the most narrow realm possible, to scientists in areas of research involving classified material and a clear national security interest. But none of these groups proved able to launch effective action to obtain this goal. The National Academy of Sciences was encumbered by its leaders' narrow interpretation of the academy's role and the institutional ties to the national security state of many of its most prominent members. The AAAS appeared more willing to tackle restrictions on civil liberties, but it lacked a tradition of political action and ultimately failed to act because of dissent from a vocal minority of its membership.

One might have expected a stronger reaction from the FAS. In the debate over the McMahon bill, the FAS had demonstrated both a commitment to political action and the ability to mobilize public support. Security investigations also produced a greater sense of urgency among the atomic scientists, whose research tended to place them under the severe scrutiny of the security clearance system. Political divisions among its membership did hamper the FAS, and it also struggled to disassociate itself from the progressive left. At the same time, however, politically moderate scientists generally argued for the need to respect the right to dissent in the Cold War political climate. In the end, though, this civil libertarian outlook could not match the political pressures of the Cold War. Impeded by fears of anticommunist persecution, the FAS adopted a tightly constrained, narrow role for itself in the fight to protect the civil liberties of scientists. The federation largely abandoned the public battle and confined itself to quiet diplomacy instead.

The FAS's constrained approach did not derive solely from fears of direct political repression or some scientists' own anticommunist attitudes. The particular form of the atomic scientists' response to security investigations also reflected the postwar institutional relationship between scientists and the AEC. World War II transformed science-state relations, bringing experts unprecedented status within the federal government. The Cold War consolidated the wartime relationship.[92] The access and influence scientists had in the AEC made informal negotiations a more attractive, less strenuous, and less risky option than public protest.

Whether to hold firmly to basic principles or settle for compromise is a problem in all decision making. But the political culture of the Cold War strongly encouraged the acceptance of pragmatism over principle. In the postwar period, the ideological tenets of liberal anticommunism combined with the administrative structures of interest group liberalism to tip decision makers' choices toward limited, practical goals. Liberal anticommunists eschewed the idealism of the progressive left in favor of what they considered to be hardheaded realism. Theirs was a sharply dichotomized worldview that defined issues in terms of hard versus soft, realism versus sentimentality, and rationality versus woolly-mindedness. Cold War liberals' conception of rationality was not value-neutral, however, but politically loaded. To them, rationality meant limited goals within severely circumscribed boundaries of legitimate political action. Sociologist C. Wright Mills, one of the few intellectuals of the era to criticize the ideological strictures of the 1950s, accused liberal anticommunists of using the mantra of realism as a substitute for their own lack of political and moral imagination.[93]

Existing institutional structures reinforced the ideological preference for pragmatism over principle. As Christopher Tomlins has argued with respect to American labor, the system of interest group liberalism created under the New Deal limited the goals of the labor movement.[94] More generally, interest group liberalism generated administrative devices that discouraged arguments based on first principles such as the protection of basic individual rights. In *The End of Liberalism*, Theodore J. Lowi's critique of the interest group liberalism that dominates modern American politics and government, Lowi discusses the dangers of informal bargaining as a means of making policy. Lowi argues that informal bargaining, or logrolling, endangers democratic institutions because it is a form of negotiation that privileges means over ends. Delegation of power to an informal bargaining process between interest groups and public agencies rather than a reliance on law, Lowi contends, tends to focus negotiations on particular decisions, rather than rules. Lowi explains:

> If by broad and undefined delegation you build your system in order to insure the logrolling type of bargain on the decision, you are very likely never to reach bargaining on the rule at all. . . . Broad delegation simply puts at two great disadvantages any client who wishes to bargain for a general rule rather than merely to logroll his case. Here is what he faces: First, the broad delegations enable the agency to co-opt the client—that is, to make him a little less unhappy the louder he complains. On top of that, the broad delegation reverses the burden of initiative and creativity, the burden of proof that a rule is needed. If the client insists on making a federal case out of his minor scrape, he must be prepared to provide the counsel and the energy to start a rule-making process himself. This means that the individual must stop his private endeavors and for a while become a creative political actor. Most behavioral research agrees that this is an unlikely exchange of roles.[95]

The end result, Lowi finds, is, "Liberalism weakens democratic institutions by opposing formal procedure with informal bargaining."[96]

Lowi's observations provide an apt description of the workings of the loyalty-security system and the relationship between the atomic scientists and the AEC. The cases of Vought, the Blewetts, and the Oak Ridge scientists all relied heavily on informal negotiations. As discussed at the end of chapter 3, the individual scientists themselves, isolated and relatively powerless, were in no position to contest their treatment on fundamental appeals to justice and basic rights, and the security system discouraged them from doing so. An individual scientist could not challenge the ideological foundations of the security clearance system without further jeopardizing his or her own chances of obtaining clearance. It was much simpler and safer to stay polite, stick to the particulars of one's own case, and avoid trying to indict the entire security clearance system. As SCLP member K. W. Ford once advised scientists facing the prospect of a hearing, "Be cooperative. Never mind underlying philosophy."[97]

The Federation of American Scientists, as an autonomous political organization, had greater freedom to act than individual scientists. But the federation followed classic interest group behavior that also kept it dependent on logrolling. Quiet diplomacy on behalf of individual scientists constituted a type of informal bargaining that resulted in, to use Lowi's language, bargaining on the decision rather than the rule. In such situations, the FAS often succeeded in attaining the immediate goal, security clearance for the scientist in question, while avoiding fundamental matters of policy.

Of course, the federation did promote procedural reforms and did not rely solely on logrolling of individual cases. But this emphasis on procedural reform ignored the political assumptions that guided loyalty and security investigations. Procedural reform addressed administrative processes rather than the underlying problem. Although procedural rights were important, as Katharine Way pointed out, the real questions involved the state's methods for determining that an individual threatened the nation's security. How was loyalty to be defined and identified? How were persons who were ideological risks to be detected? Could they be detected? What was the relationship between one's political associations and one's patriotism? In a democratic nation, who was to decide what beliefs were unsafe and constituted a potential threat to national security? To pose such questions was to challenge the very underpinnings of Cold War anticommunism. The FAS's leaders instead chose the easier and safer path of targeting process rather than Cold War ideology, and procedural rights rather than fundamental principles of civil liberties. This choice reflected not just fear of repression but also the constrained political style of liberal anticommunism and the institutional limits of postwar liberalism.[98]

The scientific profession's initial lackluster response to the dangers posed by the security clearance system and the loyalty program left scientists ill prepared when domestic anticommunism began to creep into other areas of scientific life. As the Cold War intensified and other government bodies entered discussions about science and anticommunism, the refuge of internal negotiation through quiet diplomacy started to crumble. The 1949 controversy over the AEC fellowship program would signal an ominous turning of the tide.

CHAPTER 7

Drawing the Line

The AEC, the National Academy of Sciences, and the AEC Fellowship Program, 1948–1950

• • •

The scientists' retreat from public strategies to a reliance on internal negotiations reflected part of a larger process by which domestic anticommunism became an accepted political and cultural reality in the United States. Groups once thought by American intellectuals to have resisted domestic anticommunism and the implementation of the Cold War political consensus — a Democratic presidential administration, liberal intellectuals, labor unions, and the legal profession, to mention a few — to varying degrees and for varying reasons cooperated with, capitulated to, and reconciled themselves to a postwar political order grounded on an assumption of perpetual U.S.-Soviet conflict.[1] The scientific profession, despite many scientists' discomfort with the intrusiveness of loyalty and security investigations and some scientists' outright opposition to Cold War policies, also both adjusted to and actively shaped the Cold War consensus.

Quiet diplomacy achieved initial successes in mitigating the consequences of politically based investigations of scientists throughout 1947 and 1948, but only by avoiding public discussion and the possibility of public controversy. Meanwhile, the deepening American commitment to the ideology of anticommunism left less and less room for the expression of views that either dissented from or could not be easily incorporated into Cold War political orthodoxy. The 1949 uproar over the Atomic Energy Commission's fellowship program provided scientists with a stark example of how much maneuvering room had been lost in the years after the war.

Under the Atomic Energy Act of 1946, the Atomic Energy Commission had a broad mandate to promote basic scientific and medical research related to atomic energy. In January 1948, as part of this mission, the AEC established a fellowship program for graduate and postdoctoral students in physics, biology, and medicine. Administered by the National Research Council under a contract with the AEC, the fellowship program awarded grants to almost five hundred young physicists, biologists, and medical researchers in 1948 and 1949. At the time, it was the largest program for advanced science education in the nation's history, far surpassing even the well-known Rockefeller Foundation fellowship program on which the AEC program was modeled. With the continued failure to establish the National Science Foundation, the AEC fellowship program also served as the premier federal undertaking for the promotion of advanced scientific training.[2]

From the start, the AEC and the Joint Committee on Atomic Energy clashed over whether to require security clearances for fellowship recipients, even though the vast majority of fellows were conducting unclassified research. In May 1949, when news spread that a young physicist holding a fellowship was a member of the Communist Party, the conflict erupted publicly. Open hearings into the fellowship program sparked events that led to David Lilienthal's resignation, increased congressional authority over the AEC, and a further tightening of the Cold War linkage between science and military research. The congressional imposition of FBI investigations as a condition for future fellowship recipients also laid the groundwork for the extension of political tests to unclassified research and other aspects of scientific life unrelated to national security.

Security Clearance and the Fellowship Program: First Discussions, 1948

The fellowship program addressed a variety of contradictory interests and embodied conflicting visions of science from its very beginning. For Congress, the fellowships helped to combat the perceived shortage of scientific personnel, a pressing problem given the importance of military technology to waging the Cold War. For scientists, the fellowship program's emphasis on fundamental science provided desired funding for basic research, as well as welcome educational support for graduate and postdoctoral students. For the AEC, sponsorship of a broad educational effort for younger scientists reinforced the original conception of the commission as a civilian agency with a wide mandate to promote all aspects of research and development related to atomic energy, thereby restraining the agency's ever growing Cold War preoccupation with increasing the nation's nuclear arsenal.[3] But although the fellowship program functioned in part as an attempt to recover the AEC's civilian mission, the commission's military orientation located the fellowship program on an uneasy borderline between classified and non-secret research, military and basic. Given these internal tensions, the fellowship program soon became the focal point of a pivotal conflict over the extension of loyalty investigations to scientists engaged in unclassified research that pitted scientists and the AEC against the JCAE and other conservative congressional forces.

In the spring of 1948, when the National Research Council began to organize the AEC fellowship program and screen the first applicants, the question of security clearance quickly entered the NRC's deliberations. Although the AEC granted the NRC maximum autonomy in administering the program and intended to support unclassified research in as wide a range of areas as possible, the simple fact of the AEC's sponsorship and the general popular tendency to equate atomic energy with defense needs made security clearance an obvious matter for consideration. Members of the fellowship boards, all scientists, firmly opposed a clearance requirement for fellows engaged in unclassified research. National Research Council chairman Detlev W. Bronk recognized the political delicacy of the issue, but he felt administration of the fellowships by the NRC would prevent any open political controversy by insulating the AEC from public criticism.[4]

Initial consultations with the AEC indicated that clearances would not be required. Lewis Strauss vehemently objected to the possibility that the federal government might inadvertently subsidize the education of a member of the Communist Party, but the other four AEC commissioners disagreed. At the June 17, 1948, meeting of the AEC, the commissioners reached an official decision not to require fellowship recipients to obtain clearance unless they needed access to restricted information.[5]

Given what the commissioners knew, their decision was a bold one. In the early stages of the fellowship program, the AEC had FBI reports compiled on some four dozen applicants whose research interests lay in nonsecret areas. According to the FBI, one of the applicants chosen by the NRC to receive a fellowship had once been a member of the Communist Party and possibly still was. Although access to restricted information was not at stake, it was a politically risky proposition for the AEC to knowingly grant a fellowship to a communist. In the postwar years, few organizations dared to hazard defending communists' rights in the name of abstract civil libertarian ideals.

Lilienthal was certain that it was worth the risk. He could accept the legitimacy of security investigations as a necessary evil to safeguard the nation's security, but he would not countenance political inquiries as a condition for educational assistance. It was time, he felt, to take a firm stand. Late in the day after the June 17 meeting, he wrote in his diary that the real problem could not be solved by formulating some sort of policy or procedure. Rather, the AEC commissioners needed to find "a kind of desperate courage . . . a willingness to stand up against fear and fear-begotten emotions that have swept the country, and that are being inflamed by almost every event, and by reactionary forces." Earlier in the day, he had even suggested to the other AEC commissioners that they make a bold gesture and go public with the situation. If the AEC's decision remained behind closed doors and someone else broke the story, the public criticism and likelihood of censure from both the JCAE and HUAC would be unmanageable. If the AEC had the courage to take the first step, however, Lilienthal felt, "The strength of our idea, made public, can be tested against the strength of the opposition, and the strength of our friends against that of our enemies." At the end of his diary entry, he again noted the limits of procedural reform: "But all the administrative devices in the world won't take the place of the essential; to wit, a courageous willingness to stand the gaff."[6] Lilienthal realized that improved procedures alone would not contain the spread of red-baiting. Domestic anticommunism could be fought only by confronting its ideological underpinnings and promoting the principle that the state could not legitimately inquire into individuals' political beliefs unless it had a clear national security interest. Even then, such inquiries had to be undertaken with utmost reluctance, not enthusiasm.

At a time when polls indicated that a large majority of the American public favored an outright ban of the Communist Party, Lilienthal's proposed course of action would have been difficult to take.[7] Other AEC officials sympathized with Lilienthal's sentiments, but there are few indications that they took his idea to go before the public very seriously.[8] As the AEC finalized its policy on the fellowship program throughout the summer and fall, the commission's deliberations occurred behind closed doors, just as Lilienthal had originally feared.

Meanwhile, JCAE chairman Hickenlooper insisted on the need for security investigations of all AEC fellows. On July 30, he informed Lilienthal that he felt the fellowships were intended to provide a supply of potential AEC employees, and he feared that without investigations "we may some day find ourselves in the position of having, with Government money, subsidized the education of a potential subversive who is not eligible for employment on the Atomic Energy Program."[9] Lilienthal responded to Hickenlooper at length on October 11, describing the bureaucratic structure of the fellowship program, the membership of the fellowship boards, and the commission's position on security clearance. He did not discuss the fundamental civil libertarian position he had formulated four months earlier. The reasons for his uncharacteristic reticence are unclear. In part, it may have stemmed from the deterioration of the relationship between Lilienthal and Hickenlooper; by this time, they shared only a common disdain for each other. But more significant, Lilienthal seems to have decided that the AEC's official explanation for its fellowship policy would be based on a bureaucratic rationale instead of a defense of civil liberties.

Rather than attempt to persuade Hickenlooper of the principles underlying his opposition to requiring clearance of fellows, Lilienthal instead explained that the policy was the product of arrangements made between the AEC and the National Research Council. The NRC had the responsibility for administering the fellowship program, and the AEC had full confidence in the ability of the NRC to judge applicants' scientific qualifications as well as their "moral character." While Lilienthal conceded that individuals granted fellowships would sometimes be ineligible for future AEC employment, he felt the established policy guaranteed the smooth functioning of the program. Again he placed the onus on the scientists: "It is our firm conviction that by pursuing this policy we will obtain more qualified fellows and achieve fuller cooperation from the scientific community of this country than would be the case if we adopted the principle of requiring security clearances."[10] Placing his personal beliefs aside, Lilienthal gave Hickenlooper the impression that the lack of a clearance requirement arose from the need to placate the scientific community.

Predictably, Hickenlooper was unimpressed by Lilienthal's explanation, and he pressured the AEC chairman about the FBI reports on the first group of applicants that revealed "substantial derogatory information" in several cases. He granted the AEC the administrative prerogative to carry out the fellowship program as the commission saw fit, but he objected to fellowship support for scientists who could not later be employed by the AEC for security reasons, and he found the potential use of government funds to educate members of the Communist Party intolerable.[11] Although disgruntled, however, Hickenlooper was not yet prepared to contest the AEC's authority in the matter.

Disagreements over the fellowship program would soon erupt into a critical political battle. Scientists and the AEC had already clashed with the joint committee over security clearance procedures for AEC personnel. Now they wanted to demarcate clearly the boundary of legitimate state inquiry into individuals' political beliefs and affiliations at the line between classified research, where the state had a national security interest, and unclassified research, where the state had no grounds to engage in political investigations. The AEC, however, failed to make this distinction clear to the joint committee. Free to formulate policy away from public attention, scientists and the AEC implemented a policy based on principle, but they did so without building support for their convictions within government circles. When the AEC's policy on the fellowship program suddenly became national news and widespread public knowledge, principle quickly faded in the light of political pressure.

Public Controversy: Congressional Hearings, 1949

As Lilienthal had anticipated, the AEC's fellowship program could not escape public scrutiny for long. In 1949, Hans Freistadt, a physics graduate student at the University of North Carolina at Chapel Hill, received an AEC fellowship to do research on general relativity. Freistadt happened to be an open member of the Communist Party, and he participated actively in politics at his university campus.[12] On April 22, 1949, an angry constituent wrote his senator to complain that Freistadt had been awarded an AEC fellowship even though he was an avowed communist. Senator Clyde R. Hoey (D.-N.C.) immediately wrote Lilienthal to demand an explanation. Lilienthal replied on May 5 with a response essentially identical to the one he had given Hickenlooper the previous October. On May 10, right-wing radio commentator Fulton Lewis Jr. went public with the news that Freistadt, "a professed Communist," held a sixteen-hundred-dollar AEC fellowship.[13]

An enraged Congress responded quickly. On May 12, Hoey announced on the Senate floor, "I think Senators will be amazed to know that the money of the taxpayers is being used to provide scholarships for known Communists through the Atomic Energy Commission." He then requested an investigation by the Joint Committee on Atomic Energy. Meanwhile, on the House floor, JCAE member Sterling W. Cole (R.-N.Y.) labeled the fellowship award to Freistadt "strange and incredible." The next day, Senator Pat McCarran (D.-Nev.) threatened to have the Senate Appropriations Committee revoke all fellowship funding from the AEC unless the commission agreed to adopt loyalty screening for fellowship recipients. The day after, Hickenlooper announced that he had found a second Communist Party member among the AEC fellows.[14] Within the

Congress, the AEC had no defenders. Hickenlooper's pronouncement was especially damaging, for it signaled publicly the rift between the joint committee and the AEC.

The following week, both the Joint Committee on Atomic Energy and a subcommittee of the Senate Committee on Appropriations began lengthy hearings on the AEC fellowship program. In the hearings, Lilienthal quickly backed off from the position he had advocated within the AEC. A. N. Richards and Detlev Bronk were also less than resolute in their support for the commission's policy. Press coverage at the time and later historical accounts have framed the testimony of Lilienthal, Richards, Bronk, and other architects of the fellowship program as a courageous defense of academic freedom juxtaposed against congressional objections to the use of public funds for the education of Communist Party members, whom members of Congress viewed as single-mindedly dedicated to domestic subversion.[15] A more careful reading of the hearings, however, reveals a much more contradictory, confused understanding of the issues among both sides, especially on the part of Lilienthal and his allies.

Atomic Energy Commission and NRC witnesses presented unsteady, self-contradictory arguments in both the JCAE and the Appropriations Committee hearings. Lilienthal embraced academic freedom, but he also weakened his position by conceding that the fellowship award in Freistadt's case was a poor decision. Richards and Bronk also acceded to the view that the award to Freistadt was a mistake, and they readily agreed to the imposition of a loyalty oath and noncommunist affidavit for future fellowship recipients. Lilienthal, Richards, and Bronk made little reference to the fundamental principles endangered by requiring investigations of individuals without any justified basis in national security. Only Alan Gregg, head of the AEC's advisory committee on biology and medicine and director of the medical sciences division at the Rockefeller Foundation, took an unequivocal stance against investigations of fellows' political beliefs. Lilienthal, Richards, and Bronk also failed to make clear the abuses inherent in the security clearance system. Instead, they emphasized the danger of extending political inquisitions to young, politically immature scientists, as if older scientists suffered no harm from investigations.

The position of members of Congress was also murky. Fundamentally, Hickenlooper, Senator Joseph C. O'Mahoney (D.-Wyo.), and other members of the JCAE and the Senate Appropriations Committee objected to the use of federal funds for the education of Communist Party members regardless of whether or not national security had any relevance to the debate. At the same time, however, they buttressed their arguments with a belief in the need to protect atomic secrets. When Lilienthal and other witnesses objected that political inquiries for fellows constituted the first step in extending investigations to all areas of federally sponsored education, the congressmen denied the possibility. Instead, they insisted that atomic energy was different from other areas and required special protections.

Lilienthal's initial testimony before the JCAE on May 16 revealed the limits of his public position on the fellowship program. In his opening statement, he emphasized the nonsecret nature of most of the fellows' research, and he stressed the necessity for AEC sponsorship of basic research in the absence of a National Science Foundation. National security was not at issue. The real problem, Lilienthal contended, was federal interference in education. By federal interference, however, Lilienthal did not mean the imposition of political tests, as most press and historical accounts have implied. Rather, he meant the more narrow issue of the separation of roles between the federal government and the NRC, and he stressed that the federal government should not intervene in the affairs of the NRC, "a private, nongovernmental agency" with a proven ability to administer successful fellowship programs. Although Lilienthal also mentioned "the danger of the Federal Government itself tying strings related to political and economic views and opinions to educational funds," he did not elaborate. Lilienthal implied that his primary objection to FBI investigations of AEC fellows centered on the maintenance of proper administrative boundaries in the operation of the fellowship program, not the danger posed to fundamental political freedoms.[16] From the very outset of the hearing, Lilienthal presented a public position far weaker than the stance he had taken in private the previous year.

Almost immediately, A. N. Richards inadvertently undercut the limited position that Lilienthal had articulated in his opening statement. Richards started out strong. He condemned political tests as a condition for educational assistance as "a first step . . . toward a disastrous intervention of Government in education," and he warned that such requirements hinted of "an element of educational procedure so closely akin to those by which the Russian bureaucracy controls science in that country as to be scarcely distinguishable from them." Only moments later, however, prompted by questioning from McMahon and Hickenlooper, Richards conceded that the NRC had made a poor decision in awarding

Freistadt a fellowship and expressed his opinion that belief in communism was incompatible with loyalty to the United States. A basic inconsistency plagued Richards's position. Given his conviction that there existed a category of individuals who should not be awarded fellowships on the basis of their political beliefs, he could not easily explain to the JCAE why the AEC should not take prior action to identify and screen out such individuals. As his testimony continued, Richards was hard pressed to explain why political tests for employment, such as those posed by the loyalty program, were acceptable while education remained a special area in which investigations would cause undue damage. His response only muddled the issues further. He stated that he could accept minimum loyalty standards for the fellowships, and he advocated that fellowship recipients take a loyalty oath. To go further and require fellows to sign a noncommunist affidavit, however, would be an illegitimate use of "political faith as a criterion for educational opportunities." If he had stopped there, he might have preserved some basis for continuing to oppose investigations. Instead, he also explicitly stated that a proven Communist Party member had "no business to be in the fellowship program."[17] If that was true, then why would it be so wrong to mandate a noncommunist affidavit? Internal contradictions riddled Richards's testimony with logical holes as he struggled to embrace simultaneously both anticommunism and intellectual freedom, and his confused reasoning left him ill prepared to provide a convincing defense of the AEC's policy of noninvestigation.

After questioning Richards, Senator Knowland asked Lilienthal directly whether Communist Party membership should disqualify applicants from AEC fellowships. Lilienthal displayed something of his essential objection to extending political tests beyond matters of national security when he pointed out, "There are many instances in which people of unworthy political beliefs have the kind of genius that the country needs in nonsecret aspects of development." Primarily, though, he deferred to the judgment of the NRC. He personally felt "a strong feeling of distaste" at the thought of awarding a fellowship to a communist, but there were "deeper and broader issues" at stake. In specifying what those "deeper and broader issues" actually were, however, he referred only to the problem of government interference in the NRC's administration of an educational program, rather than questions of basic political rights.[18]

Lilienthal probably emphasized the separation of administrative roles in an attempt to defuse the situation by avoiding more fundamental but politically explosive issues. Placing the onus on the NRC, however, gave the impression that the AEC was trying to avoid responsibility. In addition, Richards's testimony severely undermined Lilienthal's position. At one point, Lilienthal indicated that Congress faced a choice between accepting an occasional communist or having the NRC resign from administration of the fellowship program. But when Representative Paul J. Kilday (D.-Tex.) asked Richards if there was any possibility of NRC withdrawal, Richards replied, "When asked by the Government to perform a task which we believe we can do, which is within our competence, we are bound to do it." Kilday then specifically asked if it would be acceptable to the NRC to judge the technical qualifications of fellowship applicants, while the AEC ruled on their political status. Richards indicated such an arrangement would be acceptable, leaving Lilienthal without firm ground on which to stand.[19]

The next day, NRC chairman Detlev Bronk added to the confusion. Although Lilienthal felt Bronk was "about as good a witness as the Joint Committee has ever heard,"[20] Bronk also vacillated between commitment to academic freedom and concession to Cold War anticommunism. Early in his testimony, he indicated that the NRC had been mistaken in awarding a fellowship to Freistadt. Bronk felt that any person who failed to see a conflict between loyalty to the United States and Communist Party membership was by definition "a person who doesn't think very clearly" and was therefore "not likely to be a very good scientist."[21] Bronk thus removed the problem of political tests for otherwise competent fellowship aspirants by defining those who might be politically ineligible as scientifically incompetent. Although scientists generally insisted that political beliefs were irrelevant to scientific talent, within academia, those who sought to expel communists and left-leaning radicals from professorships resorted increasingly to the argument that adherence to communism made a person a dogmatic teacher and scholar of poor quality who had no place in the nation's colleges and universities.[22] Bronk then recanted somewhat. He confessed that he had his own prejudices against communism, which, he admitted, "does not make me a very good scientist, when I come to talk about them." Trying to speak more objectively, he suggested that there were many young scientists with intellectual potential who, with "a little more maturity," would eventually "recognize the fallacy of their point of view." Under these circumstances, Bronk suggested that the occasional radical with scientific potential should be admitted to the fellowship program "in the hope that he will become wiser as he develops." Bronk continued, "As he develops scientific competence he is also going to develop the maturity

Detlev Bronk, 1948. (National Academy of Sciences; reproduced by permission of Johns Hopkins University Archives)

in his social outlook which is going to make him not only a useful scientist but a respectable citizen of the country."[23] Thus Bronk provided a rationale for not requiring FBI investigations of AEC fellows that fit tightly into the Cold War political consensus. The AEC should risk supporting the occasional communist, not because young communists could also be good scientists who could make contributions to basic research, but because with proper guidance, young radical scientists might achieve political "maturity" and free themselves to achieve their scientific potential. Bronk's position preserved the AEC's policy only by framing it within the confines of domestic anticommunism.

Bronk's (and Richards's) emphasis on the idea of maturity is particularly important when taken within the context of postwar intellectual thought about communism. Liberal anticommunists such as Arthur M. Schlesinger Jr., Daniel Bell, and others gave little credence to the political style of the progressive left in their analyses of postwar politics, and they differentiated sharply between the realism of their own political views and what they perceived as sentimentality on the part of popular front liberals. Liberal anticommunists laid heavy emphasis on the idea of maturity. The rhetorical style of the progressive left, grounded on uncompromising principle and laden with moral conviction, was anathema to liberal anticommunists, who viewed such a political style as immature and unpragmatic. Although Bronk phrased his arguments in terms of the AEC fellows' relative youth, his basic point had less to do with their age than with the notion that the political approach of what remained of the American Left was in and of itself immature and so far out of the mainstream as to be illegitimate.

Of those involved with running the fellowship program, only J. Robert Oppenheimer and Alan Gregg expressed steadfast opposition to any form of inquiry into the political beliefs of fellowship holders. In a May 14 letter to Senator McMahon that was read at the hearing, Oppenheimer, in an atypical foray into political controversy,[24] pointed out that many basic discoveries in nuclear physics had been made by scientists who were communists or communist sympathizers. The French scientists Irène Curie and Frédéric Joliot, who discovered artificial radioactivity in 1932, provided his prime examples; Curie supported policies of the Communist Party in France, and Joliot, her husband, was a prominent member of the Communist Party. Oppenheimer contended that supporting communism was not incompatible with doing important scientific work, and in unclassified areas of research, communists could make valuable discoveries that would benefit American society.

Furthermore, Oppenheimer stressed, as Lilienthal, Richards, and Bronk had not, that the methods used to determine loyalty were "far from satisfactory." Loyalty inquiries, Oppenheimer observed, involved "secret, investigative programs which make difficult the evaluation and criticism of evidence; they take into consideration questions of opinion, sympathy, and association in a way which is profoundly repugnant to the American tradition of freedom."[25] Oppenheimer's letter attempted to remind the joint committee that the process of investigation was not risk-free but damaging and deeply disturbing. Under restricted circumstances, investigations in some form might be a necessary evil, but they were never a positive good.

Oppenheimer's letter made little impact. He did not ask to testify at the hearings, and the joint committee did not call upon him to do so. Hickenlooper politely expressed his personal admiration for Oppenheimer but quickly dismissed the scientist's position. Calling attention to massive U.S. expenditures on the Marshall Plan for Europe, foreign aid to Greece and Turkey under the Truman Doctrine, and military aid to China, all for the purpose of containing communism abroad, Hickenlooper insisted that the joint committee had a minimal responsibility to do whatever it could to contain communism at home. As Hickenlooper's reply to Oppenheimer indicates, despite Truman's personal misgivings about the postwar red scare, his anticommunist foreign policy contributed to the generation of a domestic anticommunist response.

Alan Gregg provided the strongest opposition to the JCAE. Gregg disputed the suggestion that the fellowship awards were in essence a favor from the government that should be withdrawn from individuals considered subversive. Instead, he contended that the fellowships constituted a business arrangement in which the government sought to benefit by "going into the market to find ability and talent and character." Since the government had a concrete interest in developing scientific talent, it was a mistake to narrow the applicant pool by imposing irrelevant political criteria.[26] By doing so, the government threatened its own interest in expanding the supply of scientific experts.

Investigation of political beliefs, Gregg elaborated, would discourage talented scientists from applying for AEC fellowships. He then stated explicitly that Communist Party membership had no relevance to a scientist's qualifications. Investigations, he continued, would do more harm than good, and efforts to pinpoint the "potentially subversive" were inherently misguided. Like other witnesses, Gregg suggested that there was something different about young people that made political inquiries particularly problematic. Young people were "potentially everything" and "restive about authority" by nature. A system of investigation based on the principle of government distrust would only breed "disillusionment" and "cynicism" and increase the attraction of communism as a political alternative.[27]

Gregg also feared that the requirement of an oath or FBI investigation for the AEC fellowships would eventually lead to widespread political repression in education. Once Congress made federal funds for fellowships conditional on political tests, state universities might be next, then high schools, then teachers. He asked, "Are you going to run a general witch hunt in this direction? I do not see what the limit may be to it." He then warned, "The storm will come slowly; but, like most big things, they come slowly at first and then they develop speed as they come along."[28] His prediction was on the mark. By the 1950s, loyalty investigations and requirements for loyalty oaths and noncommunist affidavits had proliferated throughout the educational system to include both state and private universities, as well as teachers in public schools.

When subjected to heated questioning by Representative Durham, Gregg held steadfast. When Durham emphasized the damage done to the AEC by the fellowship award to Freistadt, Gregg responded fervently, "I know there will be examples that will be shocking, and terrifying and all that, but, darn it, they do not amount to as big a price as you pay if you go into this fear-stricken attitude." Durham replied, "This young man has probably done as much damage as thousands." Gregg retorted, "We can do a lot more if your reaction is to make some more feel it is all a political uproar."[29] Gregg made it very plain that he believed political investigations did more harm than good. The damage done by impeding scientific research and fanning the flames of anticommunism, he contended, was far greater than the risk posed by granting the occasional communist a fellowship. Lilienthal had qualified his acceptance of communists as nonsecret fellows by deferring to the judgment of the NRC. When Richards and Bronk assented to the JCAE's position that communists should be disqualified, Lilienthal had little basis for further objections. Gregg, on the other hand, presented the absolutist position that the AEC had no rational reason to make inquiries into fellows' political beliefs under any circumstances, a position that Lilienthal had developed in private more than a year earlier but was unable or unwilling to defend publicly.

The JCAE hearings were a breeze compared with the storm Lilienthal faced in testimony before the Independent Offices subcommittee of the Senate Appropriations Committee. The senators displayed confrontational and hostile attitudes from the beginning, and Lilienthal was forced to retreat further. After the first day of hearings on May 19, Lilienthal noted in his diary, "Well, did I say this was going to be *rough*? What an understatement. It has been rough, hot and heavy, and rather a torment."[30]

From the outset, subcommittee chairman Joseph C. O'Mahoney brandished his unremitting disapproval of the AEC fellowship policy. To O'Mahoney, the Soviet Union was engaged in an international conspiracy "to propagandize the whole world and to undermine the free governments throughout the world, and by a policy of infiltration to secure positions of importance in governments everywhere."[31] Given the dangers of the world situation, O'Mahoney found the AEC fellowship policy unacceptable. Lilienthal countered by pointing out that most fellows were engaged in unclassified research and posed no possible security threat. In reply, O'Mahoney immediately objected that the issue had nothing to do with access to restricted data but with the AEC fellows and "whether they belong to organizations the principles of which advocate the overthrow of the Government by force of violence."[32] The AEC policy of noninvestigation assumed a distinction between secret and nonclassified research. O'Mahoney indicated that from the appropriations subcommittee's point of view, the distinction was irrelevant. Instead, O'Mahoney deemed that federal funds should not support any individual considered subversive, whether or not they were in any position to pose a conceivable threat to the nation.

Flustered and constantly interrupted by O'Mahoney and other subcommittee members, Lilienthal tried to lay out the principle that the federal government ought not interfere in the administration of an educational program. At the same time, however, given the lack of objection from the NRC, he told O'Mahoney he would, if necessary, accept the requirement of a loyalty oath as a condition for receiving a fellowship. Lilienthal did advise caution in the setting of conditions on educational funds, and he urged the committee not to take "a bad step that will poison the wells of American educational freedom."[33] By assenting to the loyalty oath, Lilienthal abandoned a line on political tests drawn at the boundary between classified and nonclassified research. Nonetheless, although political pressure forced him to weaken his position still further, he maintained that a line had to be drawn somewhere, for fear of an endless cascade of government mandated, politically based restrictions on individuals' participation in educational programs.

Logically, the position outlined by O'Mahoney and other members of Congress did imply a process without end. Any political, economic, or social relationship funded or mediated by the government could theoretically be made conditional on political tests. The federal loyalty program had already obliterated the distinction between jobs with national security implications, which might require special conditions, and jobs without such implications, and mandated FBI checks across the board for all government employees. The Taft-Hartley Act, passed in June 1947 over Truman's veto, required noncommunist affidavits from all union officials as a condition for recognition by the National Labor Relations Board, even though unions were nongovernmental bodies, on the grounds that the federal government had the power to place political conditions on participation in labor negotiations mediated by the federal government.[34] With the loyalty program and the Taft-Hartley Act as precedents, it became increasingly difficult to argue against the enactment of further legislation that threatened to penalize holders of unorthodox political views.

At this point, O'Mahoney made a telling remark that revealed the inconsistencies and underlying fears contained in congressional denunciations of the fellowship program. O'Mahoney vehemently rejected Lilienthal's slippery slope argument and denied that political restrictions would inevitably spread to "normal education activities." Instead, atomic energy was a special case: "This is utterly different, because here we are dealing with the activities of research in nuclear science. . . . It is an utterly different proposition when members of Congress say, 'You shall not use public funds to educate members of the Communist Party in this field of science.' It is just self-defense in the midst of the cold war."[35] Although O'Mahoney had brushed aside Lilienthal's distinction between secret and nonsecret research, underneath his objections to the fellowship award to Freistadt lay a subtext of fear about the loss of atomic secrets and the belief that there was no such thing as nonsecret research as far as atomic energy was concerned.

O'Mahoney's sentiment was not an uncommon one. Later in the appropriations hearings, Senator A. Willis Robertson (D.-Va.) also stressed the need to protect the atomic monopoly as a reason for requiring political standards for the AEC fellowships. He declared, "Every man with two grains of sense knows that, of all the secrets in the world, the one that the Communists are most anxious to get is that of the atomic bomb. . . . It is the duty of those responsible for this program . . . fully to demonstrate to us that, whether there was a technicality or no technicality, they have done everything humanly possible to protect the secret of the atomic bomb."[36] On the Senate floor, appropriations committee member Homer Ferguson (R.-Mich.) and JCAE member John W. Bricker (R.-Ohio) could see no other purpose for the AEC fellowships than to train scientists for defense research, and they both indicated that the possibility of communists in the AEC posed a far greater threat to American national security than the presence of communists in any other agency of government, such as the State Department or the armed forces.[37] The atomic scientists had failed to allay fears that there were atomic secrets that could easily be lost through espionage. Nor had they convinced political leaders and the public that the most important secret, the atomic bomb's existence, had been revealed at Hiroshima and Nagasaki and that other nations would inevitably attain nuclear weapons on their own. In short, they had failed to break the cognitive association between atomic energy and national security. Indeed, since the 1946 debate over the McMahon bill, the connection between science and national security had only deepened. This widespread belief in the union between science and security was not simply a problem of poor public relations. Although congressional conservatives' fears of losing the secret of the atomic bomb to espionage were blown out of proportion, they rightly saw a connection between scientific research and defense. In the absence of a strong National Science Foundation, more than 90 percent of funding for basic research in the physical sciences came from the military and the

AEC, and such funding served military ends. The perceived linkage between science and security was deeply rooted in material reality.

Steadfast in their belief that any program connected to atomic energy required special security measures, the senators on the appropriations subcommittee delivered unrelenting criticism of the fellowship program. Lilienthal felt cornered. On the afternoon of May 20, he wrote in his diary, "A very rough time this morning indeed. They feel they have me on the run, have tasted blood, and it will go on from here." Political pressure forced the AEC to back down. At their May 20 meeting, the AEC commissioners decided that all holders of AEC fellowships would be required to sign a loyalty oath and noncommunist affidavit, effective immediately. On May 22, a Sunday, Lilienthal reflected on the hearings of the previous week. He wrote in his diary that the ordeal was "one of the most trying and unhappy crises in all my years of controversies." Nervous and depressed, he noted that the loyalty oath requirement constituted a "retreat" that had little chance of improving his own position or that of the nation. Loyalty oaths, he wrote, "won't appease those who want to drive me out, and I know it. Nor will it allay the almost hysterical fear of Communism, change, ideas, science, etc., that underlies this." Late in the afternoon, after hearing of the suicide of former secretary of defense James Forrestal, Lilienthal reminded himself to keep the situation in perspective. "Look Dave," he wrote, "don't let this political fracas get you down. Take it as it comes."[38]

With Lilienthal already under fire, Hickenlooper dropped another bomb. The next morning, newspapers reported Hickenlooper's charges that Lilienthal was responsible for "incredible mismanagement" of the AEC. Meanwhile, HUAC's latest hearings on supposed wartime espionage at the Berkeley Radiation Laboratory, which had begun in late April, resumed on May 25 with David Bohm's testimony and continued with scattered hearing dates in June, August, and September. The hearings added to the impression that America's nuclear program was in disarray and endangered by espionage. Lengthy JCAE hearings to investigate Hickenlooper's charges began on May 26 and dragged on for three months. Despite Hickenlooper's allegations of lost documents, missing uranium, lapses in personnel security, and incompetent administrative practices, the JCAE's final report on the matter completely exonerated Lilienthal. Nevertheless, the three months occupied by the JCAE investigation virtually paralyzed day-to-day operations at the AEC as staffers anxiously prepared for the hearings. Although Lilienthal emerged clean, he was not untouched. The hearings compromised his influence, and on a

The AEC under fire at Senator Hickenlooper's "incredible mismanagement" hearings, June 6, 1949. (*Front, left to right*) Carroll L. Wilson, David E. Lilienthal, and AEC general counsel Joseph Volpe Jr.; (*rear, left to right*) AEC deputy manager Carl Shugg, AEC commissioners Sumner T. Pike, Gordon Dean, Henry DeWolf Smyth, and Lewis L. Strauss. (New York World Telegram and Sun Collection, Library of Congress; reproduced by permission of Wide World Photo)

personal level, they left him feeling tired, disheartened, and burned out. In the following months, he began to contemplate a life away from public service. On November 7 he met with President Truman and informed him of his decision to resign from the AEC chairmanship, effective February 1950.[39]

Lilienthal immediately withdrew from most daily duties at the AEC, staying on only to serve as part of a special three-person committee to advise the president about whether the United States should proceed with a crash program to develop the hydrogen bomb. Lilienthal desperately opposed the crash program. Since the Soviet atomic explosion had been detected back in September, he had heard too many military and political leaders, including Brien McMahon, now JCAE chair, talk as if war with the Soviet Union were inevitable and perhaps not too far distant. Going ahead full throttle with the hydrogen bomb, Lilienthal felt, would recklessly escalate Cold War tensions and make any kind of U.S.-Soviet accommodation virtually impossible. He tried to persuade Truman not to go ahead with the hydrogen bomb project, but the other committee members, Secretary of State Dean Acheson and Secretary of Defense Louis Johnson, held sway.[40]

The hydrogen bomb secured the AEC's final assimilation into the national security state. Lilienthal, like many of the atomic scientists, originally thought that the AEC's primary mission ought to be the promotion of the peaceful uses of atomic energy. To Lilienthal, the AEC was supposed to be a nuclear-powered Tennessee Valley Authority that would someday provide cheap and abundant energy for much of the country. By 1950, however, the AEC had deviated far from its planners' initial aspirations. After Truman decided to go ahead with the hydrogen bomb, Lilienthal lamented to State Department science adviser R. Gordon Arneson that the AEC had become "nothing more than a major contractor to the Department of Defense."[41] The transformation from New Deal to Cold War agency was complete.

Scientists and the Fellowship Compromise: Reactions

Once Richards and Bronk conceded the acceptability of a loyalty oath and noncommunist affidavit as requirements for recipients of AEC fellowships, Lilienthal had little choice but to follow along. To Richards and Bronk, the move provided a necessary compromise that forestalled the threatened congressional imposition of full-fledged FBI investigations for recipients of AEC fellowships. Other scientists were less certain if the concession afforded a proper balance between principle and expediency. Most members of the National Academy of Sciences grudgingly consented to the measure but indicated FBI investigations would be another matter entirely. Only the FAS voiced strenuous objections in public. By the summer of 1949, however, the federation lacked the influence even to begin to turn the tide. The powerlessness of the FAS to modify the fellowship program demonstrated the shortcomings of quiet diplomacy as a long-term strategy.

In late May, the academy officially accepted the loyalty oath and noncommunist affidavit as conditions for recipients of AEC fellowships. The new fellowship policy provoked relatively few reactions from the NAS. Among members of the fellowship boards, most found the oath and affidavit distasteful but acceptable as long as the AEC did not require FBI investigations and the compromise did not make FBI investigations more likely in the future.[42] Mathematician Marshall Stone, a member of the

Postdoctoral Fellowship Board in the Physical Sciences, resigned outright as a matter of principle, but his response was exceptional. Stone had earlier criticized the academy's failure to take vigorous action on the Condon case and civil liberties. One of the sons of Supreme Court chief justice Harlan Fiske Stone, he felt there was little room for compromise when dealing with matters of political freedom and human dignity. He explained, "Fundamentally, it seems to me, the imposition of political conditions upon the pursuit of scholarship, however supported, is contrary to the political principles on which our nation is founded; prejudicial to the proper development of basic research in the United States; and most difficult to contain within those limits which the proponents of the current measures appear to accept as necessary." To Stone, the oath and affidavit ran contrary to basic tenets of American democracy, and as a politically expedient gesture, they not only would fail to prevent the imposition of further restrictions but would increase the likelihood of such restrictions. Even if the oaths became no more than "a kind of empty ritual," they contained a disturbing element of coercion that created "a certain sense of humiliation and a corresponding feeling of resentment." Given such effects, Stone felt he could not lend assistance to the new fellowship policy: "I could not bring myself to continue a remote but identifiable association with the compulsion to participate [in this ritual]."[43] Stone's resignation made little impact. Roger Adams, the leading organic chemist of his generation and also a member of the Postdoctoral Fellowship Board in the Physical Sciences, simply suggested that Stone "merely be replaced by a mathematician who has no objection."[44]

Of the major scientific organizations, only the FAS objected strenuously to the oath and affidavit. At the outset of the hearings, the FAS stayed relatively quiescent, but after the AEC announced its new fellowship policy, the federation reacted quickly and stepped up its activities to a level unseen since the Oak Ridge security clearance cases. Higinbotham went to Washington and prepared for a long stay, and the FAS executive office began "working to the endurance limit" to generate press coverage, mobilize other scientific organizations, and renew old congressional contacts.[45] The stakes were high. A last-minute addendum to the May 21 FAS newsletter warned that the fellowship controversy had ominous implications for the future of the AEC, civilian control of atomic energy, plans for a National Science Foundation, and the general character of science-government relations. The newsletter observed, "This is clearly no minor skirmish." It continued, "The struggle is growing into a test of major proportions, possibly determining for some time to come the relations between government and science." Lilienthal, Bronk, and Richards, "shaken by the violence of the attack," had abandoned their previous position for lack of public support and accepted the loyalty oath and noncommunist affidavit. Now, the newsletter urged, FAS members had to act and telegram McMahon immediately to express their opposition to further restrictions.[46] On May 22 and 23, the FAS Washington office sent telegrams to the most active FAS chapters and prominent individual scientists to reiterate the main points of the newsletter and entreat scientists to wire McMahon and notify news outlets. On May 24, Higinbotham issued a memorandum to the FAS chapters emphasizing the urgency of the situation and detailing the steps taken thus far. The situation was "deadly serious," with Lilienthal in an especially tenuous position because he had "received very little support for his position from scientists, educators, and others who are alive to the implications." With the active support of scientists, however, the tide could be turned. Higinbotham wrote encouragingly, "We still have a role to play."[47]

The Washington office's May 22 telegram to other FAS chapters noted that the federation favored Lilienthal's original policy of excluding the fellowships from anticommunist political regulation. Opinions solicited from present and former members of the FAS administrative committee and the FAS advisory panel also showed strong support for the original policy. Four of the responses, including, surprisingly, that from the usually outspoken Harlow Shapley, considered the oath and affidavit a necessary expedient. Nine other replies, however, including those of David Hawkins, John P. Peters, Oscar K. Rice, Philip Morrison, and Robert R. Wilson, insisted that there be no compromise.[48] While the leaders of the National Academy of Sciences accepted the oath and affidavit as a means to forestall the more stringent requirement of FBI investigations, high-level FAS members tended to condemn the new policy as an ill-considered concession that would only hasten the implementation of further politically based restrictions on scientific research.

Over the next two and a half months, the FAS and several of its chapters openly criticized the use of political criteria as requirements for participation in the AEC fellowship program. In press releases and letters to major political figures, they opposed both the newly implemented compromise policy and the stricter alternative of FBI investigations. In late May and early June, the FAS administrative committee, the Association of Oak Ridge Engineers and Scientists, the Atomic Scientists of Chicago, and the Princeton Association of Scientists all protested the loyalty oath and noncommunist affidavit, denounced the idea of FBI investigations, and expressed support for Lilienthal in an assortment of telegrams and letters to President Truman and several senators, including McMahon, Vandenberg, and Hickenlooper, as well as in publicly released statements.[49]

The AEC's new policy failed to ease congressional pressure. On July 7, the Senate Appropriations Committee, acting on an earlier proposal from O'Mahoney, approved an amendment to the appropriations bill that required FBI investigations for all AEC fellowship recipients. Again, the FAS went into action. On July 13, in an attempt to convince the Senate to reject the amendment, the Washington Association of Scientists sent letters to forty-two senators, primarily known liberals, detailing the scientists' objections to the O'Mahoney rider. On July 17, Cornell University physicist Robert Rochlin, a longtime FAS member and former chairman of the Cornell Committee on Secrecy and Clearance, wired Bronk with a pressing appeal: "Urge you actively oppose FBI investigation of AEC fellows before Senate votes on it. This bill would set dangerous precedent." In its characteristic style, the NAS avoided the public fray, although Bronk did speak to and correspond with O'Mahoney privately to relate his reservations about the rider. On July 19, the FAS announced that it had the support of a diverse group of fifteen national organizations, including the American Association of University Professors, the American Association of University Women, the American Civil Liberties Union, Americans for Democratic Action, the National Farmers Union, and the National Women's Trade Union League of America, in its fight against the O'Mahoney rider. A statement signed by the FAS and the other organizations warned of the dangers posed by the rider, especially the threat of investigations being extended "to all Federal support of science and education and to other vitally important areas of our national life as well." When he learned of the FAS's efforts, Lilienthal allowed himself a moment of optimism. Upon receipt of a copy of the letter to the senators from the Washington Association of Scientists, he observed, "There seems to be a hopeful unanimity in the scientific and educational community."[50]

But the Senate delivered a discouraging response. On August 2, it adopted the O'Mahoney amendment by a voice vote. Only Glen H. Taylor (D.-Idaho), Henry Wallace's running mate in 1948, voted against the rider. Wayne Morse (R.-Oreg.) and Claude Pepper (D.-Fla.) voiced some concerns, but they did not join Taylor in voting no.[51]

By 1949, the FAS had lost the political influence that it had demonstrated three years earlier during the debate over the McMahon bill.

External and internal changes placed the federation in a poor position to mount an influential protest. In a conversation during the first week of congressional hearings on the AEC fellowship program, Lyman Spitzer and Oppenheimer discussed the problem the scientists faced in formulating an effective response. In his notes of the conversation, Spitzer observed that the political situation had changed markedly since the end of World War II. He commented, "Rosy days of '45 are no more. Scientists not unified. . . . Difficult for individual scientists to influence Senators (so called conservative group, e.g. Conant, Compton have no inclination to do so)." He also lamented that all the criticism of the AEC had come from the political Right. He noted, "Too bad AEC attack from right, not left, which has adequate reason to do so (and which attack might be very desirable. Too much secrecy.)."[52]

Spitzer's notes indicated that although the scientific community remained disorganized and divided, both Spitzer and Oppenheimer believed scientists still faced essentially the same issues that the FAS had stressed during the debate over civilian control of atomic energy. Spitzer highlighted the need to insist that the purpose of the fellowships was to support broad basic research in the physical sciences, not to train scientists specifically for work in the AEC. He insisted, "Must emphasize this. Try to eliminate notion of *Atomic Energy* fellowships. . . . Certainly what is needed is National Science Foundation." Without a National Science Foundation, scientists saw the AEC as an agency that could provide funding for basic research. Support from the AEC, however, only reinforced a vicious circle by strengthening the perceived connection between scientific research and military objectives. Once more, Spitzer noted, scientists had to debunk the notion of atomic secrets: "Should again point out how difficult it is to give secret away."[53] Scientists had originally envisioned the AEC as an agency devoted primarily to the development of atomic energy for peaceful purposes, and they had tried to separate the linkage between atomic energy and national security forged at Hiroshima and Nagasaki. In the years since the war, however, the congressional and public perception of the atomic bomb as the basis for American defense, and the fear that its "secret" could be easily lost through carelessness or espionage, became more powerful than ever, strengthened by the American commitment to Cold War confrontation with the Soviet Union, as well as the growth in military sponsorship of basic research.

Internal changes within the FAS also contributed to the federation's inability to influence the results of the Senate vote. The atomic scientists' relative success in the passage of the McMahon bill depended on direct access to members of Congress, well-organized press connections, and the capacity to mobilize public opinion through groups such as the National Committee on Atomic Information. But, as we have seen, throughout 1947 and 1948, the FAS gradually turned away from public tactics and grew to depend on internal negotiations with the AEC. As long as primary decision-making authority lay with the AEC, such a strategy was viable. But by the late 1940s, the JCAE took a more assertive role in the atomic energy program and began to exercise its full statutory power. Unlike Lilienthal and other AEC officials, the JCAE members insisted on a hard-line attitude toward security. Once the AEC came under open congressional attack, the atomic scientists had nowhere else to turn.[54]

Members of the National Academy of Sciences privately relayed their strident objections to the O'Mahoney amendment to the NAS leadership, but they did not act publicly to oppose its passage. After it became law, there were few reasons to believe that the NAS would do anything other than quietly accept it. Once the O'Mahoney rider actually passed and FBI investigations of fellows became mandatory, however, the leaders of the academy began to reconsider the extent of the NRC's participation in the fellowship program. Although Richards and Bronk had pledged the academy's continued unconditional administration of the program during the May 1949 hearings, the implementation of the O'Mahoney amendment led NAS members and other scientists to pressure Bronk, Richards, and the members of the fellowship boards to restrict the academy's participation in the program. After months of dragging their heels, the scientist-administrators in charge of the fellowship program were finally forced to confront the question of placing limits on their complicity in government-mandated policies of political repression.

Delayed Reaction: The National Academy of Sciences and the O'Mahoney Amendment

A delayed reaction finally set in. In the months leading up to Senate consideration of the O'Mahoney amendment, NAS members had devoted little concrete thought to what would happen in the event of its passage. After it became law, though, NAS members and other scientists began to debate the merits of restricting the academy's involvement in the AEC fellowship program.

During the May hearings before the JCAE, Bronk had indicated that the academy would accept the imposition of political criteria as conditions for fellowship recipients as long as the AEC, and not the academy, carried the responsibility for judging the political suitability of candidates. After the passage of the O'Mahoney rider, however, Bronk's reservations grew, and he began to discuss possible responses with other scientists involved in the administration of the fellowship program. In reply to an inquiry from Bronk, Henry A. Barton, director of the American Institute of Physics and chairman of the Predoctoral Board in the Physical Sciences, drafted a proposed statement for the NRC on August 26 that urged Congress to repeal the O'Mahoney amendment and hinted that as long as the amendment remained in force, scientists would not assist in the operation of the fellowship program.[55]

Other scientists pressured the NRC to withdraw or limit its participation in the fellowship program. Physicist and Nobel laureate Carl D. Anderson, who had earlier threatened to resign from his position as a fellowship board member if fellows were required to undergo FBI investigations, promptly did so on August 11. On September 6, Bronk asked Anderson to reconsider in light of other board members' discomfort with the O'Mahoney amendment; apparently reassured that the academy would take some sort of action, Anderson rescinded his resignation on September 21.[56] Mathematician A. A. Albert reported to Bronk in a September 30 letter that University of Chicago professors who were sponsors of AEC fellows "all agreed that the only effective protest against the rider requiring FBI clearance would be that of refusal by the National Research Council to administer fellowships on non-classified subjects."[57] By late September, the academy had also learned of the FAS's recommendation that it refuse to administer the nonsecret portion of the fellowship program and pressure the AEC to limit the fellowships to research in secret areas. On October 7, R. D. Stiehler, chairman of the Washington Association of Scientists, elaborated the federation's position to Bronk in writing. Although funding for unclassified research would be lost, the price was "a small one to pay for keeping science from becoming a servant of politics." He added that the loss might be compensated by the possible passage of the National Science Foundation legislation, and in that context, it was even more imperative to make clear scientists' opposition to loyalty tests as prerequisites to research support. Otherwise, "the utility of such a Foundation in advancing fundamental scientific research would be seriously curtailed."[58]

On October 4, at a joint meeting of the AEC fellowship boards, Bronk outlined four policy alternatives. The NRC could either sever all connection with the fellowship program, restrict its participation to evaluation of the scientific merits of the applicants, continue to administer the program under protest, or simply administer the program without any dissent. Eleven of the fourteen fellowship board members present, plus one absent member, favored the third alternative, administration of the program under protest. The reasons expressed for following the third option fell into two categories. Chemists Roger Adams and Charles C. Price contended that under the terms of its charter, the NAS simply had no power to withdraw from the program. Biologist Douglas Whitaker, anatomist Sam L. Clark, and biophysicist H. K. Hartline adopted the self-defeating argument that if the NRC did not run the program, another, less competent organization would. They saw little reason to launch a major protest if it ultimately would have no material effect. Since the program would go on anyway, the NRC, as the best-qualified organization, might as well administer it.[59]

Physicists Kenneth T. Bainbridge and William A. Buchta and mathematician G. A. Hedlund dissented from the majority view. Bainbridge regretted that the academy had not made a strong public statement at the time it agreed to accept the loyalty oath and noncommunist affidavit, and he felt that to go further and continue to administer the program under the loyalty clearance requirement was "a compromise we would pay very dearly for." Next, he feared, Congress would demand loyalty investigations for fellowship programs in the upcoming National Science Foundation legislation. "Drastic action" was required to avoid such an outcome. Buchta agreed: "It seems to me that it is high time we called for a real protest."[60] Both physicists felt that in compromising on the loyalty oath and noncommunist affidavit, the academy had already stretched its credibility to the limit. It could not afford to yield further.

The next day, the NAS council met to consider Bronk's four alternatives and to prepare a statement to be voted on by the membership at the academy's fall meeting in October. With the August victory of the communists in the civil war in China and Truman's September announcement that the Soviet Union had achieved an atomic explosion still freshly in mind, the linkage between science and military research and the role of science in the Cold War dominated much of the council's deliberations. According to Richards's summary of the meeting, at one point the council observed "that the state of the world is such that national security requires the broad expansion of research activities, both theoretical and practical, in all branches of science in which the solution of problems can be advanced by utilization of knowledge of nuclear energy." Concerned about the propriety of challenging the will of Congress, council members noted that the executive order that had created the NRC stipulated the duty of the NRC to promote research for defense purposes, and council members generally agreed that "participation of NRC in the AEC fellowship program would seem to fall within this obligation." In addition, they believed that the academy had little power to dissent from government policy under its 1863 charter. Council members were also anxious to maintain the cooperative nature of science-government relations. As discussed in the previous chapter, the academy's ties to the national security state generally made it unwilling to protest the loyalty-security system. In the case of the fellowship program, council members also feared lessening the NRC's influence with the government: "If, in an effort to keep faith with scientists and at the same time to register protest, we refuse to participate in the program, the ability of the Research Council to serve the Government on a broad basis would be weakened."[61]

Like the joint fellowship boards, the council also advocated Bronk's third alternative. Cold War concerns played a strong role in the decision. In the proposed draft statement on the fellowship program written at the meeting, the council stressed the national security aspects of the fellowship program and recommended its continuation. With an oblique reference to recent international events, the statement declared, "Because of the present world situation we cannot recommend that the AEC make a decision which might lessen the number of individuals who should be trained in science." In a reversal of scientists' previous insistence that the fellowships were for general training in basic research, the statement went on to suggest that in light of the Independent Offices Appropriation Act of 1950, the fellowships should be considered as preparation for future AEC employment. Again, the NAS called attention to the Cold War: "This interpretation need not lessen the number or range of fellowships since it may be expected that, should another 'emergency' occur, all AEC-trained scientists will be mobilized." The statement pledged the academy's continued assistance to the AEC, as long as the commission made it clear that the NRC's administrative responsibility for the fellowships did not include loyalty clearance, and the academy urged the AEC to work for the repeal of the O'Mahoney amendment.[62] By the standards of the draft statement, then, the council members' idea of administration of the program under protest consisted solely of a relatively weak recommendation that the AEC continue to pressure Congress to reverse its position. At the same time, the academy was ready to reinforce the perceived connection between the fellowships and Cold War military research.

In addition to their Cold War concerns, the council members also based their decision on the assumption that blanket refusal to administer the program was not an option. Subsequent legal opinions solicited in preparation for the October meeting indicated otherwise, however. Two lawyers, John Lord O'Brian, a prominent attorney in Washington, D.C., and Charles H. Bradley, the academy's regular legal adviser, concluded that the NAS charter did not obligate the academy to administer the fellowships.[63] According to reputable legal opinion, then, the NAS could freely refuse to administer the program, leaving Bronk's second alternative an open possibility despite the general acceptance of the third option by both the fellowship boards and the council.

Meanwhile, Richards distributed the draft statement to NAS members, and in the weeks before and after the October meeting, the NAS president received dozens of letters. Opinions were divided. A few scientists, such as Walter S. Adams and Harold D. Babcock, two of the foremost astronomers of their generation, had no qualms about the O'Mahoney amendment and felt that any loyal scientist should readily assent to a loyalty investigation.[64] Most members of the academy, however, expressed deep-seated objections to the requirement of FBI investigations. They differed sharply, though, over the extent of the academy's continued involvement in the fellowship program and were split more or less evenly between those who felt the NRC should either withdraw from the program or at least refuse to administer it and those who backed the position of the fellowship boards and the council.[65] Chemist Donald M. Yost categorized "the whole complex of secrecy, oaths, and clearance" as "a vast and silly waste of time, energy and money" that created "ill will within our own country and among our own people," and he recommended that the academy withdraw from the fellowship program as long as the O'Mahoney rider remained in effect. Frank B. Jewett felt that the charter required the academy to evaluate candidates but the academy "should not continue to administer A.E.C. fellowships with all that that administration implies of acquiescence in the clearance provision." Physicist P. W. Bridgman, who had earlier condemned the academy's unwillingness to address civil liberties issues, urged the NAS not to administer the unclassified portion of the fellowship program. He noted, "It seems to me that there are certain traditions in our democratic inheritance which are more precious than the exigencies of the moment created by the atomic bomb emergency." Albert Einstein and Oswald Veblen (who had urged the academy to take a strong position in the Condon case the previous year) opposed NAS administration of the fellowships and suggested that the academy recommend discontinuation of the entire program. John P. Peters, who had condemned the weakness of the academy's response to the Condon case the previous year, felt the NRC had to continue to participate in the selection of fellows so as to avoid "a completely negative attitude," but he insisted that the academy's objections be stated in "unmistakable terms" in a publicly released resolution, and he castigated the academy's earlier indecisive responses to threats against scientists' civil liberties as "deplorable." Harvard scientists George B. Kistiakowsky, Edwin C. Kemble, Paul D. Bartlett, and Julian Schwinger declared that withdrawal from the unclassified part of the program offered "the only effective warning to Congress that future similar legislation covering the Office of Naval Research, the National Institutes of Health and other agencies of Government, including eventually the National Science Foundation will meet with strong opposition from American scientists and will hurt scientific progress in the United States."[66] To these scientists, loyalty investigations posed too great a threat to civil liberties and scientific research for the academy to content itself merely with issuing a critical public statement. Protests had to be backed with action.

Other academy members favored continued administration of the fellowship program. Wendell M. Stanley, who had earlier warned the academy not to become too deeply entangled in a dispute with HUAC over the Condon case, believed the NAS should not refuse a government request that fell within the academy's capabilities, and he felt that a separation between the academy's administrative role and government administration of the clearance requirement constituted a sufficient protest. However, some scientists who supported administration of the program implied that the NAS needed to use stronger language. Lee A. DuBridge, an esteemed physicist and president of the California Institute of Technology, felt the academy had to accept the O'Mahoney amendment as "the law of the land," but he urged the NAS to make "strong representations" in opposition to the new policy. Carl D. Anderson, former AEC commissioner Robert F. Bacher, G. W. Beadle, E. T. Bell, Lee A. DuBridge, J. G. Kirkwood, Charles C. Lauritsen, and Linus Pauling, all prominent scientists at Caltech, approved the joint fellowship boards' position but recommended "a strong statement" that the O'Mahoney rider was "detrimental to the promotion of science, to democratic government, and to the national security."[67]

This general lack of agreement among academy members presented a sharp contrast to the consensus of the NAS leadership. As in the academy's debate over the Condon case, opinion was divided between the leadership and the general membership, with the latter displaying greater opposition to continued administration of the fellowship program. In the end, however, the split was not as severe as it had been in deliberations over the Condon case. The congressional threat to the fellowship program, a general education initiative that fell under the direct authority of the NRC, challenged the academy's autonomy in a way that ultimately united the NAS against assuming permanent responsibility for the fellowship program.

In response to legal advice and members' opinions received prior to the academy's fall meeting, the NAS council revised its statement regarding the AEC fellowships. The latest draft, distributed just before the fall meeting, reversed the decision of the joint fellowship boards and the council. Apparently, once the legal opinions of O'Brian and Bradley removed objections based on the charter, the council felt ready to reshape the statement in response to other academy members' protests. The new version labeled FBI investigations of fellows engaged in unclassified research "ill-advised" and expressed "grave doubts whether the continuance of the Atomic Energy Commission Fellowship Program thus restricted" served the national interest. In light of these considerations, the NAS council announced, "Since we hold these views, we believe that the National Research Council should not accept the responsibility for administering the altered fellowship program." The statement noted that the academy would evaluate the qualifications of candidates if the AEC was unable to do so, and it urged the AEC "to take all proper steps" to remove FBI investigations as a requirement for the fellowship program. The NAS fall business meeting was usually not well attended, and the 1949 meeting was no exception, but the members present adopted the new statement by an overwhelming majority.[68]

Richards had vacillated over the Condon case, but he now felt determined to resist the O'Mahoney amendment. He later wrote Oppenheimer, "I made up my mind that sooner than sign a contract with [the] AEC (as Pres. of NAS) which required FBI investigation of the non-secret fellows, I would resign."[69] Richards sent a copy of the NAS council's statement to AEC general manager Carroll Wilson on November 2. Wilson replied on November 17 with a counterproposal for a revised program that he hoped would allow the NRC to continue to administer the fellowships. The revised plan eliminated new appointments of predoctoral fellows and limited postdoctoral fellowships to scientists whose work would require access to classified information, thus preserving the boundary between the legitimate use of loyalty inquiries when there was a national security basis and the avoidance of investigations of scientists engaged in unclassified research. On October 18, Richards and M. H. Trytten, director of the NRC Office of Scientific Personnel, met with Shields Warren, director of the biology and medicine division at the AEC, and Kenneth Pitzer, director of research at the AEC, to discuss the proposal. Richards was somewhat troubled by the continued responsibility of the NAS for current predoctoral fellows engaged in unclassified research who wished to renew their fellowships under the requirements imposed by the O'Mahoney amendment. Pitzer pointed out that past promises obligated the AEC to offer those fellows the opportunity for renewal, but the AEC would phase out the predoctoral part of the fellowship program by mid-1951. The academy decided it could live with the proposal. On November 30, after consultation with the NAS council, Richards informed Wilson that the NAS would administer the program until June 30, 1951. Because of the secret nature of the revised program, however, the academy declined long-term administrative responsibility: "Since, in this postdoctoral program, considerations of classification policy, security clearance and the AEC program play so great a part, it is deemed inappropriate for the National Research Council to agree permanently to administer it." Wilson accepted the academy's understanding of its role, and on December 16 the AEC publicly announced the limits of the new fellowship program.[70]

The agreement ended the controversy over the fellowship program as far as the academy was concerned, but the AEC was not yet finished. Officials of the AEC still wanted to fund a predoctoral fellowship program, with grants to graduate students engaged primarily in unclassified research. They quickly found new sponsors. Four regional, university-based organizations agreed to administer the predoctoral program. They were Associated Universities, Inc., a consortium of nine eastern universities that ran Brookhaven National Laboratory; the Oak Ridge Institute of Nuclear Studies, Inc., a consortium of twenty-six southern universities in charge of the Oak Ridge Institute of Nuclear Studies; Argonne National Laboratory, run by the University of Chicago under an AEC contract; and the University of California at Berkeley. The Federation of American Scientists condemned the university groups' easy willingness to administer the predoctoral fellowships, but the rest of the scientific community stayed silent.[71] The issue seemed to have worn itself out.

Because the new program was based within the AEC itself, resistance to running it through the national laboratories may have seemed unnecessary or inappropriate to most scientists. Such an explanation, however, points to the disturbing postwar relationship between academia and government, as symbolized in this case by the blurred distinction between the AEC's laboratories and the nation's universities. After the collapse of the Kilgore-Magnuson bill in 1946, FAS members worried about the growth of military support for scientific research. Their fears were justified. In the postwar years, military funding came to dominate American science. The temporary transformation wrought by World War II became long term with the rise of the Cold War. Liberal-left scientists tried to chart out an alternative path for postwar science, but they failed. The proposed National Science Foundation went down in repeated defeats, while the Cold War forced the civilian AEC to devote its resources primarily to nuclear weapons research and development. Military research came to be a normal fact of scientific endeavor, particularly in the physical sciences. Secret research entered college campuses, students took classified courses and wrote classified dissertations, and entire disciplines came to be redefined around military problems. Prominent professors regularly shuttled back and forth between the universities and government advisory committees as defense intellectuals, creating, as Silvan S. Schweber has put it, "a social reality in which working on military problems was an accepted norm for scientists at universities; more than that, their involvement suggested that one aspect of being the very best was participation in defense projects."[72] The new AEC fellowship program constituted yet another instance of universities' loss of autonomy as they became entangled in the Cold War institutional web of the military-industrial-academic complex. By allowing the distinction between university and government management of the program to be blurred, participating universities evaded their responsibility to promote the open pursuit of knowledge and instead presided over a program that restricted political and academic freedom.[73]

The ease with which the O'Mahoney amendment passed through Congress reflected the triumph of Cold War ideologues in American politics. As late as 1947 and early 1948, it had still been possible to argue publicly for the political legitimacy of the American Left. By 1949, such tolerance had disappeared. The rise of the Cold War political consensus was not the product of some sort of inevitable unfolding of events, however. Rather, anticommunists gained political power as those who might have effectively opposed them knowingly or inadvertently lent their support, complied passively, or were actively suppressed.

The stark contrast between 1947 and 1949 and the narrowing of American political discourse during that period can perhaps be summarized by reference to two incidents. During the 1947 AEC confirmation hearings, allegations regarding the supposed socialistic character of the Tennessee Valley Authority and other forms of red-baiting endangered David Lilienthal's nomination. In response to a particularly ugly insinuation by Senator McKellar, Lilienthal launched into a ringing declaration about democracy, freedom, and the meaning of civil liberties. Senators, including JCAE conservatives Bricker and Knowland, rushed to shake Lilienthal's hand afterward, the news media gave the story top billing, and the statement became a public triumph.[74] Two years later, during the hearings on the AEC fellowship program, Lilienthal did not even attempt such a statement of high principle. Even if he had, there is no reason to believe that members of the JCAE and the Independent Offices subcommittee would have been at all receptive. The political space for tolerance of communists and opposition to measures that extended government-sponsored loyalty requirements no longer existed; it was a casualty of the Cold War.

The American scientific community contributed to the building of the Cold War political consensus through its general failure to speak out, especially in the crucial year of 1948. Of course scientists were not completely silent, and it is easy to find numerous examples of individual scientists who publicly denounced HUAC's attacks on scientists, the use of security clearance requirements to discourage liberal-left political action, and the nation's growing obsession with atomic security. As a group, however, scientists did not formulate strong, unequivocal organizational responses to counter the ideology of anticommunism. Scientists' organizations preferred quiet diplomacy, and not without reason. Except for cases involving HUAC, most scientists' confrontations with the federal loyalty-security apparatus occurred out of the public spotlight. Under such circumstances, direct intervention within government agencies provided an effective way to resolve individual cases without risking public controversy. But dependence on quiet diplomacy prevented efforts to address, in a forthright manner, the damage to civil liberties and political freedom wrought in the name of anticommunism. By the time the AEC fellowship program erupted into a front-page news story, scientists lacked the means to respond effectively. The border scientists wanted to draw between classified and nonsecret research in order to define the boundaries of legitimate loyalty investigations had already been compromised by the loyalty program and the Taft-Hartley Act, and there was little congressional or public sympathy for ignoring communists who received government fellowships. The Federation of American Scientists proved unable to mobilize public opinion, and the NAS only reacted after the fact. When the academy, after months of equivocation, finally took a stand and withdrew from the fellowship program in the fall of 1949, its action came far too late to have meaningful influence.

By the spring of 1949, a qualitative change had taken place in American political culture. Although they eventually shifted greater emphasis toward procedural reforms, in the early postwar years scientists had also stressed the indignities, the affronts to basic First Amendment principles, and the suppression of political freedoms inherent in loyalty and security investigations. In the 1949 debate over the AEC fellowship program, especially during congressional hearings, these concerns went virtually unspoken. When mentioned at all, they were buried underneath more provincial considerations, such as administrative prerogatives and university autonomy. The Federation of American Scientists tried to revive the debate over matters of principle, but by 1949 it had almost no hope of success. International and domestic events had strengthened antiradical nativism to the point at which dissent was virtually useless. With the Soviet-backed communist takeover of Czechoslovakia in February 1948, the summer crisis in Berlin, the resounding defeat of Henry Wallace amid heavy red-baiting in November, and the Smith Act convictions of eleven American Communist Party leaders in January 1949 on the grounds that their adherence to Marxism-Leninism alone was enough to show they were engaged in a conspiracy to overthrow the U.S. government by means of force and violence, protection of civil liberties and tolerance for left-wing political views were firmly subordinated to the Cold War political consensus. The opportunity for persuasive dissent from scientists and other influential political groups had passed in relative silence, and by the spring of 1949, even before the announcements of the communist victory in China, the Soviet acquisition of the atomic bomb, and the onset of the Korean War, Cold War anticommunism was in full force.

CHAPTER 8

Consequences

The 1950s and Beyond

• • •

By the spring of 1949, the American commitment to the Cold War and domestic anticommunism was already overwhelming. The Cold War political consensus, a political and cultural phenomenon that took perpetual U.S.-Soviet conflict for granted and invoked anticommunism as a shorthand to simplify and avoid the consideration of complex political issues, dominated American life. Dissenters existed, and critics of the Cold War persisted in their protests, but in the face of political repression and the predominance of Cold War ideology, they could not establish a significant presence in American politics, not even as a loyal opposition.[1]

In 1950, the U.S. entry into the Korean War provided a final confirmation of American leaders' determination to stop the spread of communism worldwide, and the arrests of Klaus Fuchs and Julius and Ethel Rosenberg heightened calls to put an end to subversion by the enemy within. Together, these events exacerbated domestic political repression. The federal government soon enacted stringent modifications in the loyalty program. In 1951, the Truman administration changed the standard for dismissal from "reasonable grounds . . . for belief that the person involved is disloyal to the United States" to "a reasonable doubt as to the loyalty of the person involved to the Government of the United States." In 1953, President Eisenhower, in one of the first acts of his administration, scrapped the loyalty program and replaced it with an employee security program that required all federal workers' employment to be "clearly consistent with the interests of national security." Each revision necessitated the reopening of hundreds of cases and required employees to meet ever more difficult tests to retain their positions.[2] Political tests also intruded more and more often into other areas besides federal employment. The State Department began to restrict the travel rights of suspected subversives, individual state governments began to require loyalty oaths and enforce anticommunist legislation, and the universities, labor unions, and other nongovernmental organizations purged persons who resisted the Cold War political consensus from their ranks.

The increasingly harsh political climate affected scientists in several ways. The Korean War solidified the dependence of the physical sciences on the national security state, despite the long-awaited creation of the National Science Foundation in 1950. Liberal and progressive left scientists who had assumed they could escape political repression by avoiding security clearances lost grants and consultantships. When McCarthyism entered the universities, they sometimes lost their jobs as well. In addition to playing havoc with the lives of politically active scientists, loyalty-security investigations grew increasingly disruptive of scientific life. Passport and visa restrictions not only deprived individual scientists of freedom of movement but also interfered with international scientific conferences and hindered scientists' ability to communicate with one another. The controversy over the AEC fellowship program had provided a grim foreshadowing of the shape of things to come.

Recapitulation and Denouement: The NSF Debate, 1949–1950

After the defeat of the Kilgore-Magnuson compromise bill in 1946, scientists continued to hope for the creation of a National Science Foundation. With annual monotony, they watched subsequent NSF bills fall by the wayside. As Howard A. Meyerhoff of the AAAS observed wryly in 1949, "The National Science Foundation bill seems to be in the same category as the poor—it is always with us."[3] In February 1947, two new bills were introduced to the Senate. One followed Vannevar Bush's ideas and the general scheme of the 1945 Magnuson bill; the other was identical to the Senate-amended version of the 1946 Kilgore-Magnuson compromise bill. With the Republicans in control of Congress, the former, S.R. 526, occupied legislative attention, and Senator Kilgore's proposals to reorient science policy on a liberal and more democratic basis never again received serious consideration. Senate Resolution 526 made it to the president's desk, where Truman vetoed it on the grounds that Bush's administrative structure removed the NSF from presidential authority and public accountability. A new bill, introduced March 25, 1948, contained administrative provisions satisfactory to the executive branch but died in committee in the House when Congress adjourned early for the 1948 election. Given the repeated failures, it was a wonder that the NSF was ever established. But in his January 1949 budget message, Truman placed the NSF on his agenda, and the legislative cycle started rolling again.

Although opponents of Kilgore's conception of the National Science Foundation periodically employed anti–New Deal and antiradical rhetoric, red-baiting and loyalty investigations did not surface as public controversies until the last round of the NSF debate. The relative merits and shortcomings of a progressive left political economy for science could still produce measured political discourse in the early postwar years. In 1949 and 1950, however, the latest version of the NSF had to overcome the hurdles of domestic anticommunism and the loyalty issue, even though the new NSF proposal primarily reflected Bush's ideas and lacked any fundamental reworking of the relationship between science and society. The renewed furor over the NSF reflected both the popular perception of science as the instrument of national security and the newly established conventionality of anticommunism.

In the National Patent Council, a trade association that represented small manufacturers, the new NSF legislation faced a relatively minor but significant obstacle. In the spring of 1949 House hearings over H.R. 12, the House counterpart of the latest NSF bill, S.R. 247, two key witnesses from the patent council condemned the NSF legislation in fervently anticommunist terms. Although patent reform and other progressive left initiatives had long since dropped out of plans for the NSF, former Democratic Texas congressman Fritz Lanham, speaking for the National Patent Council, claimed the NSF constituted a dangerous expansion of centralized state authority and denounced the supposed presence of subversive influences backing the science foundation. In a flurry of antiradical nativism, he suggested that "foreign regimes" were behind "this departure from our established and productive governmental policy," and, citing literature from HUAC, he attacked Harlow Shapley as one of the un-American supporters of the NSF.[4]

When hearings resumed three weeks later, John W. Anderson, president of the National Patent Council, continued where Lanham had left off. Advocates of the NSF, Anderson declared, were "coldly fanatical men, alien-inspired, believed to have as their objective the stifling of that incentive to invent and produce which has given to America its phenomenal industrial growth." Despite the vast philosophical divide between Kilgore's original legislation and H.R. 12, Anderson traced the origins of the NSF to "the long leftist shadow of Henry Wallace" and contended that the new legislation represented "the same general pattern of controls as did the Kilgore bill."[5]

Like Lanham, Anderson assaulted the NSF as the product of a foreign conspiracy. House Resolution 12, he contended, was the creation of "men at heart committed to ideologies repugnant to American traditions—men now perhaps having no choice but to follow the dictates of alien masters or be confronted with dire consequences of public disclosures of previous indiscretions." In addition to Shapley, he named Higinbotham and the FAS as the underhanded scientific influences behind the bill.[6] The federation's desire for the NSF to formulate science policy, Anderson alleged, was evidence of a plot "to invade the realms of private research and invention." He also twisted the FAS's insistence that the NSF perform only nonsecret research into a scheme to release secret information. He declared, "Thus the foundation, empowered to invade all research, governmental, academic, and commercial, would be mandated to disclose it to the world, of course including Mr. Higinbotham's conferees from Russia." Drawing on the image of national security and the need to protect atomic secrets, he added, "Mr. Higinbotham lays down, with painful frankness, his challenge to our right to retain in secret, and employ for world peace, scientific discoveries and developments for which we alone pay. An example, of course, is the atomic bomb."[7] In the changed political atmosphere of the late 1940s, any attempt to place scientific research on a basis other than defense needs could be misrepresented and distorted into a communist-inspired conspiracy.

Although the National Patent Council was the only organization to launch a full-scale assault on the 1949 NSF legislation, the Inter-Society Committee for a National Science Foundation (led by the AAAS and the FAS) feared the patent council might have a disproportionate effect in Congress.[8] In reality, however, the patent council had insufficient sway to keep the bill bottled up in committee and even less influence on the House floor. The main obstacle to the NSF bill in 1950 proved to be loyalty amendments. During the fracas over the AEC fellowships, scientists constantly expressed fears that the imposition of loyalty requirements on the AEC fellowship program would affect the National Science Foundation.[9] Such fears soon proved correct. In mid-June, when the Interstate and Foreign Commerce Committee reported out its version of the NSF bill, H.R. 4846, the committee stipulated that all recipients of scholarships or fellowships from the NSF must sign a loyalty affidavit. By 1949, after the repeated failure to establish a National Science Foundation, the scientific community was willing to accept virtually any science legislation, no matter how far short it fell from earlier expectations about making fundamental changes in the relationship between science and society. The loyalty amendments, however, forced scientists to reexamine their positions. Since most scientists had already reluctantly accepted the loyalty oath requirement for AEC fellowship recipients, they had little political ground for resisting the loyalty affidavit tacked on to H.R. 4846. Within the Inter-Society Committee for a National Science Foundation, both Dael Wolfle and Howard A. Meyerhoff reluctantly agreed that the loyalty affidavit was a necessary concession.[10] The Federation of American Scientists urged scientists to speak out against the loyalty affidavit and also observed that the bill was less than satisfactory in terms of patent reform and the lack of any specific provision for the social sciences. Nevertheless, the FAS gave the bill a tepid endorsement: "Those FAS members who have been active in support of National Science Foundation legislation do not look on H.R. 4846 with great enthusiasm. It is certainly not exactly the kind of bill the Federation would have written. But the consensus is that it will be better to have this NSF than none."[11]

The federation feared the loyalty affidavit would result in further restrictions and attempted to head off more onerous amendments. In December 1949, the Scientists' Committee on Loyalty Problems polled its sponsors. Asked whether they would support a National Science Foundation if full FBI investigations and loyalty clearance were required for fellowship holders, respondents answered a resounding "no" by a vote of 51 to 4. The SCLP quickly relayed this result to key members of Congress, indicating scientists would withdraw support from H.R. 4846 if Congress adopted security amendments beyond the loyalty affidavit.[12]

In the wake of the arrest of Klaus Fuchs on charges of wartime and postwar atomic espionage, however, House conservatives viewed security provisions as critically important. When H.R. 4846 finally made it to the House floor in late February 1950, Representative Howard W. Smith (D.-Va.) added an amendment that not only required FBI investigations of all NSF employees and scholarship recipients but also gave the FBI, not the NSF, the authority to certify the loyalty of individuals. Such a mandate exceeded, by far, the FBI's existing authority under the federal loyalty program. Smith placed the amendment explicitly in the context of the need to protect scientific secrets: "When this scientific foundation is set up it will have access to the most vital secrets of this Government. . . . If there is any one place in the Federal Government where we should undertake to protect ourselves, it is in connection with the loyalty of anybody who may be connected with or may have the opportunity to get at those most vital secrets."[13] Smith's statement reflected the extent to which many members of Congress identified all scientific research with national security by the late 1940s. Another amendment, introduced by Representative Daniel J. Flood (D.-Pa.), required foreign nationals to be cleared by the FBI before taking up any duty related to the NSF. With these added stipulations, H.R. 4846 passed the House by a 247–126 vote on March 1 and moved on to the House-Senate conference committee.

The Federation of American Scientists decided that if the security amendments were not deleted by the conference committee, "scientists should urge defeat of the bill."[14] The SCLP directly pressured conference committee members to remove the Smith amendment.[15] In a *Washington Post* editorial that Representative Barratt O'Hara (D.-Ill.) placed in the *Congressional Record*, Clifford Grobstein, vice chairman of the FAS and chairman of the Washington Association of Scientists study group on the NSF legislation, declared, "There are at least two good reasons why this legislation, now unfortunately loaded with political dynamite, is not better than none at all." First, Grobstein contended, the loyalty restrictions would force scientists to seek funds from sources other than the NSF. (Where else they were supposed to go in order to avoid loyalty requirements was not entirely clear.) More important, Grobstein pointed to the amendment's larger implications for free political discussion in American society. He wrote, "The issue raised goes beyond the Science Foundation, indeed almost relegates it to the background. It is not only as scientists, but as citizens concerned for the future of our way of life, that we protest this new height in antidemocratic measures offered to protect democracy."[16] The federation had decided it would be better to do without the NSF than to allow the further incursion of loyalty requirements into unclassified research and the general erosion of political freedom in the supposed name of national security.

Other scientists' organizations also moved to have the onerous amendments struck. Wolfle labeled the Smith amendment "about the most severe loyalty test ever considered" and coordinated the Inter-Society Committee's efforts to pressure the conference committee. Even the council of the National Academy of Sciences, which had stayed quiet months earlier when the Senate debated the O'Mahoney amendment, indicated its disapproval. In a public statement, the NAS council declared, "We are gravely disturbed . . . to see that security measures are being extended widely over the scientific life of the country, even in those areas remote from possible military application. . . . If we are concerned with these developments it is not because we ask a special privilege for scientists; it is because they cannot lose their freedom without jeopardizing the freedom of all Americans."[17] National Academy of Sciences president A. N. Richards also wrote President Truman directly and warned that the Smith amendment "would so distort the purpose of the original bill as to work serious damage to the development of science in the United States and to those persons upon whom that development depends."[18] The previous year, scientists had responded too slowly to the controversy over the AEC fellowships. This time they quickly made clear their unyielding opposition to the imposition of loyalty requirements in a foundation dedicated to basic research.

Assistance came from an unexpected quarter. The amendments were too strong even for the Justice Department and the FBI. By granting the FBI the authority to pass judgment on the loyalty of individuals, the Smith amendment proposed to violate J. Edgar Hoover's insistence on the fiction that the FBI was a neutral fact-finding agency that did not evaluate individual loyalty. The FBI also lacked the ability to investigate foreign nationals, as required by the Flood amendment. Assistant Attorney General Peyton Ford, writing on behalf of the Justice Department, the secretary of defense, and the director of the Bureau of the Budget, informed Congress that the amendments constituted "an extremely radical and undesirable change in the basic responsibilities and functions of the Federal Bureau of Investigation."[19] Given the lack of executive branch sentiment in favor of the Smith and Flood amendments and the widespread opposition of scientists, the House-Senate conference committee replaced the amendments with a modest and uncontroversial statement that no NSF employee could have access to restricted information without security clearance. In late April, the full House and Senate passed the bill, and on May 10, President Truman signed the National Science Foundation Act of 1950 into law.

Even in 1950, Cold War anticommunism had some political and institutional limits, and anticommunists' proposals were occasionally stymied. Although an important part of the history of American science, the NSF bill was a relatively minor piece of legislation in a year that included Truman's decision to develop the hydrogen bomb, the rise of Senator Joseph McCarthy and his attacks on the State Department, the U.S. commitment to increased defense spending called for under NSC-68, a secret National Security Council document, and the American entry into the Korean War. Unlike the AEC fellowship program, the NSF remained a small enough issue to avoid becoming completely embroiled in the politics of anticommunism. Without the rhetorical power of appeals to the prospect of lost atomic secrets and the support of the

Justice Department and the FBI, attempts to impose heavy loyalty and security requirements on the National Science Foundation failed. Although the loyalty affidavit remained, the final establishment of the NSF constituted a moral victory for scientists who wished to keep political tests out of basic research.

But the National Science Foundation created in 1950 fell far short of the original aspirations of either Vannevar Bush's or Harley M. Kilgore's supporters. The new foundation lacked a commitment to any of Kilgore's principles, and with a severely limited budget it could not meet Bush's vision of becoming a major supporter of pure research. A cost-conscious Congress imposed a $15 million maximum budget and provided much less in its initial outlays—only a paltry $225,000 for start-up costs in the NSF's first year and $6.3 million in fiscal year 1952. By comparison, the 1946 Kilgore-Magnuson compromise bill had projected an initial budget for the NSF of $40 million, expected to increase to the range of $150 million to $200 million after three years. As originally conceived, the NSF was supposed to be a major new federal agency that would fund the bulk of basic research in the United States, not the small, minor bureau that it actually was in its early years. Although the NSF provided significant support for the development of a few particular scientific disciplines, such as radio astronomy, the military remained the primary source for postwar funding of basic research in the physical sciences.[20]

The Federation of American Scientists greeted the passage of the National Science Foundation Act of 1950 with ambivalence. The federation observed, "The Foundation which will thus come into existence after 4 years of bitter struggle is a far cry from the hopes of many scientists. Successive compromises have weakened the original conception. The suspicions and distrust of the Cold War are woven through it. . . . Yet enough of value probably remains to earn a welcome from most scientists and to justify the hope that the Foundation may yet grow to fully fill the needs which originally suggested it." The federation took some solace in the deletion of the Smith amendment and noted, "Another case of successful political action by scientists can be chalked up." At the same time, however, the loyalty affidavit meant the NSF was "hardly free of security and secrecy." Even more worrisome, the 1950 enabling act contained a clause that allowed the new foundation to undertake research related to military needs if requested by the secretary of defense. The provision, the FAS feared, "could open the domination of the Foundation by military considerations." The restricted budget of the foundation was also a pittance compared with the $85 million per annum that the Office of Naval Research devoted to research and development. On net, "The result of four years of effort is thus an NSF limited budget-wise, open to military distortion, shot through with security provisions, under part-time management—but still preserving the original emphasis on basic research, and responsibility for overall policy formulation."

Although a disappointment compared with initial expectations, the NSF, by its very existence, contained the potential for playing a larger role in the future. The Federation of American Scientists hoped the foundation would immediately undertake "the comprehensive study of the organization, activities, economics, and limitations of post-war science in the U.S." The federation advised, "It should particularly concentrate on the effects of the Cold War—the emphasis on secrecy and military technology."[21] Despite the disappearance of Kilgore's ideas and the long series of compromises contained within the final legislation, the FAS still thought the National Science Foundation might provide a haven from the Cold War and find a way to promote the funding of basic research on grounds other than national security.

The federation's hopes went largely unfulfilled. Although the Bureau of the Budget wanted the NSF to take on a policy-planning role, NSF director Alan T. Waterman, former chief scientist and deputy chief of the Office of Naval Research, was interested only in educational and research programs. The NSF made no attempt to study the role of the Cold War in the political economy of American science. More important, with the budgetary stresses of the Korean War, the NSF could only attain significant funding by selling Congress on the importance of basic research for defense purposes. The NSF's educational programs, even before the Soviet launch of Sputnik in 1957, owed much to Cold War concerns about the supply of scientific and technical personnel.[22]

Furthermore, the military and the AEC remained by far the largest federal patrons of research in the physical sciences. In the late 1940s, many observers, including scientists and military leaders, had assumed that most basic research programs in the military would be transferred to the NSF once it was established. With the budgetary ceiling mandated by Congress, however, the NSF could not absorb large military research programs, and both the Office of Naval Research and leaders of science resisted any such transfer.[23] A 1953 amendment removed the NSF budget ceiling. Even so, by 1960 the Department of Defense and the AEC still provided 92 percent of federal support for physics research. In subfields such as solid-state physics, the proportion of military funding was even higher. Although the NSF's budget grew rapidly in the late 1950s and early 1960s, by the mid-1960s, the NSF accounted for only 10 percent of basic research funded by the government.[24] In 1952, Melba Phillips lamented the subordination of scientific research to military objectives. She noted that immediately after the war, "many [scientists] did band together for a brief period and succeeded in arousing effective public sentiment against the military control of atomic energy." The Cold War, however, had "killed the movement of protest."[25] In the context of the Cold War, liberal and progressive left scientists' hopes that the NSF could provide an alternative to military funding and the basis for a reworking of the relationship between science and society remained but a distant dream.

"Perpetual Jeopardy": Loyalty and Security Clearance in the 1950s

Meanwhile, in London, the confession and arrest of Klaus Fuchs in February 1950 transformed atomic espionage from an ominous hypothetical into a grim reality. The identification of Harry Gold, a Philadelphia industrial chemist, as Fuchs's American contact subsequently led to the arrest of David Greenglass, an army machinist who had relayed limited information about the design of the plutonium bomb to Gold and to his brother-in-law, Julius Rosenberg. Rosenberg and his wife, Ethel, were arrested in the summer of 1950, convicted of conspiracy to commit espionage in March 1951, and executed in June 1953.

The apprehension of the Rosenbergs intensified the federal government's efforts to identify potential subversives before they could do any damage. The Rosenberg case had particularly powerful repercussions for scientific personnel at the army's Signal Corps Engineering Laboratories in Fort Monmouth, New Jersey, which carried out important research on electronics, communications, radar, guided missile controls, and other military problems. The work of Ronald Radosh and Joyce Milton has made it quite clear that Julius Rosenberg headed a spy ring dedicated to sending classified scientific and technical data, including information about the Manhattan Project, advanced aircraft design, and radar, to the Soviet Union. Although no hard evidence indicated security breaches had occurred at Fort Monmouth, circumstances seemed to imply that the army installation was a likely target of the Rosenberg ring. An electrical engineering major at City College of New York (CCNY) in the 1930s, Rosenberg worked as an engineer-inspector for the U.S. Army Signal Corps from 1940 until 1945, when the army fired him after uncovering his past Communist Party membership. City College of New

York was a center of communist political activity during the depression, and Rosenberg had been a highly visible member of the CCNY branch of the Young Communist League during his college years. He eventually became a full-fledged member of the Communist Party. Sometime in late 1943, he dropped out of public political activism and moved into the underworld of Soviet espionage. Most of his recruits for his spy ring were fellow engineering alumni from CCNY, including Joel Barr, Alfred Sarant, and the Rosenbergs' codefendant, Morton Sobell. Barr worked at Fort Monmouth from 1940 to 1942. Sarant was also an engineer at the Signal Corps Engineering Laboratories in the early 1940s, where he became a close friend of Barr. Neither Barr nor Sarant was ever convicted of espionage, but their guilt is not in doubt. Both fled from the United States shortly after the Rosenbergs' arrest and eventually ended up in the Soviet Union, where they pursued careers in Soviet military research. The Venona Project intercepts, Soviet cables sent between 1942 and 1946 and decoded by U.S. intelligence in the 1940s and 1950s, also confirm the identity of Barr and Sarant as Soviet agents.[26]

In August 1953, the army launched an investigation of civilian employees at the Signal Corps Engineering Laboratories. President Eisenhower's new employee security program provided one rationale for the intensive loyalty probe, but the execution of the Rosenbergs only two months earlier, combined with signs that Senator McCarthy intended to set his own investigation of Fort Monmouth in motion, provided the main impetus for the army's actions. Over the next six months, the army suspended or revoked security clearances from some forty-odd employees of the Signal Corps laboratories, including more than thirty scientists and engineers. In October and November 1953, McCarthy also held hearings and promised to expose espionage at Fort Monmouth. He found none. Eventually, most of the employees affected by the army's investigation regained their prior status, but eight lost their positions permanently.

Specific sociological conditions contributed to the army's suspicions at Fort Monmouth. For Jewish scientists and engineers, the Signal Corps laboratories had long functioned as a haven from the discrimination that ran rampant in private industry. Almost all the employees who were suspended or relegated to working on unclassified research in the laboratories' "leper colony" were from New York City's Jewish community, and many had attended City College of New York. A few had known Julius Rosenberg or Morton Sobell in their college days. Others had relatives, friends, colleagues, or acquaintances who were Communist

Party members or had participated in communist political causes in the 1930s and early 1940s. Some had themselves attended political functions or belonged to organizations alleged to have been infiltrated or dominated by communists. These personal and political associations formed the basis for the army's fears of security breaches at the Signal Corps laboratories. According to an analysis by the FAS's Scientists' Committee on Loyalty and Security (formed at Yale University to replace the Scientists' Committee on Loyalty Problems after the latter disbanded in 1951), of nineteen cases still pending in March 1954, 85 percent of the charges leveled against the Fort Monmouth scientists concerned associations with suspect individuals or groups. Many of the charges focused on interactions with alleged Communist Party members, but they also covered liberal and progressive left affiliations. The accusations against one scientist included his active involvement in the FAS, which the army noted was "reported to have been infiltrated by Communists or Communist sympathizers," and his attendance at a conference on atomic energy where he introduced physicist Philip Morrison as a speaker. Six of the Fort Monmouth scientists were also accused of being members of the Communist Party or other communist organizations. Five of the six denied the accusations under oath; the sixth explained he had participated in communist functions only at the behest of his mother, when he was twelve or thirteen years old.[27]

Given the interests of the Rosenberg spy ring, the army had an understandable anxiety about the Signal Corps laboratories. But the Fort Monmouth probe also revealed the limitations of security clearance investigations and the quest for absolute security. On average, the persons whom the army pinpointed as security risks had been employed at the Signal Corps for more than a decade and had passed previous security checks. If any of them had nurtured aspirations toward spying, they had probably already fulfilled them. But no convincing evidence of espionage at Fort Monmouth, by either the Rosenberg ring or another group, has ever emerged.[28] How meaningful, then, was the "security risk" designation? By examining general political beliefs and associations, the loyalty-security system was supposed to identify individuals who might commit disloyal acts for ideological reasons. Certainly such individuals existed, Klaus Fuchs and Julius Rosenberg being the most notorious examples. But could they really be detected with any accuracy ahead of time? The strongest case could be made for excluding past members of the Communist Party or other communist groups from sensitive positions. Of all the known instances of wartime atomic espionage committed by Americans, the parties involved—Morris and Lona Cohen, Harry Gold, David and Ruth Greenglass, Julius and Ethel Rosenberg—were, with the exception of Harry Gold, members of the Communist Party or other communist organizations at some time in their lives.[29] But some 250,000 people passed through the American Communist Party between 1919 and 1960. The vast majority joined the party in good faith and never engaged in illegal clandestine acts.[30] Many more Americans belonged to progressive left political organizations that included communist members or held political opinions that also happened to be favored by communists. No more than the tiniest fraction of those persons were likely recruits for Soviet-sponsored subversion. The pattern of behaviors and associations that the AEC, military, and civil service used to identify loyalty and security risks was a blunt instrument, one that could not detect the "potentially disloyal" without calling all liberal-left political activity into question.

With the political tensions created by the Korean War and the Fuchs and Rosenberg cases, the standards employed by the loyalty-security system seemed to grow increasingly cavalier during the early 1950s. Scientists continued to contact the FAS with stories about being denied loyalty and security clearances for dubious reasons. Common charges included involvement in Henry Wallace's 1948 Progressive Party presidential campaign and criticism of U.S. foreign policy, especially the Korean War. In particular cases, scientists also reported being taken to task for, among other things, supporting socialized medicine, speaking before the National Council of American-Soviet Friendship, and criticizing the loyalty program. One industrial physicist complained that the only charge against him was that he had the *Daily Worker* in his home (an accusation that he refuted with affidavits from past visitors to his home, as well as a statement from his mailman). Another scientist reported to the FAS that his membership in the federation appeared responsible for his having been denied access to classified information.[31]

Even membership in the American Association for the Advancement of Science, an organization dedicated more to professional concerns than political activism, led to problems. In 1954, several scientists notified the AAAS that they had been denied security clearance by the military and the Eastern Industrial Personnel Security Board because of their AAAS membership. The loyalty-security system had never linked such membership with potential subversion before; the election of Kirtley Mather, a prominent Harvard University geologist whose political views were distinctively progressive left, as the AAAS president for 1952, followed by the election of E. U. Condon for the 1953 presidency, may have been the source of the association's newly acquired suspiciousness. One military scientist who was refused army travel funds for attendance at the AAAS annual meeting learned from the director and the executive officer of the Office of Armed Forces Information and Education that the presence of "alleged communists or communist sympathizers among the officers of the American Association for the Advancement of Science" meant no travel on government funds could be permitted.[32]

In 1948, scientists at Oak Ridge had emphasized the need for the loyalty-security system to certify that charges were demonstrably relevant to determining individuals' security status. In the 1950s, the troubling issues raised by Katharine Way and other scientists remained unresolved. Although set rules and guidelines had replaced improvised and jury-rigged proceedings, formal procedures did not guarantee fair results. Procedural reforms did nothing to promote careful consideration about the criteria used to identify security risks. The tendency of government boards to consider liberal political activity an indicator of disloyalty worsened with the consolidation of the Cold War political consensus. The frequent citation of participation in the Progressive Party and criticism of U.S. foreign policy as grounds for questioning a person's loyalty suggested that *any* dissent from mainstream political opinion could lead to one's exclusion from government employment.

Although the loyalty-security program purported to possess the capability to distinguish between loyal and potentially disloyal individuals, personnel security procedures could not even distinguish between probable security risks and individuals who were understandably angry at having their loyalty questioned. It was with good reason that the FAS advised scientists who faced hearings to maintain a polite, cooperative attitude. John Phelps of the FAS's Scientists' Committee on Loyalty and Security wrote of one case "where the charges are pretty feeble but where the guy is so incensed at the whole business that he insults the board, displays magnificent contempt for the security programs, and gives no sign of cooperating with anybody."[33] The scientist's hostile demeanor, Phelps indicated, had virtually nullified his chances of being cleared. Another scientist fired off an indignant response to his loyalty board without consulting his lawyer, to the latter's dismay. He eventually resigned his position rather than fight his case; apparently he felt that he had damaged his chances too greatly with his ill-advised reply.[34]

Given the increasingly constricted political climate of the 1950s, some scientists wondered if they should avoid the loyalty-security system altogether. Yale University physicist Ernest Pollard wrote to Samuel A. Goudsmit for advice about a student whose mother had once campaigned for a communist mayoral candidate in the 1930s and who had himself, as a college student, enthusiastically supported Henry Wallace. Would the student be eligible for classified work, or should he restrict himself to seeking future employment in nonsecret fields?[35] The scientist in question had done nothing more than exercise his political rights. Now he wondered if he had irreparably restricted his future career opportunities.

Established holders of security clearances fared no better. The situation faced by the longtime employees of the Signal Corps laboratories was not at all unusual. Scientists who already held clearances acquired little immunity to future problems. A change in regulations, a switch to another job, or the surfacing of new information could lead to an annoying and damaging reevaluation of one's eligibility. The loyalty-security system placed its subjects in what Philip M. Stern, in his book about the Oppenheimer case, called a state of "perpetual jeopardy."[36] Old charges were never settled definitively. Once cleared, an individual could confront the same accusations in subsequent loyalty-security investigations.

John and Hildred Blewett faced such a situation in January 1952, when the AEC informed them that their eligibility for security clearance and employment by the AEC were again in doubt. This time, thanks to Leland Haworth's leadership at Brookhaven, the AEC resolved their case quickly and without too much disruption of their lives. Why they faced renewed scrutiny at all is unclear. The AEC had already brought five of the six allegations against them before, including their friendship with Israel Halperin, support for the United Electrical Workers at GE, and brief membership in the Independent Citizens Committee of the Arts, Sciences, and Professions. The only new charge, which John believed was made by an unpopular and barely competent engineer who he thought was reporting casual lunchtime conversations to AEC security, concerned their opinions about the state of world politics.[37] The allegation, as stated in the AEC's notice to John, is worth quoting in full for what it reveals about the anticommunist mentality:

> It was reported that your wife, Myrtle Hildred Blewett, had stated in effect that in case of another war she would just as soon go to jail as to work on war material; that she did not consider the threat of Communism to be as great a threat to the United States as is the popular feeling existing in this country at the present time; that she has stated

> the belief that the United States should withdraw to the Western Hemisphere to avoid further conflict. It was further reported that you and your wife . . . interpret the aggression in Korea as a local problem and do not believe that the initial action was Communist inspired or Moscow-directed; also that the United States should not have attempted to keep its forces in Berlin by the use of the airlift, but should have abandoned Berlin.[38]

The perceived versus the actual reality of the communist threat, the wisdom of U.S. intervention abroad, and the limits of American power in the world were legitimate issues for public debate. The security division of the Atomic Energy Commission, however, apparently considered criticisms of domestic anticommunism and U.S. foreign policy to be signs of possible subversive intent. By the 1950s, relatively few people dared to challenge the prevailing tenets of the Cold War political consensus. The experiences of the Blewetts and other scientists suggest some of the reasons why. Freewheeling and open discussion of ideas outside the political mainstream without fear of retaliation was not a simple prospect in Cold War America.

Edward U. Condon also endured perpetual jeopardy throughout the early 1950s. Between World War II and 1954, four different boards cleared him on four separate occasions, despite HUAC's best efforts. But, as discussed in chapter 4, repeated reinvestigations eventually forced Condon from government service and industrial employment. In December 1954, after losing his latest clearance as a result of political pressure, Condon decided his never ending struggle was no longer worth the effort. He stated, "At the present time I do not feel there is any possibility of my securing a fair and independent judgment in a reconsideration of the decision by the Eastern Industrial Personnel Security Board of last July in favor of my security clearance. . . . I now am unwilling to continue a potentially indefinite series of reviews and re-reviews."[39]

Condon may have had the revocation of J. Robert Oppenheimer's security clearance in mind when he decided to end his own battle with the loyalty-security system. The AEC initially suspended Oppenheimer's clearance in December 1953. In June 1954, after extensive hearings before a Personnel Security Board, the AEC commissioners voted four to one to deny Oppenheimer access to classified information permanently.

The Oppenheimer case was, of course, the most infamous security clearance case in the history of Cold War American science. To a great extent, Oppenheimer's ordeal was sui generis, and it is simplistic to concentrate, as I do here, on the features it shared with other scientists' experiences.[40] Highly exceptional circumstances surrounded the eminent physicist's security clearance. For many years, Oppenheimer benefited from preferential treatment. As Barton J. Bernstein has pointed out, any other scientist with a similar political past would have been classified a security risk long before 1954. During the postwar period, however, Oppenheimer also compiled a long list of prominent political enemies that surpassed those of any other scientist. Cold War ideology, varied political interests, and complex individual personalities overlapped and entangled, producing what Bernstein has called "a tragic admixture . . . of the highly contingent and the nearly inevitable."[41] Procedural abnormalities not present in everyday security clearance disputes abounded in the AEC's handling of the Oppenheimer case. The prosecutorial style of the hearings and the public release of the transcript directly contravened AEC policy. Oppenheimer's month-long hearing was a show trial, rather than the neutral inquiry mandated under normal AEC procedures. Owing largely to the manipulation of AEC chairman Lewis L. Strauss, the decision against Oppenheimer was virtually guaranteed before the hearing ever began.

In its broadest outlines, however, Oppenheimer's case followed the general pattern of perpetual jeopardy that characterized other loyalty-security cases in the early 1950s. The charges against Oppenheimer in 1954 concerned, for the most part, matters that the AEC had known about when it first granted him security clearance in 1947. Oppenheimer's participation from the mid-1930s to the early 1940s in the lively social, political, and intellectual world offered by communists and the popular front in the San Francisco Bay area was old news. So were his inconsistent accounts of a 1943 incident in which a friend apprised him of a possible way to deliver scientific information to the Soviet Union. (In all versions of the story, Oppenheimer flatly discouraged any attempt to recruit him for espionage, and there is no convincing evidence that he ever violated wartime or postwar security restrictions.)[42] In 1954, the only new charge against Oppenheimer was that his policy recommendations had inhibited the development of the hydrogen bomb, a matter irrelevant to determining the likelihood that Oppenheimer might improperly disclose classified information.

Oppenheimer's political record had not changed significantly in the years between 1947 and 1954. What had changed was the nature of the U.S. nuclear program and the depth of American policymakers' commitment to a perpetual state of conflict between the United States and the Soviet Union. In 1947, Oppenheimer's distinguished wartime service still lay within recent memory, and the assets of having the acclaimed "father of the atomic bomb" at the center of postwar nuclear policy far outweighed any liabilities. His opposition to the development of the hydrogen bomb in 1949 placed him at odds with the prevailing nuclear winds, however. By the early 1950s, especially after Lewis Strauss became AEC chairman in 1953, H-bomb enthusiasts dominated the upper reaches of the AEC. Strauss, JCAE staffer William L. Borden, air force secretary Thomas K. Finletter and a coterie of air force officers, and a group of scientists led by Edward Teller all favored the hydrogen bomb and fervently believed that the United States needed ever more powerful nuclear weapons to maintain the nation's security. To them, Oppenheimer's recommendations regarding the H-bomb and other policy matters did not simply constitute instances of bad advice. They believed it possible, perhaps even probable, that Oppenheimer had not abandoned his radical political past and that his advice reflected not an honest difference of opinion but a deliberate attempt to sway the balance of military power in favor of the Soviet Union. Moreover, whatever the source of his reasoning, Oppenheimer possessed, they feared, a dangerously high level of influence over nuclear weapons policy. Although other prominent scientists, including James B. Conant, Enrico Fermi, and I. I. Rabi, also opposed the development of the hydrogen bomb, they seemed far less threatening, sharing neither Oppenheimer's flirtation with radicalism nor his public stature and persuasive eloquence. The hydrogen bomb proponents concluded that the best way to remove Oppenheimer's voice from policymaking circles was to render him effectively ineligible for security clearance—hence the revival of old charges and the predetermined verdict.[43]

The judgment against Oppenheimer suggested that no scientist could count on holding a security clearance indefinitely. From the case, scientists drew the lesson that honestly rendered advice threatened dismissal from government service if it ran contrary to conventional Cold War wisdom. In the late 1940s, scientists had stressed that unfair security clearance procedures would cause scientists to avoid government service. A few still made this claim in response to the Oppenheimer case, but most simply voiced their dismay at the AEC's decision to penalize Oppenheimer for his political views.[44] Possibly, by this time, they sensed that even if scientists assiduously avoided security clearances and government employment, they would not be free from political tests and loyalty inquiries.

"Creeping Security": Expanding Loyalty Tests in the 1950s

During the late 1940s, scientists and their organizations had tried to have political tests restricted to the narrowest realm possible, namely, to individuals requiring access to classified information. They contested this point most clearly in the debate over the AEC fellowship program. Unfortunately, by concentrating primarily on personnel security and the difference between classified and nonsecret research, scientists neglected extensive discussion of other government policies that voided the very distinction they wanted to make. Truman's loyalty program completely overturned decades-old civil service protections that prohibited questions about federal employees' political and religious opinions and associations. Truman had hoped the loyalty program would restrain domestic anticommunism and weaken groups like HUAC, but instead it strengthened them and paved the way for the extension of political tests to a wide range of areas in American life. The Taft-Hartley Act provided the next step in what the Scientists' Committee on Loyalty and Security later called "creeping security." In the short run, the noncommunist affidavit required under the act only affected labor unions. It established the principle, however, that political tests could become mandatory not just for federal employment but for any process mediated by the government. Together, the loyalty program and Taft-Hartley suggested that there were few areas in which the government could not intrude for the sake of containing communism.

During the 1950s, loyalty inquiries expanded well beyond security clearances and federal employment. Scientists with perceived or actual progressive left or communist ties faced increasing pressure from universities fearful of state-level and congressional investigating committees. They also risked having their research grants revoked by government agencies and their freedom to travel restricted by the State Department. None of these developments were entirely new. Both the universities and the State Department had on occasion discriminated against individuals for political reasons over a period of several decades. The first Cold War cases came in the late 1940s. But the full force of "creeping security" did not strike until the 1950s.

State investigating committees, HUAC, and the Senate Internal Security Subcommittee (created in 1951) gained new respectability and authority in the 1950s. They derived much of their power from the willingness of groups they investigated to exact reprisals against individual members who became uncooperative witnesses. Labor set an early tone. Anticommunists within the CIO welcomed visits from HUAC as they tried to quell left-wing influence within the unions, and they also appreciated tacit assistance from the FBI. Taft-Hartley, HUAC, and the FBI all helped to tip the political balance within the labor movement against the Left. In 1949 and 1950, the CIO completed the purge by expelling its eleven left-led unions.[45]

In the early 1950s, congressional and state investigating committees became experts in staging rituals of political purification. Descending on Hollywood, colleges and universities, public school systems, and other institutions, they targeted witnesses with communist pasts and gave them two choices. They could either "prove" they repudiated communism by cooperating with the committee, or they could accept reprisals. Willingness to discuss one's own political beliefs and activities was insufficient—proof of one's current loyalty required identifying others who had participated in the Communist Party. Many witnesses, though no longer members of the party, did not wish to endanger others by naming names, and they also disputed the investigating committees' right to subject them to political inquisitions. But the courts had ruled against the invocation of the First Amendment as a protection against unwanted political inquiry. Witnesses who did not want to cooperate with investigating committees could only avoid contempt charges by pleading the Fifth Amendment. Immunity from prosecution did not provide immunity from other forms of harassment, however. Rather than granting unfriendly witnesses the benefits of invoking a right guaranteed under the Constitution, most American colleges and universities fired professors, tenured and untenured, who refused to cooperate with HUAC and other committees on the ostensible grounds that their lack of candor was incompatible with the openness required by scholarly life. Academics also came under fire for participating in the 1948 Wallace campaign or refusing to sign loyalty oaths.[46]

Universities demonstrated their allegiance to the Cold War political consensus in other ways. In order to forestall the threat of congressional inquiries into the universities themselves, administrators initiated their own loyalty proceedings. In some cases, they also solicited information from or actively cooperated with the FBI and other government intelligence agencies in rooting out left-wing faculty members. Colleges and universities also participated in a blacklist, refusing to hire faculty members who had been penalized at other institutions because of their political commitments.[47]

Investigations by HUAC, the Senate Internal Security Subcommittee, and state-level committees had a significant effect on academic scien-

Columbia University physiologist Harry Grundfest (*left*) consults with his lawyer, Osmond K. Fraenkel (*right*), at a hearing before Senator McCarthy's Senate Internal Security Subcommittee, November 25, 1953. (New York World Telegram and Sun Collection, Library of Congress; reproduced by permission of Corbis-Bettmann)

tists. Ellen W. Schrecker estimates that professors and graduate students constituted approximately a fifth of the witnesses before congressional and state committees. A significant proportion of these, perhaps as much as half, were scientists.[48] Scores of academic scientists were subpoenaed by investigating committees or subject to other political tests by the universities; dozens lost their jobs.[49] David Bohm and the other former Berkeley scientists called before HUAC were only a few of those affected. Melba Phillips and Philip Morrison, both of whom were active in the FAS as well as progressive left politics, also suffered political persecution. Phillips lost her professorship at Brooklyn College in 1952 for invoking the Fifth Amendment when asked by the Senate Internal Security Subcommittee whether she had ever been a member of the Communist Party. Philip Morrison fought with Cornell University throughout the 1950s over the extent to which he would be allowed to engage in progressive left political activities without being subject to pressure from university administrators. Called before the Senate Internal Security Subcommittee in 1953, Morrison took the "diminished Fifth." He did not invoke the Fifth Amendment, but he talked only about his own political beliefs and refused to answer questions about the activities of his friends and acquaintances. By not using the Fifth Amendment, he saved his job but risked a contempt citation. Fortunately, he made a good impression before the committee, and no charges were brought. Even Harold Urey was vulnerable. Despite his abandonment of the progressive left for Cold War liberalism, he came under scrutiny during an investigation of the University of Chicago by the Illinois Seditious Activities Investigation Commission.[50]

The universities' acceptance of Cold War ideology endangered scientists' livelihood regardless of whether or not their research had implications for national security. Meanwhile, creeping security expanded further and began to impede scientists' freedom of movement. With increasing regularity, the State Department used political criteria to deny passports to American citizens and visas to foreign nationals throughout the 1950s. By statute, the secretary of state had broad discretion over the issuance of passports and could refuse to grant them without explanation. Before the Cold War, however, the State Department usually withheld passports only if it feared that the applicant was attempting to escape prosecution. Between 1900 and 1940, the State Department denied the right to travel abroad to political radicals in a number of instances, but not as a general rule. In the immediate postwar years, it became more common for the Passport Division to advise individuals that "travel abroad at this time would be contrary to the best interests of the United States," bureaucratic code for a passport denied because the applicant purportedly had past or present communist associations.[51]

The passage of the McCarran Act in 1950 strengthened the State Department's resolve to refuse passports and visas to politically suspect individuals. The McCarran Act, passed over Truman's veto, created the Subversive Activities Control Board to identify and register communist and communist-front organizations, and it prohibited members of registered organizations from receiving passports. The act also forbade visas for foreign nationals who were members of communist organizations or who might "engage in activities which would be prejudicial to the public interest" while in the United States.[52] Before the Subversive Activities Control Board had even registered any organizations, the State Department cited the McCarran Act to justify its growing restrictions on the travel rights of individuals who, whether or not they belonged to the

Communist Party, might "engage in activities which support the Communist movement."[53]

Attendance at international meetings and consultation with foreign colleagues were an intimate part of doing science, so the State Department's policies hit the scientific community especially hard. Several scientists had already been denied passports before, but it was the State Department's refusal to issue a passport to the renowned biochemist Linus Pauling in 1952 that alerted the scientific community to the growing restrictions on scientists' travels. Although he was not part of the Manhattan Project during the war, Pauling's concerns about the destructive power of nuclear weapons led him to become involved in the postwar atomic scientists' movement. He was also extremely active in progressive left political organizations, especially the Independent Citizens Committee of the Arts, Sciences, and Professions. He spoke out vociferously against domestic anticommunism and its repressive effects on civil liberties, and he espoused the virtues of internationalism. Like Harlow Shapley, he refused to subject political organizations to litmus tests, and he readily lent his energies and assistance to any group that had common goals, regardless of its supposed communist affiliations.[54]

In January 1952, Pauling applied to renew his passport. The Passport Division rejected his application with only the vague explanation that his "proposed travel would not be in the best interests of the United States." The State Department had no mechanism for providing due process in cases of passport denials. It did not give specific reasons for refusing to grant passports, and it had no formal appeals procedure. By communicating with State Department officials, Pauling learned that his passport was being withheld "because of the suspicion that I was a Communist, and because my anti-communist statements had not been sufficiently strong."[55] He quickly submitted affidavits to the State Department that denied any past or present communist connections. Nevertheless, the State Department again refused to allow him to travel, and Pauling had to cancel a trip to England.[56]

The Passport Division forbade Pauling from traveling overseas not out of a fear of what he might do abroad but of what he might say at home. Pauling wrote to Farrington Daniels, "I may say that my experiences with the State Department in my effort to get my passport gave me the strong impression that the State Department does not in fact fear that there is danger to the Nation in my traveling abroad, but that instead the denial of a passport to me is an effort by the Government to force me to change my political ideas or give up my activities."[57] He knew from painful personal experience that government pressure encouraged citizens to restrain their own political actions. After World War II, Pauling carefully stayed away from positions that would have required him to obtain a security clearance. In 1951, however, as a member of a California Institute of Technology faculty committee on grants, he needed a clearance in order to review classified research proposals. The Industrial Employment Review Board turned down his application. Pauling had previously withstood a high level of pressure against his political engagements, including a subpoena from the California State Investigating Committee on Education the previous year, without qualms. He now feared, however, that his outspoken politics would lead to unbearable limits on his life as a scientist. Although it turned out that his failure to obtain clearance resulted from a bureaucratic error, he decided it would be safer to keep out of politics for the time being. Pauling resigned from the National Council of the Arts, Sciences, and Professions, the American Association of Scientific Workers, and other progressive left organizations, and he avoided public comment on the Cold War and arms control for most of the next two years. He did speak publicly about his passport difficulties, but he restricted himself to temperate remarks about the facts of his case and abstained from criticizing the politically repressive effects of the State Department's policies.[58]

After considerable news coverage, in both the national and international press, the State Department relented partially, but only after a barrage of public criticism. Secretary of State Dean Acheson intervened, and in July 1952, Pauling received a passport for travel to Paris limited to the one specific trip. In subsequent years, Pauling continued to travel on restricted passports. Late in 1953, Soviet and American thermonuclear tests, combined with the AEC's initial decision to revoke Oppenheimer's clearance, drew Pauling back into the political arena. At about this time, the State Department again withdrew Pauling's passport, and unlike Acheson, Secretary of State Dulles refused to reconsider the case. Not until Pauling won the Nobel Prize in Chemistry in 1954 did the State Department permit him to travel without restrictions. Nine years later, Pauling's political activities won him an unprecedented second prize. In 1963, he was awarded the Nobel Peace Prize for his efforts to alert the world of the dangers posed by radioactive fallout from nuclear testing.[59]

As of 1958, the State Department had refused passports on political grounds to approximately six hundred persons. Probably only a small percentage were scientists, but the figure does not include all those who experienced unexpected delays in obtaining permission to travel abroad or decided not to try. It is likely that almost every scientist with a progressive left political record had difficulty obtaining a passport in the 1950s.[60] In 1952, Harlow Shapley's passport application was stalled until Shapley appealed to Senator Leverett Saltonstall (R.-Mass.) and other government officials for assistance. Shapley finally received a passport limited only to specific travel, and he wondered if he would have to endure similar problems when planning future trips.[61] In April 1953, the Passport Division notified Oswald Veblen, one of Condon's defenders in the National Academy of Sciences and a participant in a variety of progressive left causes, that it was rejecting his passport application because "it has been alleged that you were a Communist."[62] With the help of an attorney, Veblen discovered that almost all the State Department's objections to him involved statements he had made or signed, or groups that he had sponsored, including his support for the abolition of HUAC, his opposition to the McCarran Act, his signature on a statement of confidence in Israel Halperin, his membership in the FAS's Scientists' Committee on Loyalty Problems, and his sponsorship of the 1949 Waldorf conference. After a four-month delay and hundreds of dollars in legal expenses, Veblen received his passport.

Harold Urey also had problems. In August 1952, the State Department denied Urey a passport despite his abandonment of the progressive left for liberal anticommunism. In a letter to Ruth Shipley, head of the Passport Division, Urey outlined his Cold War political transformation: "I have been such an outspoken opponent of Communism that I have feared I would be called a rabid red baiter, and I would not like to have a reputation so extreme as that." He received permission to travel after only a few days of waiting, but throughout the 1950s, he had to settle for passports good only for a strictly limited duration.[63]

The fear of being denied a passport was enough to deter some American scientists from foreign travel altogether. Philip Morrison did not attempt to obtain a passport until 1960. He also avoided applying for any kind of security clearance in the postwar years. Both actions served to protect his progressive left political involvement by preventing university administrators from acquiring any proof that the government had ever reached a negative assessment of his loyalty.[64]

Passport problems sent some American scientists into exile. After losing his job at Princeton University, David Bohm found himself unemployable in the United States because of the academic blacklist. With the help of Albert Einstein, he obtained a position at the University of São Paolo. Shortly after his arrival in Brazil, however, the State Department limited Bohm's passport for return travel to the United States only. Bohm feared renewed controversy if he returned to the United States, but he did not welcome the prospect of being stranded in Brazil. Unhappy with the quality of intellectual life at the University of São Paolo and beset with physical ailments, Bohm searched for a way out. In 1954 he took Brazilian citizenship and used a Brazilian passport to go to Haifa, Israel, to take up a new position at the Technion. In 1957 he went to England on a fellowship at Bristol University. Four years later, he obtained a chair at Birkbeck College in London, where he completed a prestigious and creative career in theoretical physics.[65]

In addition to refusing passports to American scientists, the State Department also restricted the entry of foreign scientists with left-wing political ties into the United States. Even before the passage of the McCarran Act, the State Department regularly denied visas to foreigners who were members of the Communist Party or active in left-wing politics. By early 1949, many foreign scientists already sensed that the U.S. government required submission to intrusive inquiries about their general political beliefs and their opinions about U.S. foreign policy as a condition for entry.[66] The McCarran Act exacerbated the situation, and the State Department barred the entry of foreigners at a rapidly increasing rate during the 1950s. In 1952 an FAS committee estimated that at least half of all foreign scientists seeking to enter the United States experienced some sort of difficulty in obtaining a visa. Scientists from France, where the Left was particularly strong, had an especially hard time. As much as 70 to 80 percent of visa requests from French scientists were unduly delayed or refused. The *Bulletin of the Atomic Scientists* published accounts from twenty-six foreign scientists who had had problems with their visa applications. Among them were several prominent scientists, including British physicist Rudolph E. Peierls, Australian physicist M. L. Oliphant, and French biochemist Jacques Monod. Peierls thought his membership in the Atomic Scientists' Association, a British organization similar to the FAS, was the source of the eight-month delay in his visa application; Oliphant believed that the State Department refused his visa because of his criticisms of U.S. policy regarding atomic energy; and Monod learned that he was barred from the United States because of his membership in the Communist Party from 1943 to 1945, even though the party, at the time, provided a primary source of resistance against the occupation of France by Nazi Germany.[67]

The State Department's visa policies caused more than inconvenience and anguish for individual foreign scientists. They damaged the image of freedom that the United States wished to project abroad as a contrast to the Soviet Union, and they disrupted international scientific life by hindering scientists' interactions and communications. The United States lost the opportunity to host several major international meetings during the 1950s. The International Congress of Psychology, the International Congress of Genetics, the International Astronomical Union, and the International Congress of Biochemistry vowed not to meet in the United States until the State Department lifted its visa restrictions against foreign scientists.[68] By denying the United States the scientific expertise of foreign nationals, the State Department may have even endangered the security that the McCarran Act was supposedly designed to protect. In 1955, in one of the more notorious examples of the excesses of Cold War anticommunism, the United States deported Tsien Hsueshen, a Caltech professor and one of the foremost aeronautics experts in the world, because of his longtime friendship with a fellow scientist who had been a member of the American Communist Party. Tsien had made important contributions to classified military projects for years, and he wanted to become a U.S. citizen. Instead he returned to China, where he directed China's development of ballistic missiles.[69] Tsien's deportation produced one of the more interesting ironies of the postwar red scare. The Cold War fostered the notion that the United States needed superior weaponry to preserve its national security, yet anticommunist policies led to the expulsion of one of the leading authorities in rocketry from the United States to a communist country.

In addition to encroaching on travel rights, creeping security also meant the extension of loyalty tests to applicants for research grants. Under the federal loyalty program, scientific consultants serving on advisory panels and study sections already had to obtain loyalty clearance. Then in 1949, the public dispute over the AEC fellowship program forced the AEC to extend FBI investigations to fellowship recipients, regardless of whether their research was classified or nonsecret. Scientists working on secret research problems had to have security clearances. Why not some form of loyalty clearance as well for all scientists engaged in unclassified research, not just research fellows? In the 1950s, some federal agencies began to deny research grants to scientists on the basis of political criteria as well as scientific merit.

The full extent to which government agencies refused grant requests on loyalty grounds is unknown. As with passport denials, government agencies had no obligation to report their use of political criteria when denying funds to grant applicants. Often individual scientists learned the reasons they were turned down through informal communications with their funding agencies, or they surmised the use of political tests from their particular circumstances.

In the early 1950s, rumors spread throughout the scientific community that the Public Health Service, which supported biological and medical research through the National Institutes of Health, had rejected grant applications on loyalty grounds. In April 1954 Oveta Culp Hobby, secretary of the Department of Health, Education, and Welfare, confirmed previously unverified reports and admitted that the Public Health Service had, for the past two years, employed loyalty criteria as one factor in judging research proposals. Approximately thirty scientists, including Linus Pauling, either had grant applications rejected or existing grants canceled because of their actual or suspected political activities and associations.[70] Threats to revoke grants for political reasons affected not only the scientists who applied for support but also their staffs and their research institutions. Scientists traded stories about researchers whom the Public Health Service pressured to fire technicians and other members of their laboratories or risk being stripped of their grants. Occasionally, the Public Health Service threatened entire research institutions with the loss of funding if they retained certain individuals.[71]

Other agencies also participated in an unofficial blacklist. Scientists suspected the State Department's Fulbright fellowship program and the Veterans' Administration, which funded research in medicine and the behavioral sciences, of denying support to scientists whose FBI files contained derogatory information. Victor F. Weisskopf was especially disturbed by reports that a prominent physicist was refused a Fulbright fellowship in part because of his participation in the FAS.[72] In some cases, the AEC exceeded the political restrictions mandated by Congress and denied grant requests for unclassified projects on the basis of applicants' politics. In November 1953, Columbia University physiologist Harry Grundfest, a longtime activist in the American Association of Scientific Workers and other progressive left political organizations, refused to answer when the Senate Internal Security Subcommittee asked if he had ever been a member of the Communist Party. In defiance of the committee's self-defined mission as political inquisitor, Grundfest was only willing to state that he had never committed espionage. He retained his professorship at Columbia, but within weeks of his congressional testimony, both the AEC and the National Institutes of Health canceled his research grants. A future member of the National Academy of Sciences, Grundfest was highly regarded for his work in neurophysiology. He knew that the quality of his research could not be at issue. Only his refusal to cooperate with Senator McCarthy could explain the loss of funds.[73]

The National Science Foundation was the one agency that drew the line on using political criteria to restrict support for unclassified research. The NSF did reserve the option of denying grants to proven members of the Communist Party or individuals convicted of sabotage, espionage, sedition, or other criminal violations of national security. But the foundation refused to use mere suspicion or derogatory reports from the FBI as the basis for proscribing research support. According to NSF policy, actual judicial proceedings and legal judgments, not arbitrary loyalty tests, constituted the only basis for using loyalty as a criterion for the support of scientific research. Otherwise, all that mattered was the quality of the researcher and his or her proposal.[74]

The NSF's grant policy allowed at least one scientist, biochemist Martin Kamen, to shift to NSF funding and avoid serious damage to his career after the Public Health Service unjustly denied him research support on loyalty grounds.[75] With the rise of the Cold War political consensus, the NSF could not undertake a fundamental reworking of the political economy of postwar science along liberal-left lines. But the NSF did provide a small refuge from the worst excesses of Cold War political repression. By the early 1950s, this was no minor achievement. The National Science Foundation fell far short of scientists' original aspirations, but it nevertheless retained a sense of identity as an institutional alternative to Cold War science.

Between the universities, congressional and state investigating committees, the State Department, and other federal agencies, the Cold War offered scientists a multitude of possibilities for being on the receiving end of political repression. Moreover, the institutions that imposed loyalty inquiries did not act in isolation but were interconnected. As the experiences of Condon, Shapley, Pauling, and other scientists indicate, one kind of political difficulty easily spilled over to create problems in another forum. Scientists denied security clearances often had difficulty finding employment in nonclassified areas of research; scientists attacked by congressional or state committees almost inevitably had their passports revoked and were vulnerable to losing research funds. Martin Kamen's Cold War ordeal illustrates how even a relatively apolitical scientist could undergo long-term, sustained political repression.

Kamen, a specialist in radiochemistry and codiscoverer of carbon-14, spent the early years of World War II working for the Manhattan Project as a researcher at the Berkeley Radiation Laboratory, where he had been a staff member since earning his doctorate in the mid-1930s. A talented violist, he played in benefit recitals for the Joint Anti-Fascist Refugee Committee and Russian War Relief, but otherwise, unlike many of the other young Rad Lab scientists, he avoided heavy political involvement. In 1944, he met Gregory Kheifetz, a Soviet consular official, at one of these recitals. Kheifetz inquired about obtaining radioactive phosphorus for the treatment of another Soviet official who had leukemia, and Kamen agreed to put Kheifetz in contact with John Lawrence, Ernest O. Lawrence's brother, to explore the possibility. There was nothing secret about the medical application of radioisotopes; the Lawrences had begun studying the subject in the early 1930s. Grateful for Kamen's assistance, Kheifetz invited Kamen to dine at a San Francisco restaurant several days later.[76]

Thus began the so-called Fish Grotto incident, which would dog Kamen for the next decade. At dinner with Kheifetz and another Soviet official, Kamen probably discussed, among many other topics of conversation, the medical uses of radioisotopes and aspects of his prewar research, but he took care not to mention anything related to the Manhattan Project. Army intelligence officers seated nearby in the crowded, noisy restaurant, however, gained the impression that Kamen was relaying classified information. Two weeks after the dinner, Kamen was summarily dismissed from the Manhattan Project.[77]

Kamen initially had difficulty finding work, but with the intercession of Ernest O. Lawrence and Washington University chancellor Arthur H. Compton, he secured a professorship at Washington University. The fallout from dinner at the Fish Grotto continued, however. In March 1947, the State Department revoked his passport, and Kamen had to cancel plans for a professional visit to Palestine. With legal assistance, Kamen eventually learned that his supposed indiscretion while employed by the Manhattan Project was responsible for his inability to travel. The FBI also pressured Washington University to fire Kamen. With Compton's full backing, he stayed, but he had to turn down an attractive offer from the National Cancer Institute because he could not obtain loyalty clearance.[78] Then in September 1948, HUAC called on Kamen in one of its early forays into the investigation of wartime atomic espionage. In closed hearings, Kamen cooperated with the committee, answered all questions, and never had to plead the Fifth Amendment. As usual, the committee had promised shocking revelations but found none in Kamen's case. Nonetheless, the State Department continued to refuse Kamen a passport, even though the AEC had no objections to his traveling abroad. Kamen was unable to leave the United States until 1955, after years of litigation. He also lost his grant from the National Institutes of Health in 1954, but the NSF stepped in to make up the shortfall. Kamen was fortunate in that the Washington University administration supported him throughout his long struggle with Cold War anticommunism, and he was never entirely without research funds. But he endured a decade of federal discrimination and harassment from the FBI, HUAC, the State Department, and the Public Health Service, all stemming from a single, overblown wartime incident.[79]

In the late 1940s, scientists still nourished the hope that as long as they stayed away from classified research they would be able to escape loyalty inquiries. But by the 1950s, there was no place to hide. Political tests affected all areas of scientific life—government, industrial, and university employment; access to government fellowships and grants; service on advisory panels and study sections; and travel. The professional reward system of science, which accorded prestige to holders of grants and members of peer review panels, made it almost impossible for scientists of any ambition to avoid encountering some sort of loyalty test during their careers, even if they managed to find a university tolerant of progressive left political action. Fame provided some measure of protection to scientists like Pauling, Shapley, and Condon, but they still had to endure considerable harassment for their political views. For younger scientists in the throes of trying to establish stable careers, it was hard to avoid the conclusion that the best way to protect one's livelihood was to follow the dictates of the Cold War political consensus and steer clear of controversial political activities altogether.

Writing in the *Nation* in January 1953, Maurice Visscher called on scientists to take a tough stance against the Cold War suppression of civil liberties in the United States. "The paramount ethical problem facing scientists today," he declared, "is what moral stand they should take in the crisis of freedom of thought and expression." Scientists, he insisted, had to "demand to be heard" and "take the offensive rather than accept a false defensive position." Statements by individuals were insufficient—major scientific organizations such as the AAAS and the National Academy of Sciences needed to become involved. Visscher noted, "Individuals, I am convinced, can do little or nothing; only large and respected organizations can get results." Instead, the AAAS and the NAS had stood by while prominent members of both organizations, including Condon, Shapley, and Pauling, were "slandered" and "vilified." Ordinary means—"conferences between a few scholars and a few well-meaning officials"—would not be enough. Visscher observed, "A resolution of censure solemnly passed by the AAAS or the NAS would be laughed at by the paranoid fringe." Scientists had to fight McCarthyism with more than quiet diplomacy and the occasional public statement: "The only language they understand is action. The AAAS and the NAS will have to undertake an educational campaign among the people."[80]

Scientists' organizations never rediscovered the public approach that had brought the atomic scientists' movement early success. The American scientific community responded to the increased repression of the 1950s much as it had in the immediate postwar years. Scientists continued to make public statements and appeal to government officials. A few individuals went beyond conventional means and announced personal acts of conscience. In 1954, Harvard biochemist John T. Edsall surprised and inspired colleagues when he responded to the Public Health Service's grant policy by declaring that he would not accept funds from any agency that used loyalty as a factor in awarding support for unclassified research. Edsall's position won him admirers but no imitators, and concerted action at the organizational level remained elusive.[81] In some cases, scientists' associations did contemplate more vigorous and creative responses than they had in the 1940s, but the new initiatives quickly faded. In 1953, Visscher pushed for the AAAS to solicit funds from its membership and launch a large-scale educational effort with other scholarly organizations. The Federation of American Scientists backed Visscher's idea and suggested creating a coalition along the lines of the earlier Inter-Society Committee for a National Science Foundation. Higinbotham hoped that such an effort might "stiffen the spineless and educate the naive," but the new organization never got off the ground.[82] In 1954, the American Psychological Association formed the Committee on Freedom of Enquiry, which it hoped would undertake a large-scale survey of scientists and produce an in-depth study of the effects of loyalty-security requirements on intellectual freedom. Unable to secure funding for the study, the committee folded in 1956 without accomplishing anything of note.[83] Late in 1954, the Scientists' Committee on Loyalty and Security began to formulate plans to separate from the FAS and incorporate as a tax-exempt educational organization. The committee had earlier made an intensive study of the security clearance cases at Fort Monmouth in 1953 and 1954. The SCLS hoped the Scientists' Committee on Security, Inc., would not only continue its studies of the effects of personnel security policies and aid individuals in trouble but also spearhead a major educational campaign about the problems posed by the loyalty-security system. Instead, after its incorporation in 1956, the committee promptly disappeared from the historical record.[84]

In the end, it took developments at the highest levels of government to shorten the reach of loyalty inquiries. The Eisenhower administration vacillated on the protection of civil liberties in the early 1950s, but in the second half of the decade, the federal government gradually eased or removed loyalty tests in many areas. After widespread negative reaction to the AEC's handling of the Oppenheimer case, the Joint Committee on Atomic Energy began to contemplate revising the commission's personnel security procedures. In May 1956, the AEC granted security clearance applicants the long-sought right to cross-examine confidential informants in hearings, mandated that "chance or casual meetings" not be considered in evaluating political associations, and ruled that membership in organizations that appeared on the attorney general's list would not be considered if it had taken place before the listing of the organization in question. Three months later, the Eisenhower administration directed all government agencies not to deny grants or contracts for unclassified research on loyalty grounds unless criminal charges were involved.[85]

Conditions for federal employees also improved after the Supreme Court ruled in June 1956 that the employee security program did not apply to federal workers in nonsensitive positions. As a result, between 1957 and 1961, no suspensions or dismissals took place under Eisenhower's employee security program. The decision to curb the power of the employee security program was only the first of a series of Supreme Court decisions that drastically eroded the power of the government to use loyalty criteria to deny rights and privileges to citizens. Between 1956 and 1960, the Court also invalidated state sedition laws, ruled against the use of confidential informants without the right of cross-examination of accusers in federal loyalty-security proceedings, reversed the Smith Act convictions on the grounds that the act did not cover "theoretical advocacy" of violent overthrow of the government, and struck down the State Department's denial of travel rights on the basis of political belief and association.[86]

Scientists played a role in these developments. In 1954, Eisenhower asked the National Academy of Sciences to look into the effects of loyalty tests on unclassified research; the administration's 1956 decision to end such tests was based in part on the recommendation of the academy's Committee on Loyalty in Relation to Government Support of Unclassified Research.[87] In 1953, John P. Peters, a prominent Yale medical researcher who was also active in progressive left politics, lost his final administrative appeal and decided to sue the Public Health Service over the revocation of his consultantship on loyalty grounds. In 1955, the Supreme Court decided in his favor. The Court avoided ruling on the larger constitutional issues raised in the case, but *Peters v. Hobby* provided a precedent for the Court's later prohibition of the use of confidential informants in loyalty decisions.[88] Two scientists, Martin Kamen and Bruce Dayton, sued the State Department for their passports. In a major departure from its general policy not to defend individual scientists, the FAS backed both suits with moral support and financial assistance. In 1955, Kamen won a favorable ruling in federal district court, and the State Department decided not to appeal in light of the Federal Court of Appeals' recent ruling in *Schactman v. Dulles*, which stipulated that the State Department's discretionary power to issue passports did not include the power to deny passports without substantive due process.[89] In 1958, *Dayton v. Dulles* was one of three passport cases in which the Supreme Court delivered the sweeping ruling that freedom of movement constituted a liberty protected by the Fifth Amendment and that the secretary of state lacked the statutory power to restrict travel on the basis of political beliefs and associations.[90]

Although scientists participated in the process that loosened the tightest constraints that Cold War anticommunism placed on American political culture in the postwar decade, their role was not decisive. The change of heart of the Eisenhower administration and the Supreme Court had more to do with the altered political conditions of the second half of the 1950s than pressure brought to bear by scientists. The threat of communism no longer appeared as grave or imminent as it had in earlier years. U.S.-Soviet relations remained troubled but began to undergo periodic thaws. In 1955, U.S. and Soviet leaders met in Geneva in their first summit meeting since Potsdam a decade earlier, and diplomacy returned as a tool in U.S.-Soviet relations. Slowly and painfully, with repeated false starts and stalled efforts, both countries began to look for ways to ensure that their differences would not be settled with nuclear weapons. Within U.S. borders, the red scare had decimated both the Communist Party and the progressive left. Even the most ardent anticommunists had to admit that the threat of internal subversion seemed slight. The unprecedented economic prosperity of the 1950s also made it unlikely that a political and economic critique rooted in the scarcity of the Great Depression would find a new audience anytime in the near future, although a burgeoning civil rights movement would soon challenge Cold War liberals' notion that America no longer required mass movements and sweeping social change.

The late 1950s retrenchment of Cold War anticommunism was far from complete, however. Although HUAC and the Senate Internal Security Subcommittee lost much of their influence by 1960, they remained in existence until the 1970s. The FBI maintained its authority, and it resorted to rougher tactics. In 1956, the bureau initiated COINTELPRO, its program of direct intervention and harassment of the Communist Party. In the 1960s, Hoover shifted his counterintelligence priorities to civil rights organizations, black nationalist groups, and antiwar protesters, claiming that communists lurked behind all of them. The federal government did not dare rein in the power of the FBI until after Hoover's death in 1972. In 1975, the FBI admitted that it still monitored some eleven hundred organizations that it considered to be infiltrated by communists.[91] Liberal and left-wing scientists remained vulnerable, despite the decisions of the Supreme Court and the directives of the Eisenhower administration. In 1969, shocked scientists learned that the Department of Health, Education, and Welfare continued to exclude scientists from advisory panels and commissions in the Public Health Service on the basis of political criteria. The department barred some because of old charges from the early postwar years and others because they had participated in civil rights demonstrations or protested against the Vietnam War. Within the National Institutes of Health alone, an unofficial blacklist affected as many as two hundred scientists.[92]

For American scientists who suffered adverse effects from Cold War anticommunism, when vindication came at all, it was often years or decades later. The academic blacklist gradually loosened, and scientists long banished from American colleges and universities slowly trickled back in. Frank Oppenheimer and Giovanni Rossi Lomanitz, who had appeared with David Bohm before HUAC and lost their jobs in 1949, returned to academe in the early 1960s after a hiatus of more than a decade. J. Robert Oppenheimer received an apology of sorts from the government when the AEC awarded him the Fermi medal in 1963. Edward U. Condon obtained a new security clearance in 1969 when he agreed to head the air force's investigation of unidentified flying objects. The attempts to make amends spread, and in the 1970s and 1980s, universities began to welcome back scientists and other academics whom they had expelled earlier with special decrees, reinstatements, and honorary degrees.[93] Such symbolic efforts provided individuals with a measure of comfort and helped them to put their ordeals behind. For the nation as a whole, full comprehension of the meaning and legacy of the Cold War lies ahead, to be wrestled with and fought over as the Cold War becomes myth and history.

CONCLUSION

The rise of the Cold War and the split in American liberalism had long-lasting consequences for American science. Political repression not only harmed individual scientists but, more significant, generated an atmosphere of fear that made all scientists question whether political activism was worth the personal risks. As a result, anticommunism narrowed scientists' political role, as well as the range of possibilities for science policy. After experimenting successfully with mass-based politics immediately after the war, the FAS largely abandoned its public approach in response to internal ideological divisions and pressure from HUAC and the FBI. As a result, scientists sacrificed a valuable alternative source of political influence to their growing dependence on the national security state, and Americans forfeited an important avenue for serious discussion of ideas that challenged Cold War assumptions about international relations and domestic political economy. With the rise of the Cold War and the integration of scientific research into the national security state, scientists also lost their effort to reconstruct the politics of science through the Atomic Energy Commission and the National Science Foundation. Despite the hopes of its architects, the AEC quickly became a Cold War agency, and Senator Kilgore's vision of the NSF never made it through Congress. The effort to rebuild basic research on principles of New Deal political economy had no place in the political order of the Cold War, which rejected the expansion of state power for purposes other than the protection of national security. Kilgore's emphasis on lay control was also incompatible with Cold War liberalism, which denied the importance of popular politics and insisted that all social problems were amenable to nonideological negotiation through guidance by responsible elites.[1]

The Cold War left the United States with an impoverished form of liberalism that viewed policy as driven by process and banished ideals from politics. The split between liberal anticommunists and the progressive left was, as Steven M. Gillon has observed, a division between politics and vision that pitted pragmatism against idealism.[2] Liberalism cannot do without pragmatism, but it also requires higher aspirations to remain a vital force in American politics. Politics itself as a means for public debate and decision making is endangered without some measure of faith in ideas and hope for improvement. Of course, left to itself, idealism leads all too easily to a fanciful utopianism. But purely pragmatic politics decays into partisanship and cynicism, as past policy experiments become immutable dogma and once respectable positions degenerate into slogans. America's current crisis of political faith has strong roots in the intellectual poverty of the Cold War political consensus.

American science could also use an infusion of vision. The end of the Cold War and tighter research budgets have left physicists and scientists in related fields in a quandary. Too many scientists are quick to assign blame for the funding shortfalls to the postmodern critics of science and bemoan the ignorance of the general public rather than look to the historical trends that have brought the physical sciences to their post–Cold War crossroads. In the 1940s, liberal and progressive left scientists made an honest and searching attempt to wrestle with difficult questions about the tension between expertise and democracy, the social roles of scientists, and the optimal relationship between science and society. Although much of their vision for American science was rather inchoate and may not hold direct answers for current matters of science policy, their creative fervor ought to provide a resource for contemporary scientists to look outside the constraints of the Cold War political economy for science and conceive of imaginative alternatives. Whether American politics can be reopened to address pressing social and economic problems with daring and inventive proposals, however, is a more difficult question amid the present skepticism about politics itself as a vehicle for positive change.

The Cold War is very much a living legacy, and the political abuses of the postwar decade continue to jar America's collective memory. With the dissolution of the Soviet Union, the struggle to interpret the historical meaning of postwar domestic anticommunism will only grow more complex as Americans attempt to come to terms with what their nation gained and lost in fighting the Cold War. The opening of archival sources in Russia and the publication of memoirs from former Soviet officials will provide a wealth of information for scholars and enhance an already rich field of study. At the same time, the contemporary passions involved in understanding the Cold War will make the problem of interpretation a complicated and contentious undertaking.

The publication of Pavel Sudoplatov's memoirs in 1994 gives some indication of the difficulties to come. Sudoplatov, a former Soviet intelligence officer, claimed that J. Robert Oppenheimer, Enrico Fermi, Leo Szilard, and Niels Bohr all knowingly passed classified information about the atomic bomb to the Soviet Union during World War II. He also contended that American scientists' opposition to the postwar arms race was a product of Soviet influence. Historians quickly rejected and discredited Sudoplatov's claims for lack of proof, numerous factual errors, inconsistencies with the proven historical record, and flawed use of oral history, but *Special Tasks* created a minor media frenzy when it was first published. Sudoplatov's claims revived postwar myths about atomic secrets, the supposed inability of the Soviet Union and other countries to develop nuclear weapons without the aid of espionage, and the alleged disloyalty of American scientists. His allegation that scientists' public opposition to the arms race and development of the hydrogen bomb was part of a Soviet plot is particularly reminiscent of HUAC's attacks on scientists and recalls an anticommunist mind-set that could not separate honest political dissent from disloyalty and betrayal.[3]

The controversy over *Special Tasks* indicates the power the Cold War holds over the American mind and the high stakes involved in the construction of historical meaning. Reputations of individuals, some of whom are still alive, stand to be made or broken; those who are deceased have their family and friends to defend them. But much more is at issue than individual reputations. Questions about the Cold War strike at the very core of America's identity and self-image as a benevolent yet powerful nation. Judgments about U.S. conduct during the Cold War—whether loyalty and security investigations were a harsh necessity or cruelly excessive; whether Cold War foreign policy constituted a well-meaning and successful projection of American power abroad or a misguided effort to remake the world in America's image; whether the Cold War military buildup and nuclear arms race were necessary to defend the nation or constituted a reckless endangerment of the world that also strained American economic resources; in short, whether the nation, and the world, lost more than they gained—will turn on historical scholarship.

Already, in the wake of the Cold War's end, some historians have begun to rehabilitate the history of anticommunism. Between 1970 and 1990, historical scholarship almost universally condemned domestic anticommunism as a harshly repressive response to vastly overinflated fears of Soviet-sponsored subversion.[4] Recently, several historians have argued that the red scare, while often excessive, was to a significant degree a rational response to a real threat. Harvey Klehr and Ronald Radosh have argued persuasively that the notorious *Amerasia* case involved actual spying and a cover-up by the Truman administration, and

Klehr, John Earl Haynes, and Fridrikh Igorevich Firsov have begun to uncover from Russian archives detailed evidence of the Soviet Union's use of the American Communist Party as a conduit for espionage.[5] Both Haynes and Richard Gid Powers have also pointed out that anticommunism, like the Communist Party and the progressive left, was far less monolithic than its opponents made it out to be. Anticommunists' ranks ranged from socialists to John Birchers, and peerless intellectuals to outright scoundrels. Less convincingly to my mind, Powers has returned to the notion that Cold War liberals were responsible anticommunists, in contrast to the countersubversives, the virulent anticommunists who too easily drew on conspiracy theories to pursue supposed subversives.[6]

Historians who remain critical of Cold War anticommunism, myself included, will need to take the new historiography into account. Although Sudoplatov's chronicle of scientists and atomic espionage has been discredited, more creditable claims may emerge in the future. As Thomas Powers warned in his review of *Special Tasks*, "Questions of loyalty and allegiance were often raised during the cold war and many have not been settled yet. One must keep an open mind, because sometimes the incredible news is true."[7] Already it has become clear that espionage made significant contributions to the first Soviet atomic bomb, a plutonium bomb modeled after the one used at Nagasaki.[8] But historians must be cautious about allowing new revelations to put too high a gloss on America's actions during the Cold War. To conclude that the existence of atomic espionage vindicates many of the abuses committed in the name of anticommunism is to fall into the same trap of reasoning that prevailed during the postwar decade.

Atomic espionage was, no doubt, a serious matter, but it was not, as Judge Irving Kaufman told the Rosenbergs, the cause of "the Communist aggression in Korea, with . . . casualties exceeding 50,000";[9] nor was it the overwhelming blow to national security that anticommunists made it out to be in the 1940s and 1950s. Although the Soviet Union began to receive data about the British nuclear project in 1941 and Los Alamos in 1943, Stalin did not make nuclear weapons an all-out priority until after the bombings of Hiroshima and Nagasaki. The most important secret lost was the revelation of the atomic bomb's power. That, not atomic espionage, provided the major impetus for the Soviet effort. In addition, although Soviet physicists made use of stolen data, they did not need to depend on espionage. Shortly after the end of the war, physicist Peter Kapitsa had enough confidence in Soviet physics to urge Stalin to allow Soviet physicists to avoid American mistakes and find their own independent scientific and technological path to building nuclear weapons. The decision to copy the American plutonium bomb rather than develop a new Soviet bomb was a political decision, not a scientific imperative. Nor is copying a sophisticated technology as easy as it sounds. Stolen data did not provide the Soviet Union with the immediate capability to produce its own nuclear weapons. Soviet scientists had to reproduce painstakingly every piece of America data, as part of a learning process as well as to rule out the possibility of disinformation by American counterintelligence. The use of American data saved the Soviet Union, at most, one to two years in achieving its first atomic bomb.[10] It is difficult to imagine that the U.S.-Soviet strategic balance would have been significantly different had the Soviet Union acquired atomic weapons in 1950 or 1951 rather than 1949.

The atomic scientists came close to the mark with their predicted timescale of four or five years for Soviet acquisition of the bomb. They were also essentially correct in their appraisal of the options. The real choice the United States faced after the war was either to negotiate for international controls or to try to stay ahead through continuous innovation in a risky technological arms race. Security restrictions were only of secondary or tertiary importance, not primary as anticommunists believed.

It must also be recognized that the antisubversion efforts of the Truman administration, HUAC, the FBI, and other sectors of the federal government had little to do with any carefully considered analysis of the reality of the Soviet threat. Domestic political calculations, not judicious assessments of the level of danger posed by criminal subversion, guided the anticommunist response.[11] The House Committee on Un-American Activities, which did nothing without an eye toward political advantage, haphazardly pursued atomic scientists when convenient but abandoned the chase when the Joint Committee on Atomic Energy made the risks outweigh the potential gains. Hoover and the FBI systematically disregarded the law and routinely neglected the slow and careful police work necessary to prosecute real cases of criminal subversion. Instead, Hoover preferred to carry out ideological warfare and target indiscriminately the liberal-left at the expense of the Bill of Rights. The Truman administration, for its part, established the loyalty program in response to political pressure, without adequately considering the consequences.

The only area in which the government had a legitimate purpose in mandating loyalty tests was for positions that required access to secret information. But, as we have seen, the security clearance system was hardly without problems. In searching for a set of procedures that would simultaneously guarantee fairness and protect the national security, scientists and AEC officials failed to realize that loyalty screening was probabilistic in nature and thus *inherently* unfair to individuals. The personal attributes and political associations used to identify individuals with a possible ideological predisposition toward espionage were vague and inevitably cast a broad net over persons whose political activities were entirely innocent. The search for potential spies motivated by ideology was by its very nature a low-probability enterprise, one that would identify hundreds of false positives for every true positive and provide no means to differentiate between them.[12] The use of FBI files, compilations of undigested opinions and observations, compounded the problem. Individual FBI files were, at best, raw data. They could occasionally provide leads in criminal cases, but by themselves they were not reliable sources of solid evidence.[13] Much as Hoover might have claimed otherwise, FBI files simply were not designed for the task for which the security clearance system used them.

The search for ideological risks also drew attention away from other motivations for espionage. As Brent Scowcroft, national security adviser for the Bush administration, noted in 1994 with respect to the now infamous Aldrich Ames spy case, "Our focus [in security checks] for so long has been on aberrant personalities . . . either those who have a secret love for communism and this and that and the other. . . . We haven't focused as much on . . . relatively normal seeming personalities who just want to make an extra buck." Other motives, such as greed, were at least as important as ideology (and just as difficult to detect before the fact), but none received the scrutiny that ideology did.[14]

Even when some legitimate purpose is at stake, loyalty investigations always raise disturbing questions about the relationship between the citizen and the state and the balance between individual freedom and national security. Soviet espionage was a reality of the Cold War, so the effort to root out potential ideological risks was not completely unjustified. But the way it was carried out was largely futile, caused individual scientists pain and threatened their careers, and hampered politically active scientists' ability to raise publicly serious questions about the wisdom of the Cold War political consensus.

Much of the debate over the meaning of the Cold War and the balance sheet of American gains and losses will turn on an assessment of Soviet intentions. Was the Soviet Union a normal state, behaving ruthlessly but pragmatically according to the geopolitical dictates of hard-core balance-of-power politics? Were Soviet statements about the inevitable collapse of capitalism simply expressions of the historical progression predicted and demanded by Marxist theory? Or were Soviet leaders actively seeking, as anticommunists claimed, the ultimate goal of world conquest and global hegemony? My own inclination is to view Soviet intentions in a more pragmatic, limited framework, much as most scholars view China's motives today. But even if Soviet leaders' designs should prove to have been more expansive, we should not be too quick to absolve American actions during the Cold War. The sports-derived calculus of wins and losses provides a wholly inadequate vocabulary for understanding the admixture of realism, fear, idealism, contradiction, and folly that characterized the U.S. response to the dangers and challenges of the post–World War II world. The American record of individual rights violated at home and misadventure abroad in Latin America, Southeast Asia, and elsewhere, all in the name of containing the communist threat, is not a past in which to take great pride. Relief, not celebration, is the more appropriate sentiment with which to greet the post–Cold War era, and grim, tragic necessity, rather than undiluted triumph, provides a better narrative for understanding the American experience during the Cold War.

NOTES

INTRODUCTION

1. On Americans' ambivalent mood at the end of World War II, see Boyer, *By the Bomb's Early Light*, 7. For discussions of the narratives that Americans used to order their conceptions of the postwar world, see Carmichael, *Framing History*, and Engelhardt, *End of Victory Culture*.

2. For example, Caute, *Great Fear*, ch. 25, depicts scientists solely as victims of anticommunism. Boyer, *By the Bomb's Early Light*, portrays scientists primarily as dissenters from the Cold War political order, while Carmichael, *Framing History*, 25, 50–55, suggests that scientists were one of the few groups to resist vigorously the imposition of the Cold War national narrative.

3. See, for example, Forman, "Behind Quantum Electronics"; Kevles, "Cold War and Hot Physics," in part a reply to Forman; Hoch, "Crystallization of a Strategic Alliance"; Schweber, "Mutual Embrace of Science and the Military"; and Leslie, *Cold War and American Science*. I understand the phrase "national security state" to connote both a specific meaning, the complex of government agencies charged with protecting the nation's security, and Daniel Yergin's more general definition: the "unified pattern of attitudes, policies, and institutions" to be used to prepare the United States for a period of "perpetual confrontation and war" with the Soviet Union. Yergin, *Shattered Peace*, 5–6.

4. Some work along these lines has been done, but it is only just a beginning. On cases of university scientists who tried to avoid military patronage and searched for other alternatives, see Silvan S. Schweber, "Big Science in Context," in Galison and Hevly, *Big Science*, on Cornell University, and Feffer, "Atoms, Cancer, and Politics," on the University of Chicago.

5. On antiradicalism as a form of nativism and the political and economic conditions under which nativism thrives, see Higham, *Strangers in the Land*. According to Higham, under conditions of economic prosperity, social stability, and international security, differences between groups are generally tolerated, and there is political space for dissent. Although nativist rhetoric is employed under such circumstances, it does not dominate political discourse during periods of relative stability. Conditions of social stress and economic upheaval, however, especially when accompanied by intense labor protest, provide the breeding ground for nativism aimed at certain religious groups, immigrants, or promoters of radical politics; jingoist foreign policy further exacerbates domestic nativism. Higham's work covers the years from 1880 to 1920, but his framework also provides a great deal of insight into the causes of the post–World War I and post–World War II red scares.

6. For a general account of the rise of science in twentieth-century America, see Kevles, *The Physicists*. On corporate science, see McGrath, "American Scientists at War." McGrath makes the provocative argument that the science-military alliance of World War II was rooted in scientists' corporate experience during the 1920s. Of the

foundations, the greatest historical attention has been paid to the Rockefeller Foundation; Robert E. Kohler, *Partners in Science: Foundations and Natural Scientists, 1900–1945* (Chicago: University of Chicago Press, 1991), is a good place to start. For the rise of federal sponsorship of science, see A. Hunter Dupree, *Science in the Federal Government: A History of Policies and Activities to 1940* (Cambridge: Harvard University Press, Belknap Press, 1957).

7. For the most forceful statement of this argument, see David F. Noble, *America by Design: Science, Technology, and the Rise of Corporate Capitalism* (New York: Alfred A. Knopf, 1977).

8. In contrast to Noble's *America by Design*, Layton, *Revolt of the Engineers*, argues that American engineers developed a new sense of social responsibility during the early part of the twentieth century. John M. Jordan, *Machine-Age Ideology: Social Engineering and American Liberalism, 1911–1939* (Chapel Hill: University of North Carolina Press, 1995), acknowledges the aspirations of early social scientists and other liberal, technocratic thinkers but also argues persuasively that their technocratic ethos masked the conservative tendencies of their own value-laden assumptions. Many historians of the Progressive Era stress the conservative nature of Progressive reform, its agenda for social control, and its antidemocratic effects.

This is not the place to debate Progressive Era historiography. My own feeling is that historians are right to wonder whether the solutions Progressive Era reformers offered were, in fact, conservative. The sincerity of Progressive reformers' motives, however, should not be dismissed too easily. Whatever their shortcomings as nascent radicals and analysts of the causes of social ills, they did see injustice in the social and economic order of their time and aspired to improve the conditions of the poor and working class.

9. My use of political terminology is somewhat unusual and needs to be made clear. I use the term "progressive left" interchangeably with the phrase "popular front liberal," which some historians prefer. By "popular front," these historians mean not the more limited sense of the American Communist Party's political strategy from 1935 to 1939 but a general term to refer to a view of liberalism grounded on an optimistic faith in rationality and the capacity of individuals to act generously for the greater good of society, ideas embraced by a liberal-left coalition of populists, progressives, liberals, and radicals both before and after World War II. I prefer to avoid the phrase "popular front" because of the pejorative connotations it acquired after the Hitler-Stalin pact and never entirely lost. "Progressive" was the adjective the popular front liberals used to describe themselves; the historian's use of the word, however, risks confusion with the Progressive Era progressives. Hence my preference for "progressive left." See McAuliffe, *Crisis on the Left*, intro., chs. 1–3. Other historians use "popular front" in the same sense as McAuliffe: consult, for example, Halpern, *UAW Politics*, chs. 10 and 17. Richard H. Pells avoids "popular front" but uses similar characteristics in describing the contrast between progressives and Cold War liberal anticommunists. Pells, *Liberal Mind in a Conservative Age*, ch. 2.

10. Kuznick, *Beyond the Laboratory*, is the standard source on scientists and political activism in the 1930s.

11. On the rise of the science-military partnership during the Cold War, see the references cited in note 3. On Conant's career as an elite science adviser and Cold Warrior, consult Hershberg, *James B. Conant*, esp. chs. 15–31.

12. I use Robert Justin Goldstein's definition of political repression: "Political repression consists of government action which grossly discriminates against persons or organizations viewed as representing a fundamental challenge to existing power relationships or key governmental policies, because of their perceived political beliefs." Goldstein, *Political Repression in Modern America*, xvi. In the American context, political repression typically consists of the violation of rights related to the First Amendment and due process (xviii–xxi).

CHAPTER ONE

1. There are several major works on the atomic scientists' movement. Smith, *Peril and a Hope*, remains the standard source. Smith emphasizes the idealistic nature of the movement and the unanimity of scientists on atomic energy issues. Strickland, *Scientists in Politics*, differs sharply with Smith on the nature of the scientists' movement and emphasizes the schisms between scientists. Hall, "Congressional Attitudes toward Science," recognizes a split between the political activities of generally younger working scientists and those of older, more conservative scientist-administrators, and he sees the atomic scientists' movement as an expression of New Deal politics. Hodes, "Precedents for Social Responsibility," is informative on scientists' motivations and discusses the ways in which scientists balanced self-interest and idealism in order to secure support within the scientific community. Boyer, *By the Bomb's Early Light*, devotes several chapters to the scientists' movement. Boyer emphasizes the public appeal of the scientists' movement and the scientists' use of fear as a mode of persuasion. Hewlett and Anderson, *New World*, provides a legislative history of the McMahon bill. My own views on the scientists' movement are drawn in part from the differing perspectives of these works.

2. Sherwin, *World Destroyed*, chs. 4, 5, 8, and Wittner, *Struggle against the Bomb*, ch. 2.

3. Smith, *Peril and a Hope*, ch. 2. For an analysis of Leo Szilard's views on the need for a system of superpower cooperation based on enlightened self-interest, see Bess, *Realism, Utopia, and the Mushroom Cloud*, ch. 2.

4. McAuliffe, *Crisis on the Left*, 1.

5. On wartime tensions between Manhattan Project scientists and the army, see Sherwin, *World Destroyed*, ch. 2. On the May-Johnson bill and scientists' initial reactions, consult Smith, *Peril and a Hope*, 128–43, and Hall, "Congressional Attitudes toward Science," 202–14.

6. On Bush, see Reingold, "Vannevar Bush's New Deal," 302. On Oppenheimer, consult Bernstein, "Oppenheimer Loyalty-Security Case," 1396, 1401–3, 1408. James Hershberg, Conant's biographer, describes Conant as "cautiously liberal but no boat-rocker." Conant's political views, however, tended to be far removed from matters of political economy and the role of the state. He believed in academic freedom and distrusted class as a barrier in American society, but he was never quick to defend his principles when faced with opposition. Slow to embrace coeducation and racial and ethnic diversity at Harvard and less than staunch in his defense of civil liberties and free political discussion as president of Harvard, he might have just as easily been described as conservative, rather than liberal. Consult Hershberg, *James B. Conant*; the quotation is from p. 57.

7. Ronald Sullivan, "William A. Higinbotham, 84; Helped Build First Atomic Bomb," *New York Times*, November 15, 1994, D29; Smith, *Peril and a Hope*, 151, 246;

Murray J. Martin, Norwood B. Gove, Ruth M. Gove, and Agda Artna-Cohen, "Katharine Way," in *Women in Chemistry and Physics: A Biobibliographic Sourcebook*, ed. Louise S. Grinstein, Rose K. Rose, and Miriam H. Rafailovich (Westport, Conn.: Greenwood Press, 1993), 572–80; and Washington Field Office, Report, "Federation of American Scientists," February 28, 1948, File on the Federation of American Scientists, Federal Bureau of Investigation (hereafter cited as FBI-FAS), FBI HQ 100-344452-170, vol. 18.

8. On the scientists' public relations efforts, see Boyer, *By the Bomb's Early Light*, ch. 5, and Smith, *Peril and a Hope*, ch. 10, esp. 292, 306. The scientists' activities in Washington are detailed in Smith, 149–73. The number of scientists who lobbied in Washington is derived from the account in Strickland, *Scientists in Politics*, 51–52, combined with Smith, 149–73.

9. Louis Falstein, "The Men Who Made the A-Bomb," *New Republic*, November 26, 1945, 709.

10. The drafting of the McMahon bill is described in Smith, *Peril and a Hope*, 272, and Hewlett and Anderson, *New World*, 441–43. Senate Resolution 1717 is reproduced in Hewlett and Anderson, app. 1, 714–22.

11. On the formation of the Federation of Atomic Scientists and the FAS, see the early issues of the *Bulletin of the Atomic Scientists* (hereafter cited as *BAS*), especially "The Federation of Atomic Scientists," *BAS* 1 (December 10, 1945): 2, 5 (the quotation is from p. 5), and "The Atomic Scientists and the Federation of American Scientists," *BAS* 1 (February 1, 1946): 4–5.

12. Smith, *Peril and a Hope*, 227–29. The quotation is from "The Federation of Atomic Scientists: National Committee on Atomic Information," *BAS* 1 (December 24, 1945): 5.

13. Smith, *Peril and a Hope*, 366–71; Hewlett and Anderson, *New World*, 489–91.

14. Emanuel R. Freedman, "Soviet Confesses It Got Atom Data; Rebukes Canada," *New York Times*, February 21, 1946, 1, 4; "Text of Soviet Statement," *New York Times*, February 21, 1946, 4. For an initial report of the Canadian spy cases, see "Canada Seizes 22 as Spies; Atom Secrets Believed Aim," *New York Times*, February 16, 1946, 1, 6. For an extensive account of the Canadian spy scare, and the lack of traditional legal protections supplied to the defendants, see Reuben, *Atom Spy Hoax*, chs. 2–4.

15. On the Soviet Union's acquisition of uranium samples, see Herken, *Winning Weapon*, 130–31. On the general postwar faith in the atomic bomb's ability to ensure American security and the belief that the "secret of the atom" could be maintained indefinitely, consult Herken, esp. chs. 5–6; Boyer, *By the Bomb's Early Light*, esp. chs. 25–27; and Hall, "Congressional Attitudes toward Science," ch. 4.

16. On American wartime plans for the postwar world, the internal inconsistencies of those plans, and the constant rifts between the United States, Great Britain, and the Soviet Union during the war, see LaFeber, *American Age*, ch. 13, and Ambrose, *Rise to Globalism*, chs. 2–4. On perceptions of U.S. global interests late in the war and in the early postwar period, consult Leffler, "American Conception of National Security." On the Truman administration's belief in the efficacy of economic leverage and the power of atomic diplomacy, see Ambrose, 67–73, and Gaddis, *United States and the Origins of the Cold War*, 222–24 and 260–61. On the events of

early 1946 and their significance for the rise of the Cold War, see Gaddis, ch. 9, and Yergin, *Shattered Peace*, ch. 7.

17. Smith, *Peril and a Hope*, 388–89; Hall, "Congressional Attitudes toward Science," 233–34. On the scientists' perceptions that the Canadian spy scare led directly to the decline of support for the McMahon bill, see Atomic Scientists of Chicago, Minutes of the Meeting of the Executive Committee, February 19, 1946, box 2, folder 54, Rabinowitch Papers, State University of New York at Albany; Louis N. Ridenour to Philip Wylie, February 25, 1946, box 213, folder 8, Wylie Papers, Princeton University; Association of Oak Ridge Scientists, "Atomic Energy Legislation," March 4, 1946, and "Letter from an AORS Member to a Friend" [undated, probably late February or early March 1946], box 11E, folder "Atomic Energy Control," Shapley Papers, Harvard University.

18. [Eugene Rabinowitch], "Military or Civilian Control of Atomic Energy?," *BAS* 1 (March 15, 1946): 1.

19. E. U. Condon, "An Appeal to Reason," *BAS* 1 (March 15, 1946): 7.

20. "70 Thousand Letters Back the McMahon Bill," *BAS* 1 (April 15, 1946): 6; "Current Status of Domestic Legislation," *BAS* 1 (April 1, 1946): 19; Smith, *Peril and a Hope*, 396; and Bernard Weissbourd and Aaron Novick, "Report from Washington," April 11, 1946, box 2, folder 54, Rabinowitch Papers.

21. "Current Status of Domestic Legislation: An Amendment Is Amended," *BAS* 1 (April 1, 1946): 1, 19, and Smith, *Peril and a Hope*, 408.

22. Hewlett and Anderson, *New World*, 513–14.

23. Ibid., 515, and Smith, *Peril and a Hope*, 412–13.

24. [Eugene Rabinowitch], "A Dangerous Lull," *BAS* 1 (May 15, 1946): 1.

25. See, for example, the comments of Representative Dewey Short (R.-Mo.) in House, 79th Cong., 2d sess., *Congressional Record* (July 17, 1946), 92, pt. 7:9253. Similar statements by Representatives Clare Boothe Luce (R.-Conn.) and Charles H. Elston (R.-Ohio) are found on pp. 9261–63 and 9272. Jones, "Science, Scientists, and Americans," 208–22, is especially informative on congressional conservatives' willingness to concede heavy state control over the domestic applications of nuclear energy.

26. House, 79th Cong., 2d sess., *Congressional Record* (July 16, 1946), 92, pt. 7:9141.

27. Ibid., July 17, 1946, 9252.

28. Ibid., 9266.

29. Edward H. Levi, "The Atomic Energy Act: An Analysis," *BAS* 2 (September 1, 1946): 18–19. Other compromises Levi discussed were the balance between civilian and military representation, and certain aspects of the patent provisions. For the text of the act, see "The Atomic Energy Act of 1946," *BAS* 2 (August 1, 1946): 18–24.

30. Newman and Miller, *Control of Atomic Energy*, 19.

31. Maddox, "Politics of World War II Science," 22–25; Kevles, "National Science Foundation," 8–9; and Rowan, "Politics and Pure Research," 29–32. For the text of the bill, see Senate Subcommittee of the Committee on Military Affairs, *Technological Mobilization: Hearings on S. 2721*, vol. 1, 77th Cong., 2d sess., October 13, 1942, 1–4.

32. Kevles, "National Science Foundation," 10–11; Rowan, "Politics and Pure Research," 33–34; Maddox, "Politics of World War II Science," 26. For the text of

S.R. 702, see Senate Subcommittee of the Committee on Military Affairs, *Scientific and Technical Mobilization: Hearing on S. 702*, pt. 1, 78th Cong., 1st sess., March 30, 1943, 1–7.

33. Maurice B. Visscher to Melba Phillips, July 3, 1948, folder "A.A.Sc.W. Phillips, Melba," Visscher Papers, University of Minnesota. See also "Report on the Activities of the A.A.Sc.W. in 1947–8," undated, and Maurice B. Visscher to J. G. Crowther, September 9, 1948, folder "A.A.Sc.W. Phillips, Melba," Visscher Papers.

34. Rowan, "Politics and Pure Research," 34–37; Kevles, "National Science Foundation," 10–13; and Maddox, "Politics of World War II Science," 26–29. See also National Association of Manufacturers, "Shall Research Be Socialized?," May 1943, copy in Jewett File, 50.271, "S. 702 — Office of Scientific and Technical [*sic*] Mobilization," National Academy of Sciences (hereafter cited as NAS Archives); and AAAS, "Resolution on the Science Mobilization Bill (S. 702)," July 15, 1943, box 2, folder "National Science Foundation (II) 1943–1950," Borras Files, American Association for the Advancement of Science, 1st Accession.

35. Kevles, "National Science Foundation," 16–18, and England, *Patron for Pure Science*, 9–10. For Bush's report itself, see Vannevar Bush, *Science, the Endless Frontier* (Washington, D.C.: U.S. Government Printing Office, 1945).

36. Rowan, "Politics and Pure Research," 59–64. For the texts of S.R. 1297 and S.R. 1285, see Senate Subcommittee on War Mobilization of the Committee on Military Affairs, *Legislative Proposals for the Promotion of Science: The Texts of Five Bills and Excerpts from Reports*, 79th Cong., 1st sess., August 1945, 1–11. For the text of S.R. 1720, see Senate Subcommittee on War Mobilization of the Committee on Military Affairs, *National Science Foundation: Preliminary Report on Science Legislation*, 79th Cong., 1st sess., December 21, 1945, 9–18.

37. "Legislative Scene," *FAS Newsletter*, March 1, 1946.

38. On the significance of the McMahon Act's policies on patents and fissionable materials, consult Jones, "Science, Scientists, and Americans," 208–16. On the New Deal and public power, see Hawley, *New Deal and the Problem of Monopoly*, ch. 17.

39. Senate Subcommittee of the Committee on Military Affairs, *Hearings on Science Legislation (S. 1297 and Related Bills)*, pt. 2, 79th Cong., 1st sess., October 11, 1945, 143–44 (hereafter cited as *Hearings on Science Legislation*). Wallace declared, "The present lack of balance in the development of the physical and social sciences is one of the important reasons for including provision for the social sciences. . . . The further development of the social sciences may well determine whether the new and terrible forces which man has discovered through the natural and physical sciences become man's servant for enhancing his welfare or the terrible instruments for his destruction." Physicist Hugh C. Wolfe, secretary of the Association of New York Scientists, later echoed this point of view in supporting inclusion of the social sciences in the NSF legislation: "Our progress in the Physical Sciences is already far beyond that in the Social Sciences. We need intensive studies of the patterns of social behavior so that we can learn how to live together in peace." Hugh C. Wolfe to Warren G. Magnuson, July 5, 1946, box 10, folder 7, Federation of American Scientists Papers, University of Chicago (hereafter cited as FAS Papers). On FAS's support for inclusion of the social sciences, see also Wolfe to Virgil Chapman, July 2, 1946, box 10, folder 7, FAS Papers, and Clifford Grobstein to Sophie D. Aberle, January 17, 1951, box 19, folder 8, FAS Papers.

40. On Jewett's beliefs, see Reingold, "Vannevar Bush's New Deal," 302–3, 318–19. As Reingold has noted, both Jewett's and Bush's preferred president was Herbert Hoover. Certainly, Jewett's structure for postwar science was consistent with Hoover's principles. On Herbert Hoover and the meaning of voluntary associationalism, see Wilson, *Herbert Hoover*. Although Jewett was the only witness in congressional hearings to oppose the expansion of postwar government support of scientific research, there were other scientists who felt the same way. See G. N. Lewis to L. C. Dunn, May 5, 1944, folder "Kilgore Bill, 1943. Correspondence (#5)," Dunn Papers, American Philosophical Society.

41. On Bush's views on centralization, the role of experts, and the nature of political interference, see Reingold, "Vannevar Bush's New Deal," 307, 309, 323. On Bush's belief in reform of the patent system, see correspondence in ser. 2, box 9, folder "Bush, Vannevar," Bowman Papers, Johns Hopkins University, esp. Vannevar Bush to Isaiah Bowman, February 23, 1945, which notes, "The [patent] system indeed needs revision, but if it is done unwisely or unsympathetically its benefits will be lost and this will be a serious calamity." On Bush's reasons for excluding patent reform from the NSF legislation, see Vannevar Bush, "The Kilgore Bill," *Science*, December 31, 1943, 573, and *Hearings on Science Legislation*, pt. 2, October 15, 1945, 203. On Bush and the social sciences, see House Subcommittee of the Committee on Interstate and Foreign Commerce, *National Science Foundation Act: Hearings on H.R. 6448*, 79th Cong., 2d sess., May 28, 1946, 53.

42. For example, Jewett once scored Kilgore as "the victim of some starry-eyed New Deal boys." Frank B. Jewett to Vannevar Bush, November 18, 1942, Jewett File, 50.271, S-702 — Office of Scientific and Technical Mobilization. Isaiah Bowman objected that Howard Meyerhoff of the AAAS and Watson Davis of Science Service "argue and think like New Dealers of the original brand," while Homer W. Smith complained of "our friends on the left like Urey and Shapley." Isaiah Bowman to Homer W. Smith, June 7, 1946, and Homer W. Smith to Charles E. MacQuigg, July 31, 1946, both in ser. 6, box 8, folder "National Science Foundation 1946 May–December," Bowman Papers. Bush was usually more circumspect and restricted himself to noting, "Kilgore has about him a group that have strange ideas." Vannevar Bush to Frank B. Jewett, April 2, 1946, box 56, folder "Jewett, Frank B. (1946)," Bush Papers, Library of Congress. In regard to Shapley's and Urey's involvement in the Independent Citizens' Committee of the Arts, Sciences, and Professions, however, he did note that the committee had "some very left-wing individuals pulling strings in it." Vannevar Bush to Frank B. Jewett, May 7, 1946, Bush Papers, quoted in England, *Patron for Pure Science*, 46.

43. Harlow Shapley to Ward Darley, November 26, 1945, box 11E, folder "Science bills," Shapley Papers.

44. On New Deal economic thought, particularly the planning and antimonopoly wings of the New Deal, consult Brinkley, *End of Reform*, esp. chs. 2 and 3, and Hawley, *New Deal and the Problem of Monopoly*.

45. K. A. C. Elliot and Harry Grundfest, "The Science Mobilization Bill," *Science*, April 23, 1943, 376.

46. Kirtley F. Mather, Harry Grundfest, and Melba Phillips, "The Future of American Science," in Senate Subcommittee of the Committee on Education and Labor, *Wartime Health and Education: Hearings before a Subcommittee of the Committee on Educa-*

tion and Labor, pt. 7, 78th Cong., 2d sess., December 14, 15, and 16, 1944, 2285–92. The pamphlet was, incidentally, published by the United Office and Professional Workers of America (CIO), one of the left-led unions later expelled by the CIO.

47. Nathanson, *Science for Democracy*, 100. Peter J. Kuznick notes that the AASW took an interest in the relationship between patents and monopoly power several years earlier, in 1939, when the Chicago chapter of the AASW made a study of the problem. Kuznick, *Beyond the Laboratory*, 240–41.

48. L. C. Dunn, "The Opposition to the Kilgore Bill," *Science*, June 4, 1943, 511.

49. Mather, Grundfest, and Phillips, "Future of American Science," 2286.

50. AASW, Boston-Cambridge branch, "Legislation for a National Research Foundation," October 15, 1945, box 11E, folder "Science bills," Shapley Papers.

51. Handwritten note by L. C. Dunn, undated, probably 1945, folder "Kilgore and Magnuson Bills, Revised. Correspondence, Testimony, etc. (#1)," Dunn Papers.

52. Handwritten notes by Dunn, undated, probably 1945, possibly fall 1945, folder "Kilgore and Magnuson Bills, Revised. Correspondence, Testimony, etc. (#2)," Dunn Papers. See also his published essay, L. C. Dunn, "Organization and Support of Science in the United States," *Science*, November 30, 1945, 548–54, esp. 549–51.

53. Layton, *Revolt of the Engineers*, chs. 6 and 7. The quotation is from p. 159.

54. *Hearings on Science Legislation*, pt. 4 (October 29–31 and November 1–2, 1945), 978.

55. Ibid., 981.

56. Study Group, Washington Association of Scientists (Clifford Grobstein, chair), "Toward a National Science Policy?," *Science*, October 24, 1947, 385. On FAS's cooperation with Shapley on S.R. 1850, see telegram (day letter), Harlow Shapley to Lawrence Heilprin, vice chairman, Washington Association of Scientists, May 25, 1946; telegram (night letter), Lawrence Heilprin to Harlow Shapley, May 25, 1946; and telegram, Hugh C. Wolfe to Harlow Shapley, July 3, 1946, in box 11E, folder "Science bills," Shapley Papers.

57. Study Group, Washington Association of Scientists, "Toward a National Science Policy?," 385.

58. Ibid.

59. On the conflict between the technocratic ideal and democracy, John M. Jordan, *Machine-Age Ideology: Social Engineering and American Liberalism, 1911–1939* (Chapel Hill: University of North Carolina Press, 1995), is excellent. For a discussion specifically focused on scientists and the idea of "best-science elitism," see Kevles, *The Physicists*. A recent scathing critique of the elitism of expertise is found in Christopher Lasch, *The Revolt of the Elites and the Betrayal of Democracy* (New York: W. W. Norton & Co., 1995).

60. Samuel P. Hays, *Conservation and the Gospel of Efficiency: The Progressive Conservation Movement, 1890–1920* (Cambridge: Harvard University Press, 1959).

61. H. J. Muller to Curt Stern, January 16, 1947, folder "AAAS—Inter-Society Committee on Science Foundation Legislation (1947) #1," Stern Papers, American Philosophical Society.

62. Rowan, "Politics and Pure Research," 86–97.

63. On the compromise between Kilgore and Magnuson, its reception, and its collapse, see ibid., 102–8, 115–21, and England, *Patron for Pure Science*, 40–43, 47–

59. For commentary at the time on the role of the scientists' split in the defeat of S.R. 1850, see Howard A. Meyerhoff, "Obituary: National Science Foundation, 1946," *Science*, August 2, 1946, 97–98, and Talcott Parsons, "National Science Legislation, Part 1: An Historical Review," *BAS* 2 (November 1, 1946): 7–9, 31.

64. Figures are found in Kevles, *The Physicists*, 359.

65. Schweber, "Mutual Embrace of Science and the Military," 3–4, and Galison, "Physics between War and Peace," 51–65.

66. Vannevar Bush to James E. Webb, December 27, 1946, box 87, folder "National Science Foundation (Mar.–Dec. 1947) General," Bush Papers. In a letter to James Forrestal, secretary of the navy, Bush labeled criticisms suggesting that the military sought control of university research "absurd." Vannevar Bush to James Forrestal, December 11, 1946, box 87, folder "National Science Foundation (Mar.–Dec. 1947) General," Bush Papers.

67. Vannevar Bush to Karl T. Compton, box 26, folder "Compton, Karl T. (1943–46)," Bush Papers.

68. Vannevar Bush to H. Alexander Smith, December 23, 1946, box 87, folder "National Science Foundation (December 1946) General," Bush Papers.

69. Hugh C. Wolfe and Joseph H. Rush to Thomas C. Hart, July 2, 1946, and Hugh C. Wolfe to Virgil Chapman, July 2, 1946, both in box 10, folder 7, FAS Papers.

70. William A. Higinbotham to Associations, November 1, 1946, box 120, folder "FAS July–December 1946," Oppenheimer Papers, Library of Congress, and Eugene Rabinowitch, "Science, A Branch of the Military?," *BAS* 2 (November 1, 1946): 1.

71. Philip Morrison, "The Laboratory Demobilizes . . . ," *BAS* 2 (November 1, 1946): 6. Morrison's article was originally presented at a public forum sponsored by the *New York Herald Tribune* in October 1946.

72. Ibid.

73. Ibid.

74. Philip N. Powers, "A National Science Foundation?," *Science*, December 27, 1946, 614–19. The quotations in the paragraph are from pp. 614–15, 618, and 619, respectively. In a footnote on the first page of the article, Powers made it clear that he was not speaking for the ONR: "The views expressed in this article are personal and not official." For indications of Powers's other activities in support of the NSF, see David Hawkins, Philip N. Powers, and William A. Higinbotham to Associations, December 9, 1946, box 120, folder "FAS July–December 1946," Oppenheimer Papers, and "Summary Report of Informal Meeting Called by Dr. Harlow Shapley," report prepared by Powers, January 2, 1947, box 20, folder 2, FAS Papers.

75. "Association of New York Scientists" [mimeographed pamphlet], undated, early 1947, box 3, folder 9, Atomic Scientists' Printed and Near-Print Material, University of Chicago.

76. Association of Oak Ridge Engineers and Scientists to Edmund E. Day, March 27, 1947, Agencies and Departments 1947: Navy: Office of Naval Research: Naval Research Advisory Committee, NAS Archives.

77. Executive Committee, Northern California Chapter, FAS, to Members, June 28, 1947, box 49, folder 5, FAS Papers. See also Oliver Johnson, Northern California Chapter, FAS to Walter J. Murphy, editor, *Chemical and Engineering News*, March 8, 1950, box 31, folder 5, FAS Papers.

78. See, for example, Forman, "Behind Quantum Electronics," and Leslie, *Cold War and American Science*.

CHAPTER TWO

1. On the FBI's tactics from 1940 to 1950 and Hoover's need to constantly balance protection of the bureau's status and reputation with the expansion of its anticommunist efforts, see Theoharis and Cox, *The Boss*, chs. 9 and 10. William W. Keller differentiates between a political police model that describes the FBI of the postwar decade (aggressive use of information, and hostile intelligence) and an internal security state model that defines the FBI of the 1960s (use of disruptive, covert operations). Keller, *Liberals and J. Edgar Hoover*, 156.

2. Details of the visit have been ably chronicled in Smith, *Peril and a Hope*, 419–24.

3. The text of Adamson's report is taken from House, 79th Cong., 2d sess., *Congressional Record* (July 17, 1946), 92, pt. 7:9257. On Thomas's activities before the House Rules Committee, see C. P. Trussell, "Peril to Security Seen in Oak Ridge," *New York Times*, July 12, 1946, 5, and Hewlett and Anderson, *New World*, 521–22.

4. House, 79th Cong., 2d sess., *Congressional Record* (July 17, 1946), 92, pt. 7: 9257–58. The activities Adamson identified as dangerous betrayed a distrust of scientists' internationalism, a belief in the secret of the atom, and suspicion of labor unions. Given the immediate postwar context of international uncertainty and domestic economic strife, Adamson's targets reflected the combination of international tensions and domestic labor protest that were beginning to define a more general antiradical nativism.

5. Trussell, "Peril to Security Seen in Oak Ridge," 5.

6. E. E. Minett, chairman, AORES, to Brien McMahon, July 12, 1946, box 16, folder 4, Association of Oak Ridge Engineers and Scientists Papers, University of Chicago (hereafter cited as AORES Papers).

7. Telegram from L. B. Borst, W. R. Cohn, P. S. Henshaw, and E. E. Minett for AORES to the House Rules Committee, July 12, 1946, box 50, folder 11, FAS Papers, University of Chicago.

8. House, 79th Cong., 2d sess., *Congressional Record* (July 15, 1946), 92, pt. 7:9001.

9. Ibid., July 16, 1946, 9136–37.

10. Ibid., 9142.

11. Ibid., July 17, 1946, 9257. Condon's passport was revoked in response to a request from General Groves. During a brief six-week stint as assistant director of Los Alamos in 1943, Condon and Groves quickly developed a deep animosity for each other over what Condon considered excessive security restrictions. The revocation of Condon's passport reflected Groves's personal enmity, not an official judgment regarding Condon's loyalty. House, Representative Chet Holifield, "Smearing the Scientists: Attempt to Discredit Civilian Atomic-Energy Control," July 22, 1947, copy in box 69, folder 12, FAS Papers.

12. House, 79th Cong., 2d sess., *Congressional Record* (July 17, 1946), 92, pt. 7:9257–58.

13. Ibid., 9265.

14. Hewlett and Anderson, *New World*, 529–30.

15. William A. Higinbotham to the Administrative Committee, July 18, 1946, box 120, folder "FAS July–Dec. 1946," Oppenheimer Papers, Library of Congress.

16. Ibid. In a 1956 interview with Harry S. Hall, Higinbotham also noted that the husband of a second secretary, as well as Daniel Melcher, director of the NCAI, were

Communist Party members. See Hall, "Congressional Attitudes toward Science," 312 n. 2.

17. Smith, *Peril and a Hope*, 326–27.

18. William A. Higinbotham to Administrative Committee, July 18, 1946.

19. Strickland, *Scientists in Politics*, chs. 3 and 4, esp. 89–90.

20. On the evolution of Wallace's thought, see Henry A. Wallace to Harry S. Truman, July 23, 1946, in John Morton Blum, ed., *The Price of Vision: The Diary of Henry A. Wallace, 1942–1946* (Boston: Houghton Mifflin Co., 1973), 589–601. The text of Wallace's Madison Square Garden address is reprinted on pp. 661–69. Wallace released the July 23 letter publicly on September 17, 1946, five days after his Madison Square Garden address, after learning that columnist Drew Pearson planned to publish a leaked copy. For an account of the September 12 speech and its aftermath, see Graham White and John Maze, *Henry A. Wallace: His Search for a New World Order* (Chapel Hill: University of North Carolina Press, 1995), 224–40.

21. Hamby, *Beyond the New Deal*, 127–34, 159–62; Pells, *Liberal Mind in a Conservative Age*, 76–83, 96–107; Schlesinger, *Vital Center*, 146.

22. Schlesinger, *Vital Center*, 36–40, 118–23, 137. The quotation is from p. 38.

23. Pells, *Liberal Mind in a Conservative Age*, ch. 3, esp. 130–35.

24. The membership figure comes from SAC, New York, to the Director, FBI, February 17, 1947, Washington Field Office File on the Federation of American Scientists, Federal Bureau of Investigation (hereafter cited as WFO-FAS), FBI WFO 65-4736-76, sec. 3.

25. Michael Amrine to Harold C. Urey, February 7, 1948, box 8, folder 6, Urey Papers, University of California at San Diego.

26. Michael Amrine to Harold C. Urey, March 14, 1948, box 8, folder 6, Urey Papers.

27. Cuthbert Daniel to Harold Urey, June 6, 1948, box 33, folder 4, Urey Papers.

28. Ibid.

29. Harold C. Urey to Cuthbert Daniel, June 16, 1948, box 33, folder 4, Urey Papers.

30. Gene Weeks to "Fellow Member," undated, probably early 1949, box 3, folder 9, Atomic Scientists' Printed and Near-Print Material, University of Chicago.

31. "ANYS Is No More," *Newsletter* (FAS Wisconsin Chapter), March 9, 1950, box 20, folder 4, FAS Papers.

32. Hall, "Congressional Attitudes toward Science," 312 n. 2.

33. The quotations are from [Name deleted; president of the Association of Monmouth Scientists] to the membership of the association, June 21, 1947, Federal Bureau of Investigation, FBI NK 100-32133-14. See also Strickland, *Scientists in Politics*. On the collapse of the association, see Peter Kihss, "Monmouth Security Woes Antedate McCarthy Visits," *New York Times*, January 11, 1954, 14, and Peter Kihss, "Monmouth Aides Reply to Charges," *New York Times*, January 13, 1954, 14.

34. San Francisco Field Office, Report, "Federation of American Scientists (FAS); Northern California Association of Scientists (NCAS); Northern California Chapter of the Federation of American Scientists (NCCFAS)," March 1, 1948, p. 2, FBI-FAS, FBI HQ 100-344452, vol. 17.

35. El Paso Field Office, Report, "Federation of American Scientists," July 12, 1948, p. 1, FBI-FAS, FBI HQ 100-344452-215, vol. 21.

36. On Urey's activities in the 1930s, see Kuznick, *Beyond the Laboratory*, 91–93, 101–4, 182–83, 205, 225.

37. "The Talk of the Town," *New Yorker*, December 15, 1945, 23–24. On Urey's positions regarding atomic energy, see, for example, Harold C. Urey, "The Atom and Humanity," *Science*, November 2, 1945, 435–39; Harold C. Urey, "I'm a Frightened Man," *Collier's*, January 5, 1946, 14–15, 50–51; and "A Scientist's Warning," *U.S. News and World Report*, August 30, 1946, 52. "I'm a Frightened Man," one of the best-known articles of the atomic scientists' movement, was actually ghost-written by Michael Amrine.

38. Joel H. Hildebrand to Harold C. Urey, April 13, 1948, box 43, folder 6, Urey Papers.

39. Harold C. Urey to Joel H. Hildebrand, April 21, 1948, box 43, folder 6, Urey Papers.

40. On Urey's resignation from the Independent Citizens' Committee of the Arts, Sciences, and Professions, see Joel H. Hildebrand to Jo Davidson, chairman, ICCASP, October 9, 1946, and Harold C. Urey to Alice Barrows, executive director, Midwest chapter, ICCASP, November 7, 1946, both in box 45, folder 4, Urey Papers. On the demise of the Emergency Committee of Atomic Scientists, see Smith, *Peril and a Hope*, 510. On the legal dissolution of the ECAS, see also Harrison Brown to Harold C. Urey, August 16, 1951, box 29, folder 21, Urey Papers. On Urey's resignation from the United World Federalists, see Harold C. Urey to Cord Meyer Jr., November 21, 1949, box 95, folder 7, Urey Papers.

41. Harold C. Urey to Michael Amrine, October 12, 1949, box 8, folder 6, Urey Papers.

42. Harold C. Urey to Francis Biddle, January 31, 1951, box 8, folder 4, Urey Papers. In an October 12, 1949, letter to Michael Amrine, Urey elaborated on his attitude toward the Soviet Union: "I believe that it is not possible to compromise on basic principles. The basic principles of the West are founded upon the ethics and morals of Christianity and the Jewish religion, and though these are practiced most imperfectly in the West, their practice stands out in marked contrast to those of the Orient, and I regard the Russians in their whole behavior as orientals. Lie detectors work in the West because of the commandment 'Thou shalt not bear false witness against thy neighbor.' I strongly suspect that Molotov could completely wreck a lie detector machine because of a lack of adherence to this fundamental code of ethics." Box 8, folder 6, Urey Papers.

43. According to a November 1948 FBI report, Daniel had been a member of the Socialist Party from 1932 to 1936. By August 1947, however, he had been expelled: "The August 21, 1937 issue of the 'Socialist Appeal' published by the Socialist Workers Party reflects that the name [Cuthbert Daniel] appears on a list of names of persons who were expelled from the Socialist Party as being Trotskyites." The report continued, "Two informants classified [Cuthbert] DANIEL as definitely Marxist in his political and economic views and a record of a federal agency indicated that he has been expelled from the Socialist Party as a Trotskyite. Informants felt, however, that his radical views were entirely his own and that he was not affiliated with or active in any particular political group. DANIEL was described as strictly anti-Communist and anti-Russian because of his feeling that Communism was no longer pure Marxism, and that the Communist Party line was intellectually dishonest." New York Field

Office, Report, "Federation of American Scientists (FAS); Association of New York Scientists (ANYS)," November 22, 1948, p. 9, FBI-FAS, FBI HQ 100-344452-256, vol. 22, version released after referral to other government agencies. It is dangerous, of course, to rely solely on FBI reports for assessments of individuals' political identities, but this description of Daniel contains a level of detail and attention to the nuances of factionalism within the American Left that suggests more than a casual, knee-jerk analysis.

44. On the methodological challenges of working with FBI files, see also Diamond, *Compromised Campus*, 4–9.

45. For a general discussion of the FBI's investigative methods, see Donner, *Age of Surveillance*, 127–38.

46. SAC Kimball to Inspector [deleted], undated, sometime after November 27, 1946, FBI-FAS, FBI HQ 100-344452, vol. 6.

47. San Francisco Field Office, Report, "Northern California Association of Scientists," March 6, 1946, pp. 3–4, 9–13, FBI-FAS, FBI HQ 100-344452-3X2, vol. 1; San Francisco Field Office, Report, "Northern California Association of Scientists," June 12, 1946, p. 3, FBI-FAS, FBI HQ 100-344452-6X4, vol. 2.

48. Memorandum, D. M. Ladd to Director [J. Edgar Hoover], September 24, 1946, pp. 1–4, FBI-FAS, FBI HQ 100-344452, vol. 5.

49. Memorandum, SAC, San Francisco, to the Director, September 12, 1947; Harry M. Kimball, SAC to the Director, September 26, 1947; teletype, Kimball to the Director, October 2, 1947; and Hoover to Communications Section, undated (or date deleted), all in FBI-FAS, FBI HQ 100-344452, vol. 14.

50. Memorandum, SAC, Boston, to the Director [Hoover], December 5, 1946, FBI-FAS, FBI HQ 100-344452-35, vol. 6.

51. Theoharis and Cox, *The Boss*, 9–14, 174–75, 257–61.

52. Guy Hottel to Director, April 16, 1946, FBI-FAS, FBI HQ 100-344452-6X, vol. 2.

53. Washington Field Office, Report, "Changed: Federation of American Scientists (F.A.S.)," August 12, 1947, pp. 168–71, FBI-FAS, FBI HQ 100-344452-83, vol. 13.

54. The taxi driver is discussed later in the chapter. On the congressional staffer, see Memorandum, SAC, Washington Field Office, to the Director, May 10, 1954, FBI-FAS, FBI HQ 100-344452-309. This FBI document, released by the Department of Energy following an FBI referral, discusses information about the FAS provided by Corbin Allardice, executive director of the Joint Committee on Atomic Energy.

55. Diamond, *Compromised Campus*, 46. See also Donner, *Age of Surveillance*, 133–34.

56. See, for example, [Name deleted], Special Agent to Guy Hottel, August 12, 1947, WFO-FAS, FBI WFO 65-4736-136, sec. 4; SAC, Memorandum, San Francisco to the Director, March 5, 1948, FBI-FAS, FBI HQ 100-344452-76, vol. 17; and SAC, Milwaukee, to SAC, Washington Field Office, March 29, 1948, WFO-FAS, FBI WFO 65-4736, sec. 6.

57. Special Agent [name deleted] to Guy Hottel, SAC, Washington Field Office, February 28, 1946, FBI-FAS, FBI HQ 100-344452-5X, vol. 1.

58. Memorandum, D. M. Ladd to the Director, March 7, 1946, FBI-FAS, FBI HQ 100-344452-3X1, vol 1.

59. Memorandum, Special Agent [name deleted] to Guy Hottel, SAC, Washington Field Office, March 12, 1946, FBI-FAS, FBI HQ 100-344452-5X, vol. 1.

60. Memorandum, D. M. Ladd to the Director, March 7, 1946, FBI-FAS, FBI HQ 100-344452-3X1, vol. 1.

61. Ibid.

62. Memorandum, D. M. Ladd to the Director, March 11, 1946, FBI-FAS, FBI HQ 100-344452, vol. 1.

63. Special Agent [name deleted] to Guy Hottel, March 12, 1946, FBI-FAS, FBI HQ 100-344452-5X, vol. 1. See also Guy Hottel to the Director, March 20, 1946, FBI-FAS, FBI HQ 100-344452-11, vol. 1.

64. Washington Field Office, Report, "National Committee on Atomic Information; Federation of American Scientists; Federation of Atomic Scientists; National Committee for Civilian Control of Atomic Energy," January 28, 1947, pp. 3–4, FBI-FAS, FBI HQ 100-344452-39, vol. 7.

65. See, for example, Chicago Field Office, Report, "National Committee on Atomic Information; Federation of American Scientists; Federation of Atomic Scientists; National Committee for Civilian Control of Atomic Energy," October 10, 1946, FBI-FAS, FBI HQ 100-344452-22, vol. 5, and Omaha Field Office, Report, "Federation of American Scientists (F.A.S.)," December 22, 1947, p. 2, FBI-FAS, FBI HQ 100-344452-123, vol. 15.

66. San Francisco Field Office, Report, "Changed: Federation of American Scientists (FAS)[,] Northern California Association of Scientists (NCAS)[,] Northern California Chapter of the Federation of American Scientists (NCCFAS)," September 22, 1947, p. 7, FBI-FAS, FBI HQ 100-344452-96X2, vol. 14.

67. Memorandum, Guy Hottel to the Director, March 27, 1948, FBI-FAS, FBI HQ 100-344452-176, vol. 19. Several months later, another memorandum discussed an FAS member who was recontacted on June 24, 1948. Whether or not it was the same person discussed in the March 27, 1948, memo is impossible to determine because of deletions in the original documents. See Guy Hottel to the Director, July 14, 1948, FBI-FAS, FBI HQ 100-344452-21, vol. 22.

68. El Paso Field Office, Report, "Federation of American Scientists," July 12, 1948, pp. 1–2, FBI-FAS, FBI HQ 100-344452-215, vol. 21.

69. New York Field Office, Report, "Federation of American Scientists (FAS); Association of New York Scientists (A.N.Y.S.)," June 13, 1950, pp. 1 and 3, FBI-FAS, FBI HQ 100-344452-303, vol. 29. The copy of this report released by the FBI deleted all the identifying information about the informant, subject to referral by the Department of Energy. The Department of Energy released all information about the informant except for his name.

70. Diamond, *Compromised Campus*.

71. Bernstein, "Oppenheimer Loyalty-Security Case," 1394–95, 1403–5.

72. On Goudsmit's reports to Colby and to the CIA, see, for example, Samuel A. Goudsmit, "Report on Conversations with Prof. Ivar Waller," undated; Goudsmit to Walter F. Colby, December 27, 1948; Goudsmit to Colby, July 7, 1949; Goudsmit to Colby, March 20, 1950; Goudsmit to Colby, November 30, 1950; Goudsmit to Charles A. Lea, February 8, 1950 (two memorandums); Goudsmit to Lea, April 20, 1951; Goudsmit to Colby, March 31, 1952, box 7, ser. 3, folder 42, Goudsmit Papers, American Institute of Physics.

73. Smith, *Peril and a Hope*, 326, and Ralph McDonald to Albert Einstein, August 5, 1946, box 33, folder "Einstein, Prof. Albert," Records of the National Com-

mittee on Atomic Information, Library of Congress (hereafter cited as Records of the NCAI).

74. Minutes, FAS Council Meeting, June 28, 1946, report that the NCAI had by that time distributed 150,000 pieces of literature, made arrangements for more than 450 speakers, and was engaged in active fund-raising. Box 48, folder "FAS Releases," Records of the NCAI. On Melcher's relations with the scientists themselves, see, for example, Daniel Melcher to Lyle Borst, February 19, 1946, and Daniel Melcher to John Topham, February 19, 1946, both in box 47, folder "Association of Oak Ridge Scientists," Records of the NCAI. On the pay raise, see "Hello from Peggy and Dan Melcher," August 1946, box 68, folder "Daniel Melcher—Memos," Records of the NCAI.

75. Smith, *Peril and a Hope*, 327.

76. Amrine's comments, provided by an informant, are found in Washington Field Office, Report, "National Committee on Atomic Information; Federation of American Scientists; Federation of Atomic Scientists; National Committee for Civilian Control of Atomic Energy," September 17, 1946, FBI-FAS, FBI HQ 100-344452, vol. 4.

77. "Scientists of Other Nations Plead for World Wide Freedom in Science," *Atomic Information*, June 17, 1946, 9; Leo Szilard, "Letter to Stalin," *BAS* 3 (December 1947): 347–49, 376; Eugene Rabinowitch, "Working for a Miracle," *BAS* 3 (December 1947): 350. On Szilard's views on the prospects for cooperative U.S.-Soviet relations, see Michael Bess, "Peace through Cooperative Diplomacy: Leo Szilard's Vision of a Superpower Duopoly," ch. 2 in *Realism, Utopia, and the Mushroom Cloud*.

78. Daniel Melcher to William A. Higinbotham, June 21, 1946, box 64, folder "W. A. Higinbotham—Memos," and "Digest of a letter of June 21 from Amrine to Jackson," box 34, folder "Melcher, Mr. Daniel," Records of the NCAI. For Melcher's reactions to Amrine and Jackson, see Daniel Melcher to NCAI, July 6, 1946, WFO-FAS, FBI WFO 65-4736 sub C, vol. 2. On Jackson and the National Farmers Union, see Dyson, *Red Harvest*, 157, 192–93.

79. Guy Hottel, SAC, Washington Field Office, to the Director, July 23, 1946, FBI-FAS, FBI HQ 100-344452, vol. 2.

80. Higinbotham's and Rush's identities are revealed in William A. Higinbotham to Tom C. Clark, August 4, 1946, box 1, folder 7, Rush Papers, University of Chicago. Higinbotham began the letter, "Mr. Joe Rush and I told you of some of our worries about our organization and some of its associates, a week or so ago." Incidentally, Rush filed his copy of the letter under the heading "Commies, etc."

81. Washington Field Office, Report, "National Committee on Atomic Information; Federation of American Scientists; Federation of Atomic Scientists; National Committee for Civilian Control of Atomic Energy," September 17, 1946, p. 27, FBI-FAS, FBI HQ 100-344452, vol. 4. See also [name deleted], Special Agent to Guy Hottel, SAC, Washington Field Office, July 31, 1946, WFO-FAS, FBI WFO 65-4736-30, sec. 2.

82. Washington Field Office, Report, "National Committee on Atomic Information; National Committee for Civilian Control of Atomic Energy," September 17, 1946, p. 27, FBI-FAS, FBI HQ 100-344452, vol. 4.

83. Typed and handwritten notes of Daniel Melcher, July 6, 1946, and typed memorandum by Daniel Melcher, July 8, 1946, both in WFO-FAS, FBI WFO 65-4736 sub C, vol. 2; and Daniel Melcher to NCAI executive committee, August 6, 1946, box

34, folder "Melcher, Mr. Daniel," Records of the NCAI. Several FAS chapters expressed concern over the dissension within the NCAI and support for Melcher. See Waldo E. Cohn, corresponding secretary, AORES, box 17, folder 14, Atomic Scientists of Chicago Papers, University of Chicago (hereafter cited as ASC Papers); L. I. Schiff, writing for the council of the Association of Philadelphia Scientists to Ralph McDonald, August 2, 1946, box 46, folder "Association of Philadelphia Scientists," Records of the NCAI; and Jerome L. Rosenberg, acting secretary of the ANYS executive committee to Ralph MacDonald, August 9, 1946, box 45, folder "MacDonald, Dr. Ralph E.," Records of the NCAI.

84. William A. Higinbotham to Tom C. Clark, August 4, 1946.

85. On the briefcase and taxi driver incident, see H. O. Bly to D. M. Ladd, July 10, 1946; [illegible initials] Mumford to D. M. Ladd, July 11, 1946; and Guy Hottel to the Director, July 12, 1946, FBI-FAS, in FBI HQ 100-344452, vol. 2; and [name deleted], Special Agent to Guy Hottel, July 12, 1946, WFO-FAS, FBI WFO 65-4736-22, sec. 1.

86. Guy Hottel to the Director, July 13, 1946, WFO-FAS, FBI WFO 65-4736-25, sec. 1. The informant's name is deleted from the document, but the description of him as "Treasurer of the National Committee on Atomic Information" leaves no doubt as to Conway's identity.

87. On Conway's contacts with the FBI, including his acquisition of the signed statement from an FAS member, see [name deleted], Special Agent to Guy Hottel, July 23, 1946, and Special Agent [name deleted] to Guy Hottel, August 19, 1946, both in WFO-FAS, FBI WFO 65-4736, sec. 2. For the actual statement, see "Statement of [name deleted]," July 22, 1946, WFO-FAS, FBI WFO 65-4736 sub C, vol. 3. Because of deletions, these documents do not provide absolute certainty that Conway was the FBI's informant in these particular instances, but the July 13 memo from Hottel to Hoover and the general context of these documents suggest that he was probably the FBI's source throughout the crisis over Melcher.

88. Edward Scheidt, SAC, New York Field Office, to the Director, March 25, 1948, FBI-FAS, FBI HQ 100-344452-182, vol. 19.

89. Guy Hottel to the Director, April 10, 1948, FBI-FAS, FBI HQ 100-344452-179, vol. 19.

90. Chicago Field Office, Report, "National Committee on Atomic Information; Federation of American Scientists; Federation of Atomic Scientists; National Committee for Civilian Control of Atomic Energy," October 10, 1946, p. 8, FBI-FAS, FBI HQ 100-344452-22, vol. 5.

91. Ibid.

92. Daniels's identity is revealed in Office Memorandum, SAC, Chicago, to Director, FBI, December 17, 1946, FBI-FAS, FBI HQ 100-344452, vol. 6. The memorandum notes, "In November 1945, when it was determined that [two lines deleted], Dr. [Farrington] Daniels was instrumental in causing the Executive Committee of the organization to bring about her dismissal."

93. Chicago Field Office, Report, "National Committee on Atomic Information; Federation of American Scientists; Federation of Atomic Scientists; National Committee for Civilian Control of Atomic Energy," October 10, 1946, p. 8, FBI-FAS, FBI HQ 100-344452-22, vol. 5. See also untitled personal memorandum by William A. Higinbotham, March 25, 1949, copy in FBI-FAS, FBI HQ 100-344452-261, vol. 24.

The memo notes, "We checked our secretary with the FBI and G-2 and, since both reports were unfavorable, fired her."

94. Chicago Field Office, Report, "Federation of American Scientists (F.A.S.)," December 19, 1947, p. 2, FBI-FAS, FBI HQ 100-344452-127, vol. 16. The report further notes, "Similar information was also obtained from [one line deleted]."

95. Memorandum, [Name deleted] to F. J. Baumgardner, April 1, 1948, FBI-FAS, FBI HQ 100-344452-172, vol. 19. For the informer's initial approach to the FBI, see his letter: [Name deleted] to J. Edgar Hoover, March 20, 1948, FBI-FAS, FBI HQ 100-344452-171, vol. 19.

96. Edward Scheidt, SAC, New York, to the Director, March 25, 1948, FBI-FAS, FBI HQ 100-344452-182, vol. 19.

97. On the ethics of informing and Americans' attitudes toward informers, see Navasky, *Naming Names*, esp. x–xxiii.

98. Harlow Shapley to Karl Mundt, November 17, 1945, box 11B, [folder 4], Shapley Papers, Harvard University.

99. On this point, see also Richard Alan Schwartz, "What the File Tells: The FBI and Dr. Einstein," *Nation*, September 3–10, 1983, 168–73.

100. Washington Field Office, Report, "National Committee on Atomic Information; Federation of American Scientists; Federation of Atomic Scientists; National Committee for Civilian Control of Atomic Energy," July 13, 1946, p. 11, FBI-FAS, FBI HQ 100-344452-11, vol. 3.

101. SAC, San Francisco Field Office, to the Director, April 25, 1946, p. 2, FBI-FAS, HQ 100-344452-6X2, vol. 2.

102. Memorandum, D. M. Ladd to the Director, September 24, 1946, p. 1, FBI-FAS, FBI HQ 100-344452-1, vol. 5. On informants in the American Legion, see Theoharis and Cox, *The Boss*, 193–98.

103. J. Edgar Hoover to George E. Allen, May 26, 1946, Subject File—FBI Atomic Bomb, President's Secretary's File, Harry S. Truman Library (hereafter cited as PSF), from reel 20, *Harry S Truman's Office Files, 1945–1953, Part 3: Subject File* (Bethesda, Md.: University Publications of America, 1985).

104. J. Edgar Hoover to Harry Hawkins Vaughn, February 25, 1947, Subject File—FBI L, PSF, from reel 22, *Harry S Truman's Office Files, 1945–1953, Part 3*. Furry and Levinson's self-admitted Communist Party membership is noted in Kuznick, *Beyond the Laboratory*, 248. Furry's political difficulties at Harvard University in the 1950s are discussed in Schrecker, *No Ivory Tower*, 197–204.

105. J. Edgar Hoover to Harry Hawkins Vaughan, March 29, 1947, Subject File—FBI L, PSF, from reel 22, *Harry S Truman's Office Files, 1945–1953, Part 3*.

106. J. Edgar Hoover to George E. Allen, February 6, 1947, Subject File—FBI N, PSF, from reel 22, *Harry S Truman's Office Files, 1945–1953, Part 3*. Another letter on December 12 reported further activities by Shapley regarding the NSF. J. Edgar Hoover to Harry Hawkins Vaughn, December 12, 1947, Subject File—FBI N, PSF, from reel 22, *Harry S Truman's Office Files, 1945–1953, Part 3*.

107. Klehr, Haynes, and Firsov, *Secret World of American Communism*, documents the clandestine activities of the American Communist Party up to 1945; the phrase "public world of American communism" is from p. 324. Haynes, *Red Scare or Red Menace?*, also discusses the role of the Communist Party in sponsoring Soviet espionage. Even sympathetic accounts convey the dogmatic character of the party's

leadership. Isserman, *Which Side Were You On?*, depicts the party's inconsistent reactions during World War II in response to directives from Moscow; Naison, *Communists in Harlem*, leaves the impression that the most creative and popular actions of the party in the 1930s took place under the direction of local organizers contrary to the instructions of the leadership.

108. On the United Electrical Workers and other left-led CIO unions, consult Schatz, *Electrical Workers*, and Rosswurm, *CIO's Left-Led Unions*.

109. Consult O'Reilly, *Hoover and the Un-Americans*, 76, 81, on the FBI's mobilization of public opinion; Powers, *Secrecy and Power*, 282, on reports to the White House as a means of protecting the bureau; and Donner, *Age of Surveillance*, ch. 5, esp. 175–76, on the general nature of FBI investigations and the fiction that the FBI did not evaluate information. The quotations are from J. Edgar Hoover to George E. Allen, December 23, 1946, Subject File—FBI Atomic Bomb, PSF, from reel 20, *Harry S Truman's Office Files, 1945–1953, Part 3*.

110. On Soviet espionage, see Haynes, *Red Scare or Red Menace?*, esp. 55–63. On the FBI's repeated abuse of its statutory authority, see Theoharis, *Spying on Americans*, and Special Committee of the National Lawyers Guild, "Report on Certain Alleged Practices of the FBI," *Lawyers Guild Review* 10 (Winter 1950): 185–201.

111. Smith, *Peril and a Hope*, 498–502; see also Strickland, *Scientists in Politics*, 92.

112. On general declines in membership, see Smith, *Peril and a Hope*, 511. For the FBI's monitoring of the collapse of the Cambridge and Los Alamos chapters, see, for example, Boston Field Office, Report, "Federation of American Scientists," June 21, 1949, pp. 1–2, FBI-FAS, FBI HQ 100-344452-262, vol. 24, and El Paso Field Office, Report, "Federation of American Scientists," July 20, 1949, p. 1, FBI-FAS, FBI HQ 100-344452-264, vol. 24. On the disbanding of other chapters, see, for example, Detroit Field Office, Report, "Federation of American Scientists," September 7, 1949, p. 1, FBI-FAS, FBI HQ 100-344452-269, vol. 24, on the Association of University of Michigan Scientists; Philadelphia Field Office, Report, "Federation of American Scientists (FAS)," October 24, 1949, p. 1, FBI-FAS, FBI HQ 100-344452-276, vol. 24, on the Association of Philadelphia Scientists; and SAC, Buffalo to Washington Field Office, undated [late 1949], WFO-FAS, FBI WFO 65-4736-359, sec. 8, on the Rochester chapter.

113. See memorandum, R. R. Roach to A. H. Belmont, July 22, 1954, FBI-FAS, FBI HQ 100-344452, vol. 29.

114. On Urey's criticisms of the Rosenbergs' treatment, see Carmichael, *Framing History*, 91, and McAuliffe, *Crisis on the Left*, 126. Urey wrote to Michael Amrine in 1959, "I personally believe that the Rosenbergs were unjustly executed and that this whole trial, including both Rosenbergs and Sobell, must be classified along with the Tom Mooney case and the Sacco-Vanzetti case. It is curious that about every 25 years or so we have to have a public circus over unfortunate people whom we do not like." Harold C. Urey to Michael Amrine, May 18, 1959, box 8, folder 6, Urey Papers.

115. On Morrison's Communist Party membership, see Schrecker, *No Ivory Tower*, 149.

116. See chapter 6 for an example of consultation between the FAS, the AASW, and the National Council of the Arts, Sciences, and Professions. On the 1949 Waldorf conference, consult chapter 4. For details regarding Higinbotham's reservations about the Waldorf conference, see untitled personal memorandum by William A. Higinbotham, March 25, 1949, copy in FBI-FAS, FBI HQ 100-344452-261, vol. 24.

117. Washington Field Office, Report, "National Committee on Atomic Information (NCAI)," October 4, 1948, pp. 1–2, FBI-FAS, FBI HQ 100-344452-231, vol. 22, and telegram, Guy Hottel to the Director and SACs Albany, Boston, Chicago, El Paso, New York, and Newark, February 2, 1949, FBI-FAS, FBI HQ 100-344452-247, vol. 23.

118. On fears that the FAS would be targeted by HUAC, see William A. Higinbotham to Aaron Novick, October 21, 1947, box 50, folder 11, FAS Papers, and Smith, *Peril and a Hope*, 387–88.

CHAPTER THREE

1. Fried, *Nightmare in Red*, 50–54, and Roger S. Abbot, "The Federal Loyalty Program: Background and Problems," *American Political Science Review* 42 (June 1948): 486–87.

2. National Science Foundation, *Employment of Scientists and Engineers in the United States, 1950–66* (Washington, D.C.: U.S. Government Printing Office, 1968), 11–12; Walter H. Waggoner, "U.S. Acts to Speed Security Checking on Secret Projects," *New York Times*, June 19, 1949, 1. Unfortunately, Gellhorn, *Security, Loyalty, and Science*, and Brown, *Loyalty and Security*, two of the best sources for quantitative data about the loyalty-security system in the late 1940s and early 1950s, do not provide precise figures about the number of university and industrial researchers who held security clearances.

3. Hewlett and Duncan, *Atomic Shield*, 23–26.

4. The quotations and details of Rabinowitch's early life are drawn from Eugene Rabinowitch, interview by Govindjee.

5. Biographical information on Rabinowitch is drawn from Smith, *Peril and a Hope*, 22; "Eugene Rabinowitch, 1901–1973," *BAS* 29 (June 1973): 3; and Linda Greenhouse, "Dr. Eugene Rabinowitch Dies; Manhattan Project Chemist, 71," *New York Times*, May 16, 1973, 50.

6. On wartime discussions at the Metallurgical Laboratory about the wartime and postwar implications of atomic weapons, see Smith, *Peril and a Hope*, ch. 1, esp. 14–34, 41–52. On the Franck Report, see also Sherwin, *World Destroyed*, 210–15, and Wittner, *Struggle against the Bomb*, 23–26. For a copy of the Franck Report, see Smith, app. B, 560–72. For a copy of the Scientific Panel's recommendations, see Sherwin, app. M, 204–5.

7. On Rabinowitch's early involvement in the postwar scientists' movement, consult Smith, *Peril and a Hope*, 85–87, 92–93, 147, 187–88, 294–97.

8. Eugene Rabinowitch to Farrington Daniels, July 10, 1946, "Subject Files," box 7, folder "Argonne, Correspondence of Director Daniels, 1946," Daniels Papers, University of Wisconsin.

9. Ibid.

10. Farrington Daniels to Eugene Rabinowitch, July 20, 1946, "Subject Files," box 7, folder "Argonne, Correspondence of Director Daniels, 1946," Daniels Papers.

11. See, for example, Eugene Rabinowitch, "The 'Cleansing' of the AEC Fellowships," *BAS* 5 (June–July 1949): 161–62; Eugene Rabinowitch, "Atomic Spy Trials: Heretical Afterthoughts," *BAS* 7 (May 1951): 139–42, 157; and Eugene Rabinowitch, "Scientists and Loyalty," *BAS* 7 (December 1951): 354–55.

12. Eugene Rabinowitch, interview by Govindjee.

13. Eugene Rabinowitch, "Scientists and Loyalty," 355.

14. Eugene Rabinowitch, interview by Govindjee.

15. Ibid.

16. On Rabinowitch's involvement in Pugwash, see Eugene Rabinowitch, "Pugwash—History and Outlook," *BAS* 13 (September 1957): 243–48, and comments by Bernard T. Feld and Joseph Rotblat in "A Voice of Conscience Is Stilled," BAS 29 (June 1973): 4–6, and 10.

17. On the nature of research at Brookhaven, see Hewlett and Duncan, *Atomic Shield*, 433, and Seidel, "A Home for Big Science," 139. On the early history of Brookhaven, see also Needell, "Nuclear Reactors." On security clearance policy at Brookhaven, see Gellhorn, *Security, Loyalty, and Science*, 112–14. On the Blewetts and Brookhaven's clearance policy, see Morse, *In at the Beginnings*, 227–28.

18. Biographical information has been drawn from John P. Blewett, interview by Lillian Hoddeson, March 22, 1979; John P. Blewett, interview by the author; and *American Men of Science*, 10th ed., vol. 1 (New York: R. R. Bowker Co., 1949).

19. The quotation is from John P. and M. Hildred Blewett to Robert F. Bacher, May 24, 1947, RG 326, Records of the Atomic Energy Commission, National Archives (hereafter cited as Records of the AEC), Records of the Commissioners, Office Files of Robert F. Bacher, Subject Files 1947–49, box 1, folder "Brookhaven National Laboratory." On Philip Morse's meeting with Carroll Wilson, see "Conference with Carroll Wilson," January 16, 1947, app., folder 5—"Internal Memoranda and Reports," Official Files, Director's Office, Brookhaven National Laboratory (microfilm), American Institute of Physics, reel 20. John P. Blewett's later recollections are drawn from John P. Blewett, interview by Lillian Hoddeson, May 11, 1979, and John P. Blewett, interview by the author. Some additional information on the Blewetts is found in FAS Committee on Secrecy and Clearance, "Some Individual Cases of Clearance Procedures," *BAS* 4 (September 1948): 281–82. In the article, the Blewetts are identified only as the "Drs. B," but the details of the article make their identity clear.

20. John and Hildred Blewett, "Summary of Blewett Case," April 8, 1948, and "Events" [handwritten chronology by John P. Blewett], undated, probably June 1947, personal possession of John P. Blewett.

21. John P. Blewett, interview by the author.

22. John P. Blewett to W. E. Kelley, January 14, 1952, personal possession of John P. Blewett.

23. On the denial of due process rights to those accused during the Canadian spy scare, see Reuben, *Atom Spy Hoax*, 37–45. On the accusations against Halperin and his acquittal (including the *New York Times* quotation), see 31–32. Blewett's remarks are from John P. Blewett, interview by the author.

24. John P. Blewett, interview by the author, and John P. Blewett to W. E. Kelley, January 14, 1952. On the strike and its consequences, see Schatz, *Electrical Workers*, ch. 7.

25. John P. Blewett and M. Hildred Blewett to Robert F. Bacher, May 24, 1947. On the general history of the UE, see Schatz, *Electrical Workers*.

26. "Events" [handwritten chronology by John P. Blewett], undated, probably June 1947. The quotation is from John and Hildred Blewett to Robert F. Bacher, May 24, 1947.

27. John and Hildred Blewett to Robert F. Bacher, May 24, 1947.

28. John P. Blewett, interview by the author.

29. "Events" [handwritten chronology by John P. Blewett], undated, probably June 1947. The quotation is from John P. Blewett and M. Hildred Blewett to Robert F. Bacher, May 24, 1947.

30. Robert F. Bacher to John and Hildred Blewett, May 31, 1947, RG 326, Records of the AEC, Records of the Commissioners, Office Files of Robert F. Bacher, Subject Files 1947–49, box 1, folder "Brookhaven National Laboratory."

31. Lilienthal, *Journals*, 189–90.

32. AEC Meeting No. 61, June 5, 1947, and AEC Meeting No. 66, June 18, 1947, RG 326, Records of the AEC, AEC Minutes.

33. FAS Committee on Secrecy and Clearance, "Some Individual Cases of Clearance Procedures," 282. The Blewetts' comments on Higinbotham's assistance are found in John and Hildred Blewett to Melba Phillips, July 28, 1947, personal possession of John P. Blewett. Higinbotham's comments on his actions are found in William A. Higinbotham to FAS administrative committee, July 21, 1947, box 1, folder 2, Rush Papers, University of Chicago.

34. Handwritten notes by John P. Blewett, July 9, 1947, personal possession of John P. Blewett.

35. Ibid., and AEC Meeting No. 87, July 29, 1947, RG 326, Records of the AEC, AEC Minutes. The July 10 report of the ad hoc panel's initial recommendation is referred to in the July 29 minutes.

36. Robert E. Marshak to David E. Lilienthal, August 10, 1947, box 50, folder 2, FAS Papers, University of Chicago.

37. Ibid.

38. Herbert C. Pollack to the United States Atomic Energy Commission, August 21, 1947, RG 326, Records of the AEC, Records of the Commissioners, Office Files of Robert F. Bacher, Subject File, 1947–49, box 4, folder "Security."

39. FAS Committee on Secrecy and Clearance, "Some Individual Cases of Clearance Procedures," 282. The "living on charity" quotation is from Morse, *In at the Beginnings*, 228. On the restriction of the Blewetts to nonsecret work, see Morse, 227–28, and Sumner T. Pike to Bourke B. Hickenlooper, October 24, 1947, RG 326, Records of the AEC, Records of the Office of the Chairman, Office Files of David E. Lilienthal, Joint Atomic Energy Committee Correspondence, January 1947–June 1950, box 1, folder "Joint Committee Correspondence, '47 July through December."

40. On John Blewett's work at Brookhaven in the 1940s and early 1950s, consult John P. Blewett, interview by Lillian Hoddeson, May 11, 1979, and Hewlett and Duncan, *Atomic Shield*, 236, 251, 500. Hewlett and Duncan make no mention of Hildred Blewett, but Morse recalls that both John and Hildred made important contributions to the cosmotron. Morse, *In at the Beginnings*, 228. In the May 11, 1979, interview by Lillian Hoddeson, John Blewett notes that Hildred did mathematical analysis for the accelerator group. Further information on the Blewetts' careers is drawn from *American Men and Women of Science*, 17th ed., vol. 1 (New York: R. R. Bowker Co., 1989).

41. Philip M. Morse to E. Newton Harvey, January 19, 1948, box 6, folder "Brookhaven [3]," Morse Papers, Massachusetts Institute of Technology.

42. Leslie, *Cold War and American Science*, esp. 27, 30–31, 38, 61, 95, 128–29, 147–48.

43. Robert H. Vought to Robert F. Bacher, January 23, 1947, box 47, folder 1, FAS Papers.

44. Robert H. Vought, interview by the author. Where not otherwise noted, all quotations and other biographical information about Vought are from this interview.

45. Résumé of Robert H. Vought, prepared in the early 1970s, personal possession of Robert H. Vought; Robert H. Vought, interview by the author.

46. Robert H. Vought, interview by the author; résumé of Robert H. Vought; R. H. Vought, "Molecular Dissociation by Electron Bombardment: A Study of $SiCl_4$," *Physical Review*, January 15, 1947, 93–101.

47. Robert H. Vought to Vernon K. Schumann, November 5, 1946, box 47, folder 1, FAS Papers; Robert H. Vought, interview by the author. Although Jim Crow is popularly thought of as restricted to the southern states, Vought recalls discriminatory practices, such as the refusal of many restaurants to admit African American patrons, as common in 1940s Philadelphia.

48. John P. Blewett, interview by the author.

49. Gellhorn, *Security, Loyalty, and Science*, 114–15.

50. Robert H. Vought, interview by the author, and Robert H. Vought to Robert F. Bacher, January 23, 1947.

51. Robert H. Vought to Robert F. Bacher, January 23, 1947.

52. Ibid.

53. Ibid.

54. Robert H. Vought to Vernon K. Schumann, November 5, 1946.

55. On anticommunists' linkage of civil rights and subversion, see Caute, *Great Fear*, 166–68, and Whitfield, *Culture of the Cold War*, 20–23. On Hoover and the FBI, see Powers, *Secrecy and Power*, 323–25. Athan Theoharis and John Stuart Cox note that "the FBI tapped and bugged virtually every civil rights organization challenging racial segregation or seeking to promote equal rights for black Americans." Theoharis and Cox, *The Boss*, 10.

56. Robert H. Vought to Vernon K. Schumann, November 5, 1946.

57. Robert H. Vought to Robert F. Bacher, January 23, 1947.

58. Ibid.

59. Robert H. Vought to C. Guy Suits, June 9, 1947, box 47, folder 1, FAS Papers.

60. Robert H. Vought to Robert S. Rochlin, June 12, 1947, box 47, folder 1, FAS Papers.

61. Robert H. Vought to Robert F. Bacher, August 22, 1947, box 47, folder 1, FAS Papers. In a letter to R. L. Meier, Vought identified his unnamed source as a faculty member of the University of Pennsylvania with whom he had spoken at the June meeting of the Montreal Physical Society. Robert H. Vought to R. L. Meier, March 9, 1948, box 47, folder 1, FAS Papers. Vought told Meier he received no reply from Bacher other than a routine acknowledgment from Bacher's secretary while the AEC commissioner was away.

62. Robert H. Vought to R. L. Meier, March 9, 1948.

63. Ibid.

64. R. L. Meier to Robert H. Vought, March 18, 1948, box 47, folder 1, FAS Papers.

65. Harold A. Gauper to Carroll L. Wilson, July 24, 1948, personal possession of Robert H. Vought.

66. Ibid.

67. FAS Committee on Secrecy and Clearance, "Some Individual Cases of Clearance Procedures," 283. The identities of the individuals involved in the cases were kept anonymous in the article, but the details of the case of "Dr. D" make Vought's identity clear.

68. "Proceedings of [the] Personnel Security Board, Office of Schenectady Operations, United States Atomic Energy Commission, in the Matter of Robert Howard Vought," October 4, 1948, 3–4, personal possession of Robert H. Vought (hereafter cited as "In the Matter of Robert Howard Vought").

69. V. P. Keay to D. M. Ladd, September 23, 1948, FBI-FAS, FBI HQ 100-344452-229, vol. 22. The memo, which also discussed the Blewetts' case, was prompted by the appearance of "Some Individual Cases of Clearance Procedures" in the September 1948 issue of *BAS*.

70. Robert H. Vought, interview by the author.

71. For example, one affidavit presented at the hearing, although it attested firmly to Vought's good character, read in part: "Mr. Howard E. Vought . . . disclosed to me that Bob was having difficulty getting clearance to work on certain projects, because he had been a member, sometime ago, of an organization for the betterment of the conditions of the negro society. This organization, as many other humanitarian ones, appealed to the finest type of American manhood. It was extremely unfortunate for many sincere young men who joined these organizations, that the organizations were infiltrated with, and probably started by subversive elements for their own selfish purposes." Tonks took pains to point out that the person who wrote the affidavit had no direct knowledge of UPAC, that the affidavit could not be taken as evidence that UPAC was a dangerous, subversive organization with hidden purposes, and that the affidavit did not imply Vought was in any way ignorant or careless in his decision to join UPAC. "In the Matter of Robert Howard Vought," 43–44.

72. Robert H. Vought, interview by the author.

73. "In the Matter of Robert Howard Vought," 6–16, 82–97.

74. In 1951, the Supreme Court found that since the attorney general's listings were made without notice and the opportunity for a hearing, they violated due process. See *Joint Anti-Fascist Refugee Committee v. McGrath, Attorney General, et al.*, 341 U.S. 123 (1950). Nevertheless, federal and state loyalty-security boards continued to use the attorney general's list to identify suspect groups. The House Committee on Un-American Activities, and sometimes other investigatory bodies, also used HUAC's more lengthy list of supposed subversive organizations.

75. "In the Matter of Robert Howard Vought," 24.

76. Ibid., Exhibits No. 3–No. 14.

77. Robert H. Vought, interview by the author.

78. "In the Matter of Robert Howard Vought," 75.

79. Ibid., 183–84.

80. Ibid., 93, 95, 96, 100–104, 113, 115–16.

81. Robert H. Vought, interview by the author.

82. Gellhorn, *Security, Loyalty, and Science*, 220.

83. John E. Gingrich to Robert H. Vought, February 3, 1949, personal possession of Robert H. Vought.

84. Navasky, *Naming Names*.

85. FAS Committee on Secrecy and Clearance, "Some Individual Cases of Clearance Procedures," 283, 285.

86. John P. Blewett, interview by Lillian Hoddeson, May 11, 1979. Such isolation was probably not uncommon. According to Ellen W. Schrecker, university scientists subjected to the academic blacklist of the 1950s often spoke of being abandoned by friends and colleagues. Schrecker, *No Ivory Tower*, 299–301.

87. M. Hildred Blewett to the author, September 28, 1993, personal possession of the author; John and Hildred Blewett, "Summary of the Blewett Case," undated, probably late 1948, personal possession of John P. Blewett.

88. John P. Blewett, interview by the author.

89. For more on this point, see Selma R. Williams's accounts of Jessica Davidson's and Allan Seale's hearings under the federal loyalty program. Her account gives the impression that the primary difference in the two cases was that Davidson vigorously protested her treatment, whereas Seale was a much more cooperative witness. Davidson was denied loyalty clearance, but Seale was cleared. Williams, *Red-Listed*, chs. 1–4.

90. Ellsworth C. Dougherty to David E. Lilienthal, June 21, 1948, box 50, folder 2, FAS Papers.

91. R. L. Meier to Ellsworth C. Dougherty, August 13, 1948, box 50, folder 2, FAS Papers.

CHAPTER FOUR

1. On the reasons for Rankin's interest in Shapley, see "Transcription of Conversation between Dr. Shapley, Mr. [Zechariah] Chaffee, Mr. [Ernie] Adamson [HUAC counsel], and Mr. Yont, Saturday morning, October 26 [1946]," box 11B, [folder 4]; Harlow Shapley to Ernie Adamson, October 28, 1946, box 11B, [folder 1]; and Ernie Adamson to Harlow Shapley, October 29, 1946, box 11B, [folder 4], all in Shapley Papers, Harvard University; "Dr. Shapley Assails Subpoena as Politics," *New York Times*, November 5, 1946, 22; John D. Morris, "Rankin Clashes with Dr. Shapley," *New York Times*, November 15, 1946, 16; and Shapley, *Rugged Ways*.

2. On the CIO-PAC, the National Citizens Political Action Committee, and the Independent Citizens' Committee, see Hamby, *Beyond the New Deal*, 34–35. On the CIO's postwar withdrawal from the popular front coalition, see Halpern, *UAW Politics*, 134–35. The coalition of Republicans and southern Democrats had already gained power by the late 1930s; see Patterson, *Congressional Conservatism*.

3. "Testimony of Dr. Harlow Shapley before the Committee on Un-American Activities," November 14, 1946, box 11B, [folder 1], Shapley Papers.

4. John D. Morris, "Rankin Clashes with Dr. Shapley," 16; and "Dr. Shapley's Statement to the Press," [November 14, 1946], box 11B, [folder 1], Shapley Papers.

5. Biographical material on Harlow Shapley has been drawn primarily from Charles Coulston Gillespie, ed. in chief, *Dictionary of Scientific Biography* (New York: Charles Scribner's Sons, 1975), s.v. "Shapley, Harlow," by Owen Gingerich; Bart J. Bok, "Harlow Shapley," in *Biographical Memoirs of the National Academy of Sciences* (Washington, D.C.: National Academy of Sciences, 1978), 49:241–59; and Shapley's own reminiscences in Shapley, *Rugged Ways*.

6. The quotation from Struve is from *Dictionary of Scientific Biography*, 12:346; the quotation from Bok is in Bok, "Harlow Shapley," 244.

7. Kuznick, *Beyond the Laboratory*. On Shapley's political activities in the 1930s, consult pp. 173, 184, 188, 219, 220, 224, 231–32, 235.

8. On the purpose of the Joint Anti-Fascist Refugee Committee, see *Joint Anti-Fascist Refugee Committee v. McGrath, Attorney General, et al.*, 341 U.S. 123 (1950), 130–32. A photograph of leading members of the Independent Voters Committee, including Shapley, with President Roosevelt at the White House is found in Shapley, *Rugged Ways*, photograph section.

9. Carlton Brown, "The Two Lives of Dr. Shapley," *Science Illustrated* 4 (January 1949): 41.

10. H. Walton Cloke, "Scientists Demand World Mutual Aid," *New York Times*, October 22, 1946, 27.

11. William L. Laurence, "Shapley Attacks Pride of Nations," *New York Times*, December 30, 1946, 8.

12. Morris Kaplan, "Shapley Assails Policy on Greece," *New York Times*, September 10, 1947, 13.

13. Harlow Shapley, "Why Amend the Golden Rule?," *American Scholar* 17 (Spring 1948): 138.

14. "Pepper Asks Fight to Save Liberties," *New York Times*, October 27, 1947, 22.

15. The quotation is from "Movies Will Oust Ten Men Cited for Contempt of Congress," *New York Times*, November 26, 1947, 27. On the Committee of One Thousand, see box 11E, folder "misc. political—also some fan," Shapley Papers.

16. Brown, "The Two Lives of Dr. Shapley," 43, 65. On Shapley's reactions to the 1930s purge of Soviet astronomers, see also Kuznick, *Beyond the Laboratory*, 142–43.

17. House Committee on Un-American Activities, *Review of the Scientific and Cultural Conference for World Peace*, House Report No. 1954, 81st Cong., 2d sess., April 19, 1949 (original release date), April 26, 1950 (ordered to be printed), esp. 18, and House Committee on Un-American Activities, *Report on the Communist "Peace" Offensive*, House Report No. 378, 82d Cong., 1st sess., April 1, 1951.

18. Carr, *House Committee on Un-American Activities*, 323–57. The quotation is from p. 351.

19. Oswald Veblen to Harlow Shapley, March 17, 1949, box 25, folder "National Council of the Arts, Sciences, and Professions 1948–49," Veblen Papers, Library of Congress, relates Veblen's regrets that he would not be able to attend the conference. Lieberman, " 'Does That Make Peace a Bad Word?,' " 204, notes that Einstein was a sponsor, but the partial list of sponsors on the call to the conference does not list Einstein's name, although he was a member at large of the National Council of the Arts, Sciences, and Professions. "Forward to Peace: Call to a Cultural and Scientific Conference for World Peace," [February 1949], box 2, folder 130, Morrison Papers, Massachusetts Institute of Technology. Einstein's attendance seems unlikely, since his name is not mentioned in press accounts of the conference. Weisskopf originally planned to participate in the conference but backed out because he felt too many of the invited speakers reflected "a heavy emphasis on those groups who constantly condoned the Russian attacks on freedom," and he believed sensationalist press coverage would make a serious discussion of U.S.-Soviet relations impossible. Victor F. Weisskopf to Harlow Shapley, March 11, 1949, box 2, folder 109, Morrison Papers.

20. House Committee on Un-American Activities, *Review of the Scientific and Cultural Conference for World Peace*, 12, 15.

21. Charles Grutzner, "Police Lift All Restrictions on Culture Meeting Pickets,"

New York Times, March 25, 1949, 1; "Pickets Denounce Soviet, Communism," *New York Times*, March 26, 1949, 1, 3. The quotation, from Sidney Hook, is found in William R. Conklin, "Soviet Is Attacked at Counter Rally," *New York Times*, March 27, 1949, 1. For a detailed discussion of the Waldorf conference, see Lieberman, " 'Does That Make Peace a Bad Word?' "

22. Conklin, "Soviet Is Attacked at Counter Rally," 1, 46, details liberal anticommunists' objections to the conference. Conference organizers' version of the meeting is found in Daniel S. Gillmor, ed., *Speaking of Peace: The Widely-Discussed Cultural and Scientific Conference for World Peace* (New York: National Council of the Arts, Sciences, and Professions, 1949), copy in box 24, folder "Loyalty Cases Printed Matter," Veblen Papers. Their version of the gathering is backed by the *New York Times*'s accounts of the conference. See Richard H. Parke, "Culture Sessions Center on Conflict of East and West," *New York Times*, March 27, 1949, 1, 45; "Keynoters Assess East-West Blame," *New York Times*, March 27, 1949, 46; and Charles Grutzner, " 'Action' Unit Set Up for Parley Goals," *New York Times*, March 28, 1949, 3.

23. Harlow Shapley to Emil Lengyel, March 12, 1949, box 25, folder "National Council of the Arts, Sciences, and Professions," Veblen Papers.

24. On the goals of the Cultural and Scientific Conference for World Peace, see "Forward to Peace: Call to a Cultural and Scientific Conference for World Peace," [February 1949], box 2, folder 130, Morrison Papers, and "Text of Resolutions Presented at Plenary Session, Cultural and Scientific Conference for World Peace," March 27, 1949, box 2, folder 110, Morrison Papers.

25. Diamond, *Compromised Campus*, 36, 38.

26. Oshinsky, *Conspiracy So Immense*, 109, 125–26, and Reeves, *Joe McCarthy*, 230, 256, 258.

27. "Senate Confirms Smith and Conant," *New York Times*, February 7, 1953, 6, and Reeves, *Joe McCarthy*, 467–68. Taft possibly questioned Conant about Shapley in order to placate McCarthy, a vehement opponent of Conant's appointment. Taft had earlier dissuaded McCarthy from delivering a speech to the Senate opposing Conant's confirmation.

28. Harlow Shapley to Robert A. Taft, February 11, 1953, box 11B, [folder 7], Shapley Papers. An earlier draft of the letter was much harsher in its language: Harlow Shapley to Robert A. Taft, February 9, 1953, box 11B, [folder 7], Shapley Papers. Why Shapley toned down the original is unclear; perhaps he simply sensed that the incident would blow over if he did nothing to inflame the situation.

29. Harlow Shapley to Herbert Wilson, March 28, 1950, box 11C, folder "misc. political—also some fan," Shapley Papers.

30. Quoted in Carr, *House Committee on Un-American Activities*, 38.

31. William Odlin Jr., "Condon Duped into Sponsoring Commie-Front Outfit's Dinner," *Washington Times-Herald*, March 23, 1947, 1, A-2; William Odlin Jr., "Condon Facing U.S. Probe into Soviet Society Affiliation," *Washington Times-Herald*, March 25, 1947, 2.

32. J. Parnell Thomas, "Russia Grabs Our Inventions," *American Magazine* 143 (June 1947): 19; J. Parnell Thomas, "Reds in Our Atom-Bomb Plants," *Liberty* (June 1947), included in the *Congressional Record* by Representative John McDowell of Pennsylvania, House, 80th Cong., 1st sess., *Congressional Record* (June 9, 1947), 93, pt. 11:A2729.

33. James Walter, "House Unit to Quiz Dr. Condon on Reds' A-Bomb 'Know-How,'" *Washington Times-Herald*, July 17, 1947, 2.

34. House, Representative Chet Holifield, "Smearing the Scientists: Attempt to Discredit Civilian Atomic-Energy Control," July 22, 1947, copy in box 69, folder 12, FAS Papers, University of Chicago.

35. House Committee on Un-American Activities, Special Subcommittee on National Security, *Report to the Full Committee of the Special Subcommittee on National Security of the Committee on Un-American Activities*, 80th Cong., 2d sess., March 1, 1948, 1 (hereafter cited as *Report to Full Committee*).

36. Biographical information has been drawn primarily from Philip M. Morse, "Edward Uhler Condon," in *Biographical Memoirs of the National Academy of Sciences* (Washington, D.C.: National Academy of Sciences, 1976), 48:125–51; Britten and Odabasi, *Topics in Modern Physics*, esp. pref., vii–xx, and Edward U. Condon, interviews by Charles Weiner, October 17, 1967, April 27, 1968, and September 11–12, 1973.

37. Morse, "Edward Uhler Condon," 129.

38. Ibid., 126.

39. On Condon's participation in the Lincoln's Birthday Committee for Democracy and Intellectual Freedom activities, consult Kuznick, *Beyond the Laboratory*, 190, and the committee's press release, February 5, 1939, box 11A, folder "ACDIF [American Committee for Democracy and Intellectual Freedom]," Shapley Papers. A search of the voluminous materials in the Shapley Papers dealing with the National Council of Soviet-American Friendship, Inc., the American Committee for Democracy and Intellectual Freedom, and the Independent Citizens Committee of the Arts, Sciences, and Professions did not indicate any high-level activity on Condon's part in these organizations.

40. Edward U. Condon to J. Robert Oppenheimer, April 26, 1943, Oppenheimer Papers, Library of Congress.

41. E. U. Condon, "Science and Our Future," *Science*, April 5, 1946, 417.

42. For Condon's views on secrecy and internationalism in science, consult his articles and speeches: "Science and Our Future," cited in the previous note; "Science and International Cooperation," *BAS*, May 15, 1946, 8–11; "Is War Research Science?," *Saturday Review of Literature*, June 15, 1946, 6; "Science and Security," *Science*, June 25, 1948, 659–65; "Science, Secrecy, Security," *Harper's Magazine* 200 (February 1949): 58–63; and "Reflections on Government," *BAS* 5 (June–July 1949): 179–81. On his criticisms of the loyalty and security system and the problems faced by scientists in the 1950s, see "Problems of Scientists," *Science News Letter*, April 12, 1952, 230, and "Scientists and the Federal Government," *BAS* 8 (August 1952): 179–82.

43. Carlson, "J. Parnell Thomas and the House Committee on Un-American Activities," abstract, pages not numbered.

44. Ibid., 7, 9.

45. Thomas, "Russia Grabs Our Inventions," 18.

46. Thomas, "Reds in Our Atom-Bomb Plants," A2729.

47. Carr, *House Committee on Un-American Activities*, 131–53.

48. House Committee on Un-American Activities, *Testimony of Dr. Edward U. Condon: Hearing before the Committee on Un-American Activities*, 82d Cong, 2d sess., Septem-

ber 5, 1952, 3848 (hereafter cited as *Testimony of Dr. Edward U. Condon*). The AEC furnished a similar explanation to the Joint Committee on Atomic Energy in 1948. AEC to Bourke B. Hickenlooper, March 10, 1948, box 1205, folder "E. U. Condon," RG 326, Records of the AEC, Secretariat Files, National Archives.

49. "Louis Welborn," "The Ordeal of Dr. Condon," *Harper's Magazine* 200 (January 1949): 50. "Louis Welborn," according to the article, was "the pen name of a correspondent of a well-known news organization which prefers that members of its staff not be identified with the presentation of their individual views on controversial subjects." I have not been able to identify him (or her).

50. President Truman's 1947 loyalty order and the dismissal standards under E.O. 9835 are described in Gellhorn, *Security, Loyalty, and Science*, 129–33. For the November 1947 list, see 13 Federal Register 1471, 1473 (March 20, 1948). On the origins and mechanics of the attorney general's list, consult Bontecou, *Federal Loyalty-Security Program*, ch. 5. The original list of eighty-two organizations grew to almost two hundred by 1950. The House Committee on Un-American Activities had a similar list that classified more than six hundred organizations as Communist, but HUAC's list did not have the legal ramifications of the attorney general's list. Bontecou, 171.

It should be noted that after a massive drop in membership and support, the National Council of American-Soviet Friendship, the Joint Anti-Fascist Refugee Committee, and the International Workers Order sued for redress. The Supreme Court, in a five-to-three decision, ruled that the basis for the attorney general's listings was arbitrary and therefore unconstitutional and remanded the case to the district court, leaving the district court to determine the reasonability of the attorney general's assessments of the listed organizations. See *Joint Anti-Fascist Refugee Committee v. McGrath, Attorney General, et al.*, 341 US 123 (1950), and Bontecou, *Federal Loyalty-Security Program*, 231–33.

51. The charge concerning the address and telephone number was made explicit in the 1952 Condon hearing: *Testimony of Dr. Edward U. Condon*, 3888.

52. Minutes of the Executive Committee, May 28, 1946, box "American P, no. 2—Ando," American-Soviet Science Society, folder 7, Condon Papers, American Philosophical Society. On the address change, see Gerald Oster to E. U. Condon, March 25, 1948, box "American P, no. 2—Ando," American-Soviet Science Society, folder 10, Condon Papers.

53. *Report to Full Committee*, 6.

54. E. U. Condon, "Science and Our Future," 415–17.

55. William S. White, "Soviet Spy Links Laid to Dr. Condon, High Federal Aide," *New York Times*, March 2, 1948, 3.

56. "Condon's Letter Professing Loyalty," *New York Times*, March 5, 1948, 4.

57. William S. White, "Subpoena Seeks Data on Dr. Condon," *New York Times*, March 4, 1948, 8. On the furor over whether or not the Commerce Department board had made its decision after March 1 and predated it, see William S. White, "Commerce Report on Condon Scored," *New York Times*, March 11, 1948, 11. Condon later recalled he believed the decision had been predated. Edward U. Condon, interview by Charles Weiner, September 11–12, 1973, transcript, 187–88.

58. Press release, March 3, 1948, box 6, folder 8, Emergency Committee of Atomic Scientists Papers, University of Chicago.

59. Press release, March 4, 1948, box 69, folder 12, FAS Papers.

60. FAS letter, March 6, 1948, copy in file on Edward U. Condon, American Institute of Physics Public Information Office Files, American Institute of Physics.

61. Condon file, PSF, Harry S. Truman Library.

62. American Physical Society press release, March 4, 1948, Oppenheimer Papers; "Princeton Groups Support Condon," *New York Times*, March 5, 1948, 4; "Scientists Criticize Attack on Dr. Condon," *New York Times*, April 26, 1948, 5; "Academy Issues Statement Criticizing Condon Attack," *Science News*, May 15, 1948, 312.

63. House, 80th Cong., 2d sess., *Congressional Record* (March 8, 1948), 94, pt. 10:A1464.

64. Ibid., (March 9, 1948), 94, pt. 2:2407.

65. Ibid., 2412.

66. Ibid., 2405–6.

67. *Report to Full Committee*, 4.

68. Goodman, *The Committee*, 235.

69. William C. Foster to Harry S. Truman, April 27, 1948; memorandum, attorney general's office to Harry S. Truman, May 3, 1948, Condon file, PSF.

70. See, for example, House, Representative John E. Rankin of Mississippi speaking, 80th Cong., 2d sess., *Congressional Record* (March 10, 1948), 94, pt. 2:2476.

71. Samuel A. Tower, "Atomic Committee Takes Up Charges against Dr. Condon," *New York Times*, March 7, 1948, 17.

72. "Condon Data Issue Pushed by Reece," *New York Times*, May 2, 1948, 32.

73. Condon to William C. Foster, May 5, 1948, PSF.

74. Both Secretary of Commerce Charles Sawyer and Acting Attorney General Phillip B. Perlman suggested Truman not deliver the speech because six weeks had gone by since the passage of H.R. 522 and the issue seemed "quiescent." Phillip B. Perlman to Clark Clifford, June 2, 1948, PSF; Charles Sawyer to Clark Clifford, June 2, 1948, file on Dr. Edward U. Condon, Office of the Secretary, box 1086, file 104475, RG 40, General Records of the Department of Commerce, National Archives.

75. House, 79th Cong., 2d sess., *Congressional Record* (July 17, 1946), 92, pt. 7:9257.

76. See, for example, Washington Association of Scientists, "The Condon Case and Its Implications," undated, approximately March 8, 1948, box 69, folder 12, FAS Papers, and House, Representative Helen Gahagan Douglas of California speaking, 80th Cong., 2d sess., *Congressional Record* (April 14, 1948), 94, pt. 4:4464. Representative Chet Holifield conveyed similar suspicions: speech before the House of Representatives, Representative Chet Holifield, "Sabotage of American Science: The Full Meaning of the Attacks on Dr. Condon," March 9, 1948, copy in box 69, folder 12, FAS Papers. According to Condon's recollections, most of Holifield's speech was written by Condon's assistant, Hugh Odishaw: Edward U. Condon, interview by Charles Weiner, September 11–12, 1973, transcript, 191.

77. White, "Soviet Spy Links Laid to Dr. Condon, High Federal Aide," 3. In later years, Condon continued to believe that his support for the McMahon bill was one of the primary causes of his troubles. See letter, E. U. Condon to J. Robert Oppenheimer, January 22, 1954, Oppenheimer Papers. In a draft of his unfinished autobiography, probably written sometime in the 1960s, Condon also cited his support of

the McMahon bill as one of the motivations for HUAC's persecution of him. Draft of Condon autobiography, p. 10, box "Atomic Energy, no. 5—Autobiography, no. 1," Autobiography, folder 11, Condon Papers.

78. R. L. Meier to member associations and members at large, March 10, 1948, box 2, folder 3, Washington Association of Scientists Papers, University of Chicago.

79. Washington Association of Scientists, "The Condon Case and Its Implications," undated, approximately March 8, 1948, box 69, folder 12, FAS Papers.

80. Owen J. Roberts to Carroll L. Wilson, June 7, 1948, box 1205, folder "E. U. Condon," RG 326, Records of the AEC, Secretariat Files.

81. Carroll L. Wilson to Donald M. Cunningham, secretary, Personnel Security Review Board, June 17, 1948, box 1205, folder "E. U. Condon," RG 326, Records of the AEC, Secretariat Files.

82. "Memorandum of Decision," July 15, 1948, box 1205, folder on Edward U. Condon, RG 326, Records of the AEC, Secretariat Files.

83. "Condon Is Cleared By Atomic Energy Commission," *BAS* 4 (August 1948): 226.

84. For details on Condon's experiences after 1948, see Wang, "Science, Security, and the Cold War," 260–69.

85. Atomic Scientists of Chicago poll responses, undated, approximately April 1948, box 19, folder 4, ASC Papers, University of Chicago.

86. Frederick Seitz, *On the Frontier: My Life in Science* (New York: AIP Press, 1994), 182–83.

87. Bernstein, " 'J. Robert Oppenheimer,' " 250, and Stern, *Oppenheimer Case*, ch. 12.

CHAPTER FIVE

1. I have borrowed the use of the term "quiet diplomacy" in the context of science-government relations from Balogh, *Chain Reaction*, 40.

2. Hewlett and Duncan, *Atomic Shield*, 18–19, 66–68, 224–25.

3. Gellhorn, *Security, Loyalty, and Science*, 81.

4. Hewlett and Duncan, *Atomic Shield*, 23–26.

5. J. Parnell Thomas, "Reds in Our Atom-Bomb Plants," *Liberty* (June 1947), included in the *Congressional Record* by Representative John McDowell of Pennsylvania, House, 80th Cong., 1st sess., *Congressional Record* (June 9, 1947), 93, pt. 11:A2729.

6. On Lilienthal's suspicions, see Lilienthal, *Journals*, 203. In a February 1948 conversation, AEC commissioner Sumner T. Pike also related to Condon the story that Groves had taken the files, a story that Condon found believable. See pp. 46–50 of Condon's typed autobiography in box "Atomic Energy, no. 5—Autobiography, no. 1," folder "Autobiography #12," Condon Papers, American Philosophical Society. On the FBI's role, see Louis B. Nichols to Clyde Tolson, March 18, 1947, and Louis B. Nichols to Clyde Tolson, March 26, 1947, in O'Reilly, *FBI File*, reel 3.

7. Thomas, "Reds in Our Atom-Bomb Plants," A2729, A2730.

8. Ibid., A2730.

9. Ibid., A2730, A2731.

10. Hewlett and Duncan, *Atomic Shield*, 89.

11. Lilienthal, *Journals*, 189.

12. Hewlett and Duncan, *Atomic Shield*, 91, 92.

13. Ibid., 92–95, and James Walter, "House Unit to Quiz Dr. Condon on Reds' A-Bomb 'Know-How,'" *Washington Times-Herald*, July 17, 1947, 2.

14. RG 128, JCAE Executive Session Transcripts, box 1, Transcripts of Executive Meetings 1947–48, Document #969, July 14, 1947; Document #970, July 17, 1947; Document #972, July 22, 1947; and AEC Meeting No. 92, August 6, 1947, RG 326, Records of the AEC, AEC Minutes, National Archives.

15. AORES, "Facts about the Recent 'Security' Suspensions," undated, probably September 7, 1947, box 65, folder 10, FAS Papers, University of Chicago.

16. G. O. Robinson to E. R. Trapnell, Office of Information, Washington, October 6, 1947, RG 326, Records of the AEC, Records of the Office of the Chairman, Office Files of David E. Lilienthal, Subject Files, 1946–50, box 10, folder "Oak Ridge—Correspondence, 1947–1948."

17. John H. Bull to the AEC, August 15, 1947, RG 326, Records of the AEC, Records of the Commissioners, Office Files of Robert F. Bacher, Subject Files, 1947–49, box 1, folder "Clinton Laboratories."

18. AORES, "Some Light on the Commission's Views," September 7, 1947, box 65, folder 10, FAS Papers.

19. Alvin M. Weinberg to Fletcher Waller, September 5, 1947, RG 326, Records of the AEC, Records of the Commissioners, Office Files of Robert F. Bacher, Subject File, 1947–49, box 4, folder "Security."

20. Eugene P. Wigner to James B. Fisk, September 3, 1947, RG 326, Records of the AEC, Records of the Commissioners, Office Files of Robert F. Bacher, Subject File 1947–49, box 1, folder "Clinton Laboratories."

21. Alvin M. Weinberg to Fletcher Waller, September 29, 1947, RG 326, Records of the AEC, Records of the Commissioners, Office Files of Robert F. Bacher, Subject File 1947–49, box 4, folder "Security."

22. Unidentified FAS council member to Leonard Schiff, March 26, 1947, box 50, folder 2, FAS Papers. The AEC was also paying attention to the security clearance issue: AEC Meeting No. 43, May 5, 1947, RG 326, Records of the AEC, AEC Minutes.

23. Joseph H. Rush to the AEC, June 11, 1947, box 47, folder 1, FAS Papers.

24. AORES, "Memorandum Prepared by Chicago Group on Loyalty Procedures Affecting Employees of Atomic Energy Commission and Its Contractors," box 15, folder 6, AORES Papers, University of Chicago.

25. William A. Higinbotham to Robert S. Rochlin, October 1, 1947, box 47, folder 1, FAS Papers. See also R. L. Meier to G. H. Goertzel, October 6, 1947, box 50, folder 2, FAS Papers.

26. Philip Morrison, Melba Phillips, and Arthur Roberts, "Draft Statement on Military Secrecy and Security," March 15, 1947, box 120, folder "Federation of Scientists 1947," Oppenheimer Papers, Library of Congress.

27. Oscar K. Rice to Joseph H. Rush, April 2, 1947; Victor F. Weisskopf to William A. Higinbotham, April 10, 1947; and William A. Higinbotham to Philip Morrison, April 30, 1947, all in box 50, folder 2, FAS Papers.

28. Joseph H. Rush to Oscar K. Rice, April 9, 1947, box 50, folder 2, FAS Papers.

29. William A. Higinbotham to Robert S. Rochlin, November 26, 1947, box 47, folder 1, FAS Papers. Several recipients of the Cornell questionnaire informed the FBI of the Cornell committee's attempts to gather data about how the security clearance system operated. See, for example, Memorandum, SAC, Boston, to the Director,

FBI, December 3, 1947, FBI-FAS, FBI HQ 100-344452-104, vol. 15; Memorandum, SAC, Albany, to the Director, FBI, March 18, 1948, FBI-FAS, FBI HQ 100-344452, vol. 17; and "Report of Mr. Harry P. Sparks on the Federation of American Scientists," March 25, 1948, FBI HQ 100-344452, copy released by the Freedom of Information/Privacy Office, U.S. Army and Security Command. In response to these reports, the FBI monitored the distribution of the Cornell questionnaire closely and actively consulted the Justice Department about whether or not several items in the questionnaire violated espionage statutes. The Justice Department informed the FBI that if the questionnaire was distributed without willful intent to harm the United States, no statute had been violated. The FBI then discontinued its investigation of the Cornell Committee on Secrecy and Clearance. T. Vincent Quinn, Assistant Attorney General, Criminal Division, to the Director, FBI, April 20, 1948, FBI HQ 100-344452-182, copy released by the Criminal Division of the Justice Department, and the Director, FBI to SAC, Albany, May 3, 1948, WFO-FAS, FBI WFO 65-4736, sec. 6. Whether Higinbotham knew of the FBI's interest in the Cornell committee is unclear.

30. R. L. Meier to Robert S. Rochlin, February 13, 1948, box 47, folder 1, FAS Papers.

31. Katharine Way to Beth Olds, [late May 1948], box 11, folder 14, ASC Papers, University of Chicago.

32. Beth Olds to Katharine Way, June 2, 1948, box 11, folder 14, ASC Papers.

33. R. L. Meier to member associations, February 7, 1948, box 50, folder 2, FAS Papers.

34. R. L. Meier to member associations and members at large, March 24, 1948, Oppenheimer Papers.

35. Political scientists describe such a three-way relationship as an iron triangle. In an iron triangle relationship, government programs derive their support from an interest group (in this case, scientists), a congressional oversight committee (the JCAE), and an administrative agency (the AEC). Each supporter acts out of self-interest, and policies emerge out of negotiations between them. See Balogh, *Chain Reaction*, 62–64.

36. On security through concealment versus security through achievement, see Gellhorn, *Security, Loyalty, and Science*, ch. 1, esp. 17–18. On the general attitudes of congressional conservatives, see Hall, "Congressional Attitudes toward Science."

37. Lilienthal, *Journals*, 140. For a detailed account of the confirmation hearings, see Hewlett and Duncan, *Atomic Shield*, 1–14, 49–53, and Hall, "Congressional Attitudes toward Science," ch. 7.

38. Lilienthal, *Journals*, 141.

39. Ibid., 166.

40. Ibid., 168.

41. Ibid., 233.

42. Balogh, *Chain Reaction*, 67; Hewlett and Duncan, *Atomic Shield*, 324.

43. Lilienthal, *Journals*, 227.

44. Ibid., 262. See also Hewlett and Duncan, *Atomic Shield*, 324.

45. Lilienthal, *Journals*, 263.

46. For the voluminous correspondence between Hickenlooper and the AEC on personnel security, consult RG 326, Records of the AEC, Records of the Office of the Chairman, Office Files of David E. Lilienthal, Joint Atomic Energy Committee Corre-

spondence, January 1947–June 1950, box 1, folders "Joint Committee Correspondence, '47 July through December" and "Correspondence with Bourke Hickenlooper January–June 1948"; and RG 128, Records of the Joint Committee on Atomic Energy, 1946–77, National Archives (hereafter cited as Records of the JCAE), General Correspondence Files, box 74, folder "AEC—Security Review Board."

47. W. W. Waymack, "Four Years under Law," *BAS* 7 (February 1951): 54.

48. Balogh, *Chain Reaction*, 67–73.

49. RG 128, JCAE Executive Session Transcripts, box 1, Transcripts of Executive Meetings 1947–48, Document #988, March 8, 1948. The Hickenlooper quotation is from p. 147, and the quotation from McMahon is from p. 149. For previous JCAE discussions of the Condon case, see Document #986, March 2, 1948, and Document #987, March 5, 1948, also in RG 128, JCAE Executive Session Transcripts, box 1.

50. RG 128, JCAE Executive Session Transcripts, box 1, Transcripts of Executive Meetings 1947–48, Document #989, March 31, 1948, 155, 156.

51. Ibid., 159.

52. Ibid., 161.

53. On the JCAE's dissatisfactions with the AEC, see RG 128, JCAE Executive Session Transcripts, box 1, Transcripts of Executive Meetings 1947–48, Document #990, April 1, 1948, 181, 203–4 (hereafter cited as Executive Session Transcript #990). On Hickenlooper's reassurances, see p. 183. On Hickenlooper and Bricker's complaints about Vought and the Blewetts, see pp. 196–97. On the JCAE's fears of action by HUAC, see pp. 198–200.

54. Executive Session Transcript #990, 206.

55. On the revocation of Goldsmith's clearance, see AEC Meeting No. 92, August 6, 1947, RG 326, Records of the AEC, AEC Minutes. See also "Memorandum for Files," August 28, 1947, Philip Morse to Carroll L. Wilson, September 3, 1947, and L. R. Thiesmeyer, "Memorandum for the Files," September 15, 1947, in app., folder 9, Official Files, Director's Office, Brookhaven National Laboratory, American Institute of Physics (microfilm), reel 20. Goldsmith was not one of the two Oak Ridge chemists whose clearance was revoked; he was working at Brookhaven at the time.

56. Knoxville Field Office, Report, "Federation of American Scientists," May 13, 1948, FBI-FAS, FBI HQ 100-344452-211, vol. 21. The information about Rice appears on pp. 25–26; for Cohn, see pp. 7–10. A slip of the censor's pen revealed Rice and Cohn's identities. The document almost certainly contains information about the other Oak Ridge scientists whom the AEC and JCAE suspected of being "security risks," but their identities remain concealed under black ink.

57. Committee on Secrecy and Clearance, "Some Individual Cases of Clearance Procedures," *BAS* 4 (September 1948): 281. An excerpt from an army document released under the Freedom of Information Act following a referral from the FBI suggests that the army was highly suspicious of the FAS's motives, especially in the late 1940s. The document observed, "The FAS has been severely criticized by private individuals and groups because of its attacks on the government's loyalty and security program and the fact that many of its past officers have records of affiliations with communist front organizations. . . . The record bears out the assertion that FAS officers have records of subversive affiliations." The document went on to name specifically J. Robert Oppenheimer, Harlow Shapley, Edward U. Condon, John P. Peters, and Harold C. Urey as "past FAS officers [who] have engaged in Communist

front activity." The document went on to note that FAS had become more moderate on security issues and generally less active since 1950. Colonel John A. Cleveland Jr., Assistant Chief of Staff, G-2, Department of the Army, date of document unavailable, but definitely after 1955, title of document unavailable (the Freedom of Information Act release included pp. 6–9 only).

58. Lilienthal, *Journals*, 328.

59. The quotations are from ibid., 399 and 547, respectively.

60. On the rumors, and the FBI's attention to them, see Memorandum, SAC, Boston, to the Director, FBI, April 3, 1948, FBI-FAS, FBI HQ 100-344452-173, vol. 19, and Guy Hottel to the Director, FBI, August 20, 1948, WFO-FAS, FBI WFO 65-4736-256, sec. 6.

61. AORES, "Security Bulletin," vol. 2, no. 1, April 17, 1948, box 65, folder 10, FAS Papers, and "AEC Interim Procedures for Local Security Boards," BAS 4 (July 1948): 198. Although the Interim Procedure was not announced publicly until May 20, it was adopted as AEC policy on April 15. See Sumner T. Pike to Bourke B. Hickenlooper, April 19, 1948, RG 128, Records of the JCAE, Correspondence Files, box 72, folder "AEC Security General 1947–1948."

62. AORES, "Security Bulletin," April 17, 1948.

63. Katharine Way to Beth Olds, May 18, 1948, box 11, folder 14, ASC Papers.

64. The four articles, all written by Stephen White and published in the *New York Herald Tribune*, are: "2 Atomic Scientists Suspended, Many More Face Loyalty Board," May 19, 1948, 1, 35; "Oak Ridge Sunk in Gloom over Loyalty Inquiry," May 20, 1948, 5; "Why Morale Sags at Oak Ridge," May 24, 1948, 18; "Senior Scientist Rolls Depleted at Oak Ridge," May 28, 1948, 11.

65. "2 Atomic Scientists Suspended, Many More Face Loyalty Board," 35.

66. "Oak Ridge Sunk in Gloom over Loyalty Inquiry."

67. "2 Atomic Scientists Suspended, Many More Face Loyalty Board," 35.

68. "Why Morale Sags at Oak Ridge."

69. "Senior Scientist Rolls Depleted at Oak Ridge."

70. "Scientists' Statement," *New York Herald Tribune*, May 20, 1948, 5.

71. "Atomic Data Barred from 2 At Oak Ridge," *New York Times*, May 20, 1948, 22; "Two Scientists at Oak Ridge Suspended in Loyalty Probe," *Washington Post*, May 19, 1948, 3; "'Demoralized': Oak Ridge Resents AEC Investigation," *Washington Post*, May 20, 1948, 7; "Loyalty Check Change Asked," *Washington Post*, May 23, 1948, 8.

72. Katharine Way to Beth Olds, [late May 1948], box 11, folder 14, ASC Papers. See also James S. Stewart to the *New York Herald Tribune*, May 23, 1948, box 65, folder 10, FAS Papers.

73. R. L. Meier to AORES Executive Committee, [late May 1948], box 65, folder 10, FAS Papers.

74. James S. Stewart to R. L. Meier, May 25, 1948, box 65, folder 10, FAS Papers.

75. AORES, "Long Range Security Policy," undated, no later than May 1948, box 15, folder 6, AORES Papers.

76. Carroll L. Wilson to James S. Stewart, June 16, 1948, RG 326, Records of the AEC, Records of the Office of the Chairman, Office Files of David E. Lilienthal, Subject Files, 1946–50, box 10, folder "Oak Ridge Correspondence 1947–48."

77. Katharine Way and Henri A. Levy to Owen J. Roberts, June 12, 1948, box 11, folder 14, ASC Papers.

78. AORES, "Security Bulletin, Vol. II, No. 3: Position of the Roberts Board," July 1, 1948, box 11, folder 14, ASC Papers. The quotation from Roberts is taken from this source.

79. Beth Olds to Katharine Way, June 18, 1948, box 11, folder 14, ASC Papers.

80. Richard B. Gehman, "Oak Ridge Witch Hunt," *New Republic*, July 5, 1948, 12–14. T. H. Davies, " 'Security Risk' Cases: A Vexed Question," *BAS* 4 (July 1948): 193–94. In the same issue, *BAS* also published an abridged version of the *New York Herald Tribune* articles by Stephen White and a list of the charges in the five Oak Ridge cases.

81. R. L. Meier to Henri A. Levy, June 4, 1948, box 65, folder 10, FAS Papers.

82. Henri A. Levy, letter to the editor, June 16, 1948, file copy in box 16, folder 2, AORES Papers. Published in "Letters to the Editor," *Chemical and Engineering News*, July 19, 1948, 2158.

83. "Eight Scientists Protest Thomas Committee's Methods," *BAS* 4 (October 1948): 290, 320.

84. "Text of President Truman's Talk to Scientists," *New York Times*, September 14, 1948, 24. Apparently Condon encouraged Truman to speak at the AAAS centennial meeting, and he and his assistant Hugh Odishaw wrote an early draft of the speech. See Dael Wolfle, "The Centennial Annual Meeting, Starring Harry Truman and Civil Liberties," *Science*, October 6, 1989, 130–31, and, for Condon's own recollections, Edward U. Condon, interview by Charles Weiner, September 11–12, 1973, transcript, 196–98.

85. Quoted in C. P. Trussell, "House Body Plans to Expose Details of Atomic Spying," *New York Times*, September 18, 1948, 8.

86. Ibid., 1.

87. "Statement of the Association of Oak Ridge Engineers and Scientists," September 14, 1948, box 65, folder 10, FAS Papers.

88. R. L. Meier, "Position of Scientists," *New York Times*, September 23, 1948, 28.

89. L. W. Nordheim to Bourke B. Hickenlooper, September 16, 1948, box 11, folder 14, ASC Papers.

90. Ray W. Stoughton to Bourke B. Hickenlooper, September 17, 1948, box 11, folder 14, ASC Papers.

91. Ray W. Stoughton to Fellow-Association Members, September 18, 1948, box 1, folder 8, Rabinowitch Papers, State University of New York at Albany.

92. Alvin M. Weinberg to David E. Lilienthal, September 20, 1948, RG 326, Records of the AEC, Records of the Office of the Chairman, Office Files of David E. Lilienthal, Subject Files, 1946–50, box 10, folder "Oak Ridge Correspondence 1947–48."

93. Ray W. Stoughton to David E. Lilienthal, September 24, 1948, RG 326, Records of the AEC, Records of the Office of the Chairman, Office Files of David E. Lilienthal, Subject Files, 1946–50, box 10, folder "Oak Ridge Correspondence 1947–48." Copies of the letter were also sent to other AEC officials, including Owen J. Roberts and Carroll L. Wilson.

94. Bourke B. Hickenlooper to David E. Lilienthal, September 14, 1948, RG 326, Records of the AEC, Records of the Office of the Chairman, Office Files of David E. Lilienthal, Joint Atomic Energy Committee Correspondence, January 1947–June 1950, box 1, folder "Joint Committee Correspondence, '48, July thru September."

95. R. L. Meier to Ray W. Stoughton, September 30, 1948, box 65, folder 10, FAS Papers.

96. David E. Lilienthal to Bourke B. Hickenlooper, September 30, 1948, RG 326, Records of the AEC, Records of the Office of the Chairman, Office Files of David E. Lilienthal, Joint Atomic Energy Committee Correspondence, January 1947–June 1950, box 1, folder "Joint Committee Correspondence, '48, July thru September."

97. David E. Lilienthal to Bourke B. Hickenlooper, October 15, 1948, RG 326, Records of the AEC, Records of the Office of the Chairman, Office Files of David E. Lilienthal, Joint Atomic Energy Committee Correspondence, January 1947–June 1950, box 1, folder "Joint Committee Correspondence, '48, July thru September."

98. Ray W. Stoughton to R. L. Meier, October 12, 1948, box 65, folder 10, FAS Papers.

99. Data was collected using [List of senior chemists and physicists who resigned from Oak Ridge National Laboratory], October 12, 1948, box 65, folder 10, FAS Papers, and *American Men of Science*, 9th ed., vol. 1 (New York: R. R. Bowker Co., 1955). In a few instances, scientists who were not listed in the ninth edition were traced using the 1949 eighth edition. Three other scientists not on the FAS list also left, but all three were almost certainly covered by the loyalty-security system in their new jobs. One went to the National Bureau of Standards, another went to Argonne National Laboratory, and the third went to work in the atomic energy research department of North American Aviation, Inc.

100. Clark D. Ahlberg and John C. Honey, "The Scientist's Attitude toward Government Employment," *Science*, May 4, 1951, 506, 507.

101. National Science Foundation, *Employment of Scientists and Engineers in the United States, 1950–66* (Washington, D.C.: U.S. Government Printing Office, 1968), 20–21.

102. On this point, consult Freeland, *Truman Doctrine*.

103. Ray W. Stoughton to David E. Lilienthal, September 24, 1948.

104. Paul C. Tompkins to the FAS, August 20, 1948, box 65, folder 10, FAS Papers. An FBI report that mentioned Katharine Way's political activities took note of her involvement in the atomic scientists' movement, her interest in the Southern Conference for Human Welfare, and her acquaintance with an acquaintance of one of the defendants in the Canadian spy cases. Knoxville Field Office, Report, "Federation of American Scientists," May 13, 1948, pp. 35–36, FBI-FAS, FBI HQ 100-344452-211, vol. 21.

105. [List of senior chemists and physicists who resigned from Oak Ridge National Laboratory], October 12, 1948.

106. Memorandum, "Security and Clearance," from ASC executive committee to present and prospective ASC members, December 9, 1948, box 11, folder 5, Urey Papers, University of California at San Diego.

CHAPTER SIX

1. "A Statement by Members and the Council of the National Academy of Sciences Concerning a National Danger," [distributed March 10, 1948], Congress 1948: Committees: Un-American Activities: Condon Case: NAS Statement: Preparation, NAS Archives.

2. A. N. Richards to Members of the Council of the National Academy of Sciences, March 10, 1948, Congress 1948: Committees: Un-American Activities: Condon Case: NAS Statement: Preparation, NAS Archives.

3. "Conversation with Dr. Bush on March 11, 1948 about the proposed statement," NAS: Congress 1948: Committees: Un-American Activities: Condon Case: NAS Statement: Preparation, NAS Archives.

4. Richards to members of the NAS council, March 16, 1948, NAS: Congress 1948: Committees: Un-American Activities: Condon Case: NAS Statement: Preparation, NAS Archives.

5. Frank B. Jewett, "E. U. Condon—Conversation November 29 [1947]," Congress 1947: Committees: Un-American Activities: 1947, NAS Archives.

6. "A Statement by Members of the National Academy of Sciences Concerning a National Danger," March 31, 1948, Lawrence Papers, University of California at Berkeley.

7. Vannevar Bush to A. N. Richards, March 20, 1948, NAS: Congress 1948: Committees: Un-American Activities: Condon Case: NAS Statement: Preparation, NAS Archives. Frank B. Jewett's incorrect prediction of Bush's reaction is found in Frank B. Jewett to A. N. Richards, March 17, 1948, NAS: Congress 1948: Committees: Un-American Activities: Condon Case: NAS Statement: Preparation, NAS Archives.

8. Rebecca S. Lowen, *Creating the Cold War University: The Transformation of Stanford* (Berkeley and Los Angeles: University of California Press, 1997), 93.

9. A. N. Richards to members of the National Academy of Sciences, March 31, 1948, Lawrence Papers.

10. A. N. Richards to members of the NAS council, April 7, 1948, NAS: Congress 1948: Committees: Un-American Activities: Condon Case: NAS Statement: Consideration by NAS Membership, NAS Archives.

11. A. N. Richards to Frank B. Jewett, April 18, 1948, NAS: Congress 1948: Committees: Un-American Activities: Condon Case: NAS Statement: Consideration by NAS Membership, NAS Archives.

12. [List of members approving and disapproving the Condon statement, April or early May 1948], NAS: Congress 1948: Committees: Un-American Activities: Condon Case: NAS Statement: Consideration by NAS Membership, NAS Archives.

13. L. C. Dunn to A. N. Richards, April 2, 1948, box "Columbia #2-DA," folder "Condon Case Materials," Dunn Papers, American Philosophical Society.

14. "Members who do not approve statement re Thomas Committee," April 14, 1948, NAS: Congress 1948: Committees: Un-American Activities: Condon Case: NAS Statement: Consideration by NAS Membership, NAS Archives.

15. [List of members approving and disapproving the Condon statement, April or early May 1948].

16. A. N. Richards to NAS members, April 20, 1948, Lawrence Papers.

17. A. N. Richards to J. Parnell Thomas, April 30, 1948, copy in the Lawrence Papers, notes that "many of the members did not approve the decision not to publish the statement." I do not have further quantification.

18. I. I. Rabi to A. N. Richards, April 8, 1948, NAS: Congress 1948: Committees: Un-American Activities: Condon Case: NAS Statement: Consideration by NAS Membership, NAS Archives.

19. Marshall H. Stone to A. N. Richards, May 3, 1948, NAS: Congress 1948: Committees: Un-American Activities: Condon Case: NAS Statement: Consideration by NAS Membership, NAS Archives.

20. John P. Peters to A. N. Richards, April 23, 1948, NAS: Congress 1948: Com-

mittees: Un-American Activities: Condon Case: NAS Statement: Consideration by NAS Membership, NAS Archives.

21. Transcript of NAS business session, April 27, 1948, p. 46, Organization 1948: NAS: Meetings: Annual: Business Sessions: Transcript, NAS Archives.

22. Ibid., 48.

23. Ibid., 49.

24. Ibid., 53–54.

25. Ibid., 54–55. J. Robert Oppenheimer to A. N. Richards, March 15, 1948, Oppenheimer Papers, Library of Congress.

26. Transcript of NAS business session, April 27, 1948, p. 56.

27. Ibid., 61, 62.

28. Press release, National Academy of Sciences, May 3, 1948, copy in the Lawrence Papers.

29. Richards reported, "So far as we can tell, no newspaper has seen fit to make the slightest allusion to the release which was given out on May 3." A. N. Richards to Frank B. Jewett, May 7, 1948, NAS: Congress 1948: Committees: Un-American Activities: Condon Case: NAS Statement: Press Release, NAS Archives. *Science News*, at least, reported the statement. See "Academy Issues Statement Criticizing Condon Attack," *Science News*, May 15, 1948, 312.

30. E. U. Condon to Martin J. Kamen, May 24, 1948, box "K-Las," Kamen, Martin J., folder 1, Condon Papers, American Philosophical Society.

31. In 1948, when the academy debated its response to the Condon case, the members of the NAS council were A. N. Richards, Luther P. Eisenhart, F. E. Wright, Detlev Bronk, Jerome C. Hunsaker, W. A. Noyes, Walter R. Miles, I. I. Rabi, Wendell Stanley, John T. Tate, and D. D. Van Slyke. Harlow Shapley, Oswald Veblen, and a few other scientists who could be considered progressive left had served on the council in the 1920s and 1930s, but progressive left scientists were noticeably absent from influential positions within the academy during the 1940s.

32. On Rabi's distrust of anticommunism, see John S. Rigden, *Rabi: Scientist and Citizen* (New York: Basic Books, 1987), 254. Not only did Rabi support Condon, but he also offered John Blewett a position at Columbia University when his chances for clearance at Brookhaven looked dim, and in 1954 he was one of Oppenheimer's staunchest supporters. In a well-known incident several weeks after the Oppenheimer hearings, he refused to shake Edward Teller's hand and castigated Teller for his damaging testimony against Oppenheimer. Rigden, 230.

33. For Jewett's general observations on the Condon case, see Frank B. Jewett to Karl T. Compton, March 29, 1948, box 56, folder "Jewett, Frank B. (1947–49)," Bush Papers, Library of Congress. On Jewett's efforts to assist Condon in 1947 by contacting New Jersey Republicans, see Frank B. Jewett to George W. Merck, June 24, 1947; Arthur Vanderbilt to Frank B. Jewett, June 30, 1947, box "United States, no. 2," folder "U.S. Cong., House, Special Comm. on Un-Am. Activities, folder 4," Condon Papers; and Frank B. Jewett to Edward U. Condon, July 9, 1947, box "United States, no. 2," folder "U.S. Cong., House, Special Comm. on Un-Am. Activities, folder 6," Condon Papers. Merck was treasurer of the Republican Party in New Jersey, and Vanderbilt was state Republican leader.

34. Vannevar Bush to Frank B. Jewett, April 2, 1948, box 56, folder "Jewett, Frank B. (1947–49)," Bush Papers.

35. "Scientists Score Thomas Actions," *New York Times*, March 12, 1948, 19.

36. Minutes of council meeting, June 10, 1948, quoted in A. N. Richards, "Committee on Civil Liberties," February 3, 1949, Administration: Organization: NAS: Committee on Civil Liberties: 1949, NAS Archives.

37. A. N. Richards to Marshall H. Stone, February 25, 1949, Administration: Organization: NAS: Committee on Civil Liberties: 1949, NAS Archives.

38. "Draft of a Proposed Statement to be Issued by the Council of the National Academy of Sciences," undated, but sent with cover letter dated January 13, 1949, Administration: Organization: NAS: Committee on Civil Liberties: 1949, NAS Archives.

39. On discussion of dissemination of the statement, see "Minutes of the Meeting of the Council of the Academy," February 3, 1949, Administration: Organization: NAS: Committee on Civil Liberties: 1949, NAS Archives. Truman's reply is discussed in Hershberg, *James B. Conant*, 438, 867 n. 67.

40. J. Robert Oppenheimer to A. N. Richards, February 10, 1949, Administration: Organization: NAS: Committee on Civil Liberties: 1949, NAS Archives.

41. P. W. Bridgman to Alfred N. Richards, February 23, 1949, Administration: Organization: NAS: Committee on Civil Liberties: 1949, NAS Archives.

42. Quoted in Cochrane, *National Academy of Sciences*, 473.

43. Frank B. Jewett, "The Academy—Its Charter, Its Functions, and Relations to Government," November 17, 1947, published in *Proceedings of the National Academy of Sciences*, April 15, 1962, 481–90. The quotations are drawn from pp. 482, 483, and 484.

44. Cochrane, *National Academy of Sciences*, 473.

45. On the government advisory positions held by these scientists, and the purposes of the various committees on which they served, see, for example, Cochrane, *National Academy of Sciences*, ch. 14, esp. 435–46, 459–63; Kevles, "Scientists, the Military, and the Control of Postwar Defense Research," and Kevles, "Cold War and Hot Physics," esp. 246–47.

46. On Bush's reservations about the military patronage of science, see Reingold, "Vannevar Bush's New Deal." On Conant's disillusionment with his work on the GAC, his opposition to the development of the hydrogen bomb, and his fears of military domination of science, see Hershberg, *James B. Conant*, esp. chs. 24 and 28.

47. Annual membership figures for the AAAS are found in Wolfle, *Renewing a Scientific Society*, 10. Peter J. Kuznick sees the AAAS Science and Society movement in the 1930s as evidence of political activism, but the activities Kuznick examines were primarily educational forums and general discussions of the social responsibility of science. The discussions were not translated into explicit political action directed at specific policy changes. Kuznick, *Beyond the Laboratory*, ch. 3.

48. "Minutes of Meeting of the Council, A.A.A.S.," December 28, 1947, Board/Council Files, AAAS, and Maurice B. Visscher, "A Half Century in Science and Society," *Annual Review of Physiology* 31 (1969): 12.

49. Maurice B. Visscher to Harlow Shapley, March 17, 1948, box 46, folder 11, FAS Papers, University of Chicago.

50. "Report of the Special Committee on the Civil Liberties of Scientists," December 18, 1948, p. 5, box 2, folder "Civil Liberties: Report of Special Committee on the Civil Liberties of Scientists," Borras Files, 2d Accession, AAAS.

51. Ibid., 74–77.

52. Ibid., 73.

53. Maurice B. Visscher to Philip Bard, Robert E. Cushman, R. L. Meier, and James R. Newman, January 11, 1949, box 28, folder 10, FAS Papers.

54. Robert E. Cushman to Maurice B. Visscher, January 14, 1949, box 28, folder 10, FAS Papers.

55. R. L. Meier to Maurice B. Visscher, Philip Bard, James R. Newman, and Robert E. Cushman, January 18, 1949, box 28, folder 10, FAS Papers.

56. Maurice B. Visscher to Members of the AAAS Committee on Civil Liberties, February 24, 1949, box 28, folder 10, FAS Papers.

57. Visscher, "A Half Century in Science and Society," 12–13.

58. Kirtley F. Mather and Howard A. Meyerhoff, for the Executive Committee, to Members of the Council, March 17, 1949, and "Report of the Special Committee on the Civil Liberties of Scientists" [copy of questionnaire, with tallied results], undated, approximately April 1949, Board/Council Files, folder "Minutes of the Executive Committee—July 7, 1949 and Agenda of the Executive Committee—April 23–24, 1949."

59. "Minutes of the Executive Committee," April 23–24, 1949, Board/Council Files, folder "Treasurer's Accounts—December 31, 1949, and Minutes of the Executive Committee—April 23–24, 1949"; "Minutes of the meeting of the Executive Committee," July 7, 1949, Board/Council Files, folder "Minutes of the Executive Committee—July 7, 1949 and Agenda of the Executive Committee—April 23–24, 1949"; "Publications Committee, Agenda, Meeting of July 6, 1949," Board/Council Files, folder "Minutes of the Publications Committee—July 6, 1949, August 20, 1949, and October 7, 1949"; "Civil Liberties of Scientists," *Science*, August 19, 1949, 177–79.

60. For Shapley's comment, see SAC, Boston, to the Director, FBI, December 12, 1948, WFO-FAS, FBI WFO 65-4736-296, sec. 7. On Visscher's observations, see Visscher, "A Half Century in Science and Society," 12. In a 1972 letter, Visscher recalled, "There was quite a controversy within the Council and Executive Board of the AAAS over the publication of this Report," but he did not provide further details. Maurice B. Visscher to William Bevan, Executive Officer, AAAS, December 14, 1972, box 2, folder "Civil Liberties—3/73; 1952–," Borras Files, 2d Accession.

61. "AAAS Preliminary Agenda, Meeting of the Executive Committee," October 14–15, 1950, folder "Minutes of the Executive Committee—June 24–25, 1950; October 14–15, 1950," Board/Council Files.

62. "AAPG Action on the Visscher Report," *Science*, June 9, 1950, 638.

63. "AAAS Preliminary Agenda, Meeting of the Executive Committee," October 14–15, 1950.

64. Throughout the early 1950s, the AAAS took steps toward expanding its political role, but its plans never came to fruition. In 1951 the AAAS engaged in an extensive reexamination of the organization's purpose and formulated the "Arden House" proposals, which included an initiative to address science-government relations. The Arden House program was eventually reduced to professional and educational activities. See box 1, folder "Arden House Statement, 1951–1954," Borras Files, 2d Accession, and related materials in Executive Committee and Board of Directors minutes, American Association for the Advancement of Science. In his term as AAAS president in 1953, E. U. Condon sought to expand political coverage

in *Science*; his effort deteriorated into an unpleasant dispute with AAAS administrative secretary Howard A. Meyerhoff, who ultimately resigned as a result. See folder "Minutes of the Board of Directors—March 16, 1953" and folder "Minutes of the Board of Directors—May 2, 1953," Board/Council Files.

65. R. L. Meier and J. M. Lowenstein to Associations and Members at Large, June 10, 1948, box 7, folder "Federation of American Scientists," Morse Papers, Massachusetts Institute of Technology.

66. Minutes of the FAS Council, May 2, 1948, box 11A, folder "Federation of American Scientists," Shapley Papers, Harvard University; R. L. Meier and J. M. Lowenstein to Associations and Members at Large, June 10, 1948; William A. Higinbotham, Robert R. Bush, and Executive Secretariat to Member Associations and Members at Large, September 17, 1948, box 40, ser. 9, folder 1, Goudsmit Papers, American Institute of Physics; Minutes of Council Meeting, FAS, December 16, 1948, box 11A, folder "Federation of American Scientists," Shapley Papers.

67. "Scientists Committee on Loyalty Problems: Interim Report," undated, probably October 1948, box "Science, no. 2—Security, no. 1," "Scientists' Committee on Loyalty Problems," folder 1, Condon Papers. The other founding members were David Bohm, R. J. Britten, R. R. Bush, Albert Einstein, Luther P. Eisenhart, Samuel A. Goudsmit, Donald R. Hamilton, M. Stanley Livingston, Stuart Mudd, Henry D. Smyth, Oswald Veblen, Arthur S. Wightman, and Irving Wolff.

68. [Confidential AORES list of senior chemists and physicists resigning from Oak Ridge], October 12, 1948, box 65, folder 10, FAS Papers.

69. Minutes of the Scientists' Committee on Loyalty Problems, September 25, 1948, box "Science, no. 2—Security, no. 1," "Scientists' Committee on Loyalty Problems," folder 1, Condon Papers. For a description of the SCLP's statement of purpose, see "Prospectus: Scientists' Committee on Loyalty Problems," box 28, folder "Scientists' Committee on Loyalty," Spitzer Papers, Princeton University.

70. SCLP, "Report on Activities," March 18, 1949, box 69, folder 9, FAS Papers.

71. For an example of Goudsmit's information-gathering efforts on behalf of the SCLP, see R. R. Bush to Herbert F. Berger, April 8, 1949, and Samuel A. Goudsmit to William A. Higinbotham, December 27, 1949, box 40, ser. 9, folder 2, Goudsmit Papers. On his attitude toward left-wing scientists, see, for example, Walter F. Colby, Director of Intelligence, AEC, to Samuel A. Goudsmit, March 16, 1950, and Samuel A. Goudsmit to Walter F. Colby, March 20, 1950, box 10, folder III-42, Goudsmit Papers. Colby noted in his letter that Goudsmit had "a bright eye for villains" and that a scientist they had discussed previously "is reported to be a secret member of the American Communist Party and his connections and relatives in [the] UK all have the wrong color." He added, "What did you do about his letter?" Goudsmit replied, "I smelled a rat, and the SCLP (Scientists' Committee on Loyalty Problems), on my advice, did not pay any attention to his request to prepare an article for 'Nature.'" Although the exact context of the exchange is not clear, Goudsmit's animosity toward the scientist in question, as well as his willingness to have the SCLP act based on that animosity, is perfectly apparent. On Goudsmit's passing information to AEC intelligence and the CIA, see chapter 2, n. 72.

72. SCLP, "Proposed Policy Statement," November 17, 1948, box 23, folder 5, FAS Papers.

73. Ibid.

74. Katharine Way, "Comments on Proposed Policy Statement of Scientists' Committee on Loyalty Problems," December 1, 1948, box 15, folder 6, AORES Papers, University of Chicago. The receipt of responses from Rabinowitch and the others is noted in SCLP, Minutes of the Meeting of December 11, 1948, box 3, folder 9, Coryell Papers, University of Chicago.

75. Katharine Way, "Comments on Proposed Policy Statement of Scientists' Committee on Loyalty Problems."

76. Ibid.

77. Ibid.

78. Ibid.

79. Lyman Spitzer Jr. to Arthur S. Wightman, December 3, 1948, box 28, folder "Scientists' Committee on Loyalty," Spitzer Papers.

80. Ibid.

81. SCLP, Minutes of the Meeting of December 11, 1948, box 3, folder 9, Coryell Papers.

82. Ibid.

83. SCLP, Minutes of the Meeting of February 12, 1949, box 29, folder "Scientists' Committee on Loyalty," Spitzer Papers.

84. For details on the Berkeley Radiation Laboratory hearings, consult Caute, *Great Fear*, 466–69; Schrecker, *No Ivory Tower*, ch. 5, esp. 126–48; and Russell B. Olwell, "Princeton, David Bohm, and the Cold War: A Case Study in McCarthyism and Physics," unpublished ms., personal possession of the author. For the hearings themselves, see House Committee on Un-American Activities, *Hearings Regarding Communist Infiltration of Radiation Laboratory and Atomic Bomb Project at the University of California—Vol. 1*, 81st Cong., 1st sess., April 22, 26, May 25, and June 10, 14, 1949.

85. Quoted in Schrecker, *No Ivory Tower*, 135.

86. Samuel A. Goudsmit to Lyman Spitzer Jr., June 2, 1949, box 28, folder "Scientists' Committee on Loyalty," Spitzer Papers.

87. Ibid.

88. Lyman Spitzer Jr. to Hugh C. Wolfe, June 6, 1949, box 28, folder "Scientists' Committee on Loyalty," Spitzer Papers.

89. Lyman Spitzer Jr. to Robert E. Marshak, July 5, 1949, box 28, folder "Scientists' Committee on Loyalty," Spitzer Papers.

90. Athan Theoharis, "The Politics of Scholarship: Liberals, Anti-Communism, and McCarthyism," in Griffith and Theoharis, *The Specter*, 270.

91. Clifford Grobstein and Alan H. Shapley, "The Scientists' Organizations in 1950," *Bulletin of the Atomic Scientists* 7 (January 1951): 24. On the AEC's policy change, see press release, "AEC Revises Hearing Procedures for Security Clearance of Personnel," September 19, 1950, RG 128, Records of the JCAE, Correspondence Files, box 72, folder "AEC Security General 1949–1950," National Archives.

92. Balogh, *Chain Reaction*, ch. 2, esp. 21–29. See also Jackson Lears, "A Matter of Taste: Corporate Culture Hegemony in a Mass-Consumption Society," in May, *Recasting America*, 50–51.

93. On the dichotomy liberal anticommunists drew between themselves and popular front liberals, see Kenneth O'Reilly, "Liberal Values, the Cold War, and American Intellectuals: The Trauma of the Alger Hiss Case, 1950–1978," in Theoharis, *Beyond the Hiss Case*, 310–313. For Mills's critique, see Mills, *Causes of World War Three*, 16–17. On Mills himself, see Pells, *Liberal Mind in a Conservative Age*, 249–61.

94. Tomlins, *State and the Unions.*

95. Lowi, *End of Liberalism*, 108.

96. Ibid., 297.

97. "Summary of talk by K. W. Ford before WAS," May 23, 1950, box 23, folder 5, FAS Papers.

98. Norman L. Rosenberg, "Gideon's Trumpet: Sounding the Retreat from Legal Realism," in May, *Recasting America*, 107–24, finds a similar transition in American legal thought during the early Cold War years. Rosenberg documents a subtle rhetorical shift from a staunch commitment to free speech and political freedom to a more strict, legalistic emphasis on the part of legal theorists. He holds FBI repression, and not interest group liberalism, responsible for the shift, however.

CHAPTER SEVEN

1. In the 1950s, most American intellectuals blamed the rise of domestic anticommunism on right-wing demagogues, especially the infamous Joseph McCarthy, and viewed anticommunism as an irrational, anti-intellectual expression of mass-based politics that was grounded in the status anxieties of a newly prosperous middle class. To these scholars, most of whom were liberal anticommunists, liberals were the embattled victims of McCarthyism. Anticommunism by itself was not problematic, but the United States was endangered by its excesses—irresponsible anticommunists like Joseph McCarthy, as opposed to responsible anticommunists like themselves. See, for example, the essays in Bell, *Radical Right*; Hofstadter, *Paranoid Style in American Politics*; and Schlesinger, *Vital Center.*

Beginning in the 1970s, historians who studied the postwar red scare began to hold the Truman administration and political elites responsible for the rise of Cold War anticommunism, rather than an anxiety-ridden, anti-intellectual middle class. See, for example, Freeland, *Truman Doctrine*, and Theoharis, *Seeds of Repression*, on the Truman administration; McAuliffe, *Crisis on the Left*, and Pells, *Liberal Mind in a Conservative Age*, on liberals and intellectuals; Levenstein, *Communism, Anticommunism, and the CIO*, Rosswurm, *CIO's Left-Led Unions*, and Halpern, *UAW Politics*, on the labor movement; and, on the legal profession, McAuliffe, *Crisis on the Left*, ch. 7, Mary S. McAuliffe, "The Politics of Civil Liberties: The American Civil Liberties Union during the McCarthy Years," in Griffith and Theoharis, *The Specter*, 154–70, and Norman L. Rosenberg, "Gideon's Trumpet: Sounding the Retreat from Legal Realism," in May, *Recasting America*, 107–24.

2. By way of comparison, between 1919 and 1950, the Rockefeller Foundation awarded 1,107 National Research Fellowships for research in the natural sciences, an average of fewer than forty per year. Raymond Fosdick, *The Story of the Rockefeller Foundation* (New York: Harper & Brothers, Publishers, 1952), 146.

3. On the AEC's institutional mission, see Hershberg, *James B. Conant*, chs. 17–19, esp. 312, 331–33, 341, 351, and Bernstein, "Four Physicists and the Bomb," 251–56.

4. Draft [of minutes of May 1 joint meeting between the AEC Predoctoral and Postdoctoral Boards in the Physical Sciences], May 3, 1948, Fellowships 1948: AEC-NRC Fellowship Boards: Postdoctoral: General, NAS Archives; "From notes of meeting of 5/5/48, AEC Postdoctoral Fellowship Board in Medical Sciences," Fellowships 1948: AEC-NRC Fellowship Boards: Postdoctoral: Medical Sciences, NAS Archives; and Detlev W. Bronk to Carroll L. Wilson, July 27, 1948, box 28, folder "AEC

Fellowship General," RG 128, Records of the JCAE, General Correspondence, National Archives.

5. Draft [of minutes of May 1 joint meeting between the AEC Predoctoral and Postdoctoral Boards in the Physical Sciences], May 3, 1948; Lewis L. Strauss, "Memorandum for the Secretary," July 2, 1948, box 1220, folder "Fellowship Program," RG 326, Records of the AEC, Secretariat Files, National Archives; Minutes of First Meeting, AEC Postdoctoral Fellowship Board in the Medical Sciences, May 5, 1948, p. 14, Fellowships 1949: AEC-NRC Fellowship Boards: Postdoctoral: Medical Sciences, NAS Archives; Hewlett and Duncan, *Atomic Shield*, 341; and AEC 4/6, "Extension of Security Clearance to Fellowships," July 16, 1948, box 1220, folder "Fellowship Program," RG 326, Records of the AEC, Secretariat Files.

6. Lilienthal, *Journals*, 189.

7. A December 1947 Gallup poll indicated 62 percent of the American public favored outlawing the Communist Party, while only 23 percent were opposed. An April 1947 poll indicated 67 percent of the American public believed communists should not be allowed to hold government jobs, and an April 1948 poll showed 80 percent of Americans believed Communist Party members should not be allowed to work in defense-related industries. "Curb Red Activity, Voters Urge," *Atlanta Constitution*, May 23, 1948, 3-D.

8. In his June 18, 1948, diary entry, Lilienthal recorded that Carroll L. Wilson, James B. Fisk, AEC field manager Walter J. Williams, and AEC personnel director Fletcher B. Waller all "welcomed warmly" his ideas about the need for the commission to take courageous action, and Fisk "said it would have a good effect all through the enterprise just to know that such things were being said." The entry gives no indication, however, that any of these officials were ready to take action on Lilienthal's ideas. Lilienthal, *Journals*, 362.

9. Bourke B. Hickenlooper to David E. Lilienthal, July 30, 1948, RG 128, Records of the JCAE, General Correspondence, box 28, folder "AEC Fellowship General."

10. David E. Lilienthal to Bourke B. Hickenlooper, October 11, 1948, RG 128, Records of the JCAE, General Correspondence, box 28, folder "AEC Fellowship General."

11. Bourke B. Hickenlooper to David E. Lilienthal, January 12, 1949, RG 128, Records of the JCAE, General Correspondence, box 28, folder "AEC Fellowship General." See also RG 128, JCAE Executive Session Transcripts, box 2, Transcripts of Executive Meetings January–June 1949, Document #472 (1), pp. 6–9.

12. On Freistadt's political activities and his open willingness to discuss his membership in the Communist Party, see his testimony in Joint Committee on Atomic Energy, *Atomic Energy Commission Fellowship Program*, 81st Cong., 1st sess., May 18, 1949, 107–34 (hereafter cited as *Atomic Energy Commission Fellowship Program*).

13. The angry constituent's letter is J. R. Cherry Jr. to Clyde R. Hoey, April 22, 1949, RG 128, Records of the JCAE, General Correspondence, box 29, folder "AEC Fellowships Freistadt, Hans." Lilienthal's reply to Hoey is reproduced in Senate Subcommittee of the Committee on Appropriations, *Independent Offices Appropriation Bill for 1950*, 81st Cong., 1st sess., May 19, 1949, 248–51 (hereafter cited as *Independent Offices Appropriation Bill*). The Fulton Lewis broadcast is reproduced in *Independent Offices Appropriation Bill*, May 24, 1949, 622–23.

14. "U.S. Atomic Fellowship Held by a Naturalized Communist," *New York Times*,

May 13, 1949, 1, 15; "AEC Study Fund Cut Weighed in Senate," *New York Times*, May 14, 1949, 11; "Senator Says Another Communist Holds a U.S. Atomic Scholarship," *New York Times*, May 15, 1949, 1, 30.

15. See, for example, Harold B. Hinton, "Lilienthal Backs Fellowship Plan," *New York Times*, May 17, 1949, 23; Gellhorn, *Security, Loyalty, and Science*, 189–90; Hall, "Congressional Attitudes toward Science," 363–68; Caute, *Great Fear*, 464; and Balogh, *Chain Reaction*, 72.

16. *Atomic Energy Commission Fellowship Program*, May 16, 1949, 1–11, esp. 2, 3, 9. The quotations are from p. 9.

17. Ibid., 11–30. The quotations are from pp. 23, 29, and 31, respectively.

18. Ibid., 32–39. The quotations are found on pp. 37 and 39, respectively.

19. Ibid., 37, 41–42. The quotation is on p. 42.

20. Lilienthal, *Journals*, 531.

21. *Atomic Energy Commission Fellowship Program*, May 17, 1949, 60. A few days later, Bronk repeated this reasoning before the Independent Offices subcommittee of the Senate Appropriations Committee: "I do not think that a person who claims that he can be loyal to the United States and be loyal to the Communist Party has the intellectual qualities which would make him a good scientific man. . . . I don't believe he has the clarity of thought to make him a good scientific prospect." *Independent Offices Appropriation Bill*, May 20, 1949, 371.

22. Schrecker, *No Ivory Tower*, 106–12. For general accounts of political tests in universities, consult Schrecker; Hershberg, *James B. Conant*, chs. 21–23; and Diamond, *Compromised Campus*.

23. *Atomic Energy Commission Fellowship Program*, May 17, 1949, 61.

24. Bernstein, "Oppenheimer Loyalty-Security Case," 1403, notes that Oppenheimer's statement on the AEC fellowship program constituted exceptional behavior. Bernstein contends that Oppenheimer was not usually overly concerned about civil liberties during the late 1940s. Of course, Oppenheimer did express dissenting political opinions at times throughout his postwar career, most noticeably in the debate over the hydrogen bomb.

25. *Atomic Energy Commission Fellowship Program*, May 17, 1949, 88–90.

26. Ibid., 93.

27. Ibid., 94, 95.

28. Ibid., 97.

29. Ibid., 105.

30. Lilienthal, *Journals*, 531.

31. *Independent Offices Appropriation Bill*, May 19, 1949, 253.

32. Ibid., 256.

33. Ibid., 260–75. The quotation is from p. 273.

34. On the history of labor policy and the evolution of labor unions as quasi-public, rather than purely private, institutions, see Tomlins, *State and the Unions*.

35. *Independent Offices Appropriation Bill*, 277.

36. Ibid., May 24, 1949, 577.

37. Senate, 81st Cong, 1st sess., *Congressional Record* (May 18, 1949), 95, pt. 5:6418–19.

38. Lilienthal, *Journals*, 531, contains the May 20 diary entry. For the AEC's change in policy, see RG 326, Records of the AEC, Minutes of the Meetings, Meeting

No. 272, May 20, 1949, and Walter H. Waggoner, "Atom Board Orders Loyalty Oaths to Bar Reds from Study Awards," *New York Times*, May 22, 1949, 1. For Lilienthal's May 22 diary entry, see Lilienthal, *Journals*, 532–33.

39. Thomas and Northrup, *Atomic Energy and Congress*, ch. 3, esp. 79–85; Hewlett and Duncan, *Atomic Shield*, 358–61; and Lilienthal, *Journals*, 533–94. For the hearings, see Joint Committee on Atomic Energy, *Investigation into the United States Atomic Energy Project*, 81st Cong., 1st sess., pts. 1–23, 1949. Although the JCAE investigation was the immediate cause of Lilienthal's resignation, Thomas and Northrup note that the dispute over the fellowship program "more than any other single issue, brought on the Congressional loss of confidence in Mr. Lilienthal." Thomas and Northrup, *Atomic Energy and Congress*, 65.

40. Lilienthal, *Journals*, 580–85, 588–90, 594–95, 620–34.

41. Quoted in Herken, *Winning Weapon*, 320.

42. See, for example, Carl D. Anderson to C. J. Lapp, June 13, 1949, NAS: Fellowships 1949: AEC-NRC Fellowship Boards: Security Clearance: NAS: Council Statements: First: General, NAS Archives.

43. Marshall H. Stone to Detlev W. Bronk, May 31, 1949, NAS: Fellowships 1949: AEC-NRC Fellowship Boards: Security Clearance: NAS: Council Statements: First: General, NAS Archives. For an example of Stone's ability to think in constitutional terms, see Marshall H. Stone to A. N. Richards, February 18, 1949, Administration: Organization: Committee on Civil Liberties: 1949, NAS Archives. Harlan Fiske Stone generally advocated judicial self-restraint, but he also developed the preferred freedoms doctrine, the principle that certain rights, such as free speech, deserve special protection. *U.S. v. Carolene Products Company*, 304 U.S. 144, 152–53 n. 4 (1938). Marshall Stone similarly placed civil liberties in a preferred position.

44. Roger Adams to C. J. Lapp, June 14, 1949, Fellowships 1949: AEC-NRC Fellowship Boards: General, NAS Archives.

45. Memorandum, Executive Secretariat and R. R. Bush to FAS chapters and members at large, May 28, 1949, box 22, folder "Federation of American Scientists 1948–49," Veblen Papers, Library of Congress.

46. "Supplement to FAS Newsletter," May 21, 1949, Institutions: Associations: Individuals 1949: Federation of American Scientists, NAS Archives.

47. William A. Higinbotham and the FAS Executive Secretariat to FAS chapters, May 24, 1949, box 11E, [folder 5], Shapley Papers, Harvard University. Harold C. Urey, incidentally, took a dim view of the federation's efforts. In response to a wire from Alan H. Shapley, he replied tersely, "Believe FAS can do nothing. Prefer to work independently." Harold C. Urey to Alan H. Shapley, May 25, 1949, box 33, folder 4, Urey Papers, University of California at San Diego.

48. William A. Higinbotham and the FAS Executive Secretariat to FAS chapters, May 24, 1949. Hawkins, Peters, Rice, and Morrison were all associated with the progressive left. Hawkins, like Morrison, was a former member of the Communist Party. In December 1950, he testified before HUAC that he had been a party member from 1938 to 1943, but he cited the First and Fourth Amendments in his refusal to testify about his colleagues. He was a tenured professor of the philosophy of science at the University of Colorado at the time of his testimony. The president and regents of the university appointed a committee to reevaluate Hawkins's status, but the Committee on Faculty Privilege and Tenure found no grounds to revoke his

tenure, and the regents voted four to one to retain him. Caute, *Great Fear*, 467–68; Goodman, *The Committee*, 277; and Schrecker, *No Ivory Tower*, 35, 39, 60, 249–50.

49. See "Atom Award Checks by FBI Are Opposed," *New York Times*, May 30, 1949, 2; Letter "To Members of PAS," May 30, 1949, and Princeton Association of Scientists to members of the JCAE, June 19, 1949, box 29, folder "Princeton Association of Scientists," Spitzer Papers, Princeton University; "Lilienthal Backed by Science Group," *New York Times*, May 31, 1949, 14; and "Lilienthal Is Defended," *New York Times*, June 4, 1949, 3.

50. Washington Association of Scientists to members of the Senate, July 13, 1949, with a list of senators sent the letter, box 2, folder 7, Washington Association of Scientists Papers, University of Chicago; Robert Rochlin to Detlev W. Bronk, July 17, 1949, and Detlev W. Bronk to Joseph C. O'Mahoney, July 20, 1949, Fellowships 1949: AEC-NRC Fellowship Boards: Security American Scientists, NAS Archives; "Plan to Investigate Atom Fellows Hit," *New York Times*, July 20, 1949, 4; David E. Lilienthal to Robert D. Stiehler, July 19, 1949, box 2, folder 7, Washington Association of Scientists Papers.

51. Robert C. Albright, "Senate Votes Curb on AEC Fellowships," *Washington Post*, August 3, 1949, 1, 2.

52. "Points Made in Conversation with J.R.O.," undated, probably the week of May 16, 1949, box 28, folder "Scientists' Committee on Loyalty," Spitzer Papers.

53. Ibid.

54. On the evolution of the JCAE's role in the atomic energy program, see Green and Rosenthal, *Government of the Atom*, ch. 1. Green and Rosenthal observe that while the JCAE initially adopted a "relatively passive role, functioning primarily in a 'watchdog,' or overseer, capacity," by late 1949 it "assumed vigorous leadership in the national atomic-energy program" (5).

55. Henry A. Barton, "Draft: Proposed Statement of Position," August 26, 1949, NAS: Fellowships 1949: AEC-NRC Fellowship Boards: Security Clearance: NAS: Council Statements: First: General, NAS Archives.

56. Carl Anderson to Detlev W. Bronk, August 11, 1949, Bronk to Anderson, September 6, 1949, and Anderson to Bronk, September 21, 1949, Fellowships 1949: AEC-NRC Fellowship Boards: Security Clearance: NAS: Council Statements: First: General, NAS Archives.

57. A. A. Albert to Detlev Bronk, September 30, 1949, Fellowships 1949: AEC-NRC Fellowship honor. In 1987, Brooklyn College faculty members passed a resolution NAS Archives.

58. R. R. Bush to Administrative Committee, FAS, July 28, 1949, Fellowships 1949: AEC-NRC Fellowship Boards: Security and that these could be solved by knowledgeable experts rather than by maNAS: Fellowships 1949: AEC-NRC Fellowship Boards: Security Clearance: NAS: Council Statements: First: General; and R. D. Stiehler to Detlev Bronk, October 7, 1949, Fellowships 1949: AEC-NRC Fellowship Boards: Security Clearance: General, all in NAS Archives. For the FAS's position, see also "O'Mahoney Rider Passes Senate," *FAS Newsletter*, August 9, 1949, 1, 2.

59. "Meeting of the AEC Fellowship Boards," October 4, 1949, Fellowships 1949: AEC-NRC Fellowship Boards: General; and Charles C. Price to M. H. Trytten, October 1, 1949, Fellowships 1949: AEC-NRC Fellowship Boards: Security Clearance: NAS: Council Statements: First: General, NAS Archives.

60. "Excerpts from notes taken at joint meeting of AEC Fellowships Boards," October 4, 1949, NAS: Fellowships 1949: AEC-NRC Fellowship Boards: Security Clearance: NAS: Council Statements: First: General, NAS Archives.

61. A. N. Richards to Members of the NAS Council, October 12, 1949, Fellowships 1949: AEC-NRC Fellowship Boards: Security Clearance: NAS: Council Statements: First: General, NAS Archives.

62. "Proposed Draft of an Answer to the AEC Request, to be Considered by the Academy, Oct. 26" [undated, probably October 4, 1949], Fellowships 1949: AEC-NRC Fellowship Boards: Security Clearance: NAS: Council Statements: First: General, NAS Archives.

63. John Lord O'Brian to J. Robert Oppenheimer, October 20, 1949, and Charles H. Bradley to A. N. Richards, October 19, 1949, Fellowships 1949: AEC-NRC Fellowship Boards: Security Clearance: NAS: Council Statements: Second, NAS Archives.

64. Walter S. Adams to A. N. Richards, October 20, 1949, and Harold D. Babcock to A. N. Richards, October 19, 1949, Fellowships 1949: AEC-NRC Fellowship Boards: Security Clearance: NAS: Council Statements: First: Comments by NAS Members, NAS Archives.

65. The files of the National Academy of Sciences show responses from fifty-four scientists. Eleven felt the NRC should withdraw completely from the program, thirteen recommended against administration of the program, twenty-six supported administration under protest, and four favored doing nothing. If a June 10 telegram from University of Chicago scientists recommending against administration of the program is included in the sample, the number opposing administration by the NRC rises to twenty-six. For the letters, consult the following files: Fellowships 1949: AEC-NRC Fellowship Boards: Security Clearance: NAS Council Statements: First Comments by NAS Members; and Fellowships 1949: AEC-NRC Fellowships Boards: Security Clearance: NAS Council Statements: First: General, NAS Archives. The June 10 telegram is in Fellowships 1949: AEC-NRC Fellowship Boards: Security Clearance: Proposed, NAS Archives.

66. Donald M. Yost to A. N. Richards, October 19, 1949; Frank B. Jewett to A. N. Richards, October 21, 1949; P. W. Bridgman to A. N. Richards, October 25, 1949; Albert Einstein and Oswald Veblen to A. N. Richards, October 19, 1949; John P. Peters to A. N. Richards, October 21, 1949; G. B. Kistiakowsky, Edwin C. Kemble, Paul D. Bartlett, and Julian Schwinger to A. N. Richards, October 21, 1949; all in Fellowships 1949: AEC-NRC Fellowship Boards: Security Clearance: NAS: Council Statements: First: Comments by NAS Members, NAS Archives. Peters's position was somewhat ambiguous. He advocated that the academy continue to participate in the selection of fellows, but the tone of his letter suggests that he, like Jewett, did not favor full administrative responsibility for the program on the part of the NAS.

67. Lee A. DuBridge to A. N. Richards, October 21, 1949, and Carl D. Anderson et al. to A. N. Richards, October 27, 1949, Fellowships 1949: AEC-NRC Fellowship Boards: Security Clearance: NAS: Council Statements: First: Comments by NAS Members, NAS Archives.

68. Minutes of the Business Session, October 24, 1949, and October 26, 1949, National Academy of Sciences: 1949 Autumn Meeting: Minutes of the Business Sessions, NAS Archives. For a transcript of the October 24 session, see "National

Academy of Sciences: Business Meeting, October 24, 1949," Organization 1949: NAS: Meetings: Autumn: Business Sessions: Transcript, NAS Archives.

69. A. N. Richards to J. Robert Oppenheimer, January 9, 1950, box 134, folder "NAS—AEC Fellowship Program," Oppenheimer Papers, Library of Congress.

70. A. N. Richards to Carroll L. Wilson, November 2, 1949, Fellowships 1949: AEC-NRC Fellowship Boards: Security Clearance: Resultant Modification of Program, NAS Archives; Carroll L. Wilson to A. N. Richards, November 17, 1949, RG 128, Records of the JCAE, General Correspondence, box 28, folder "AEC Fellowship General"; "Notes of Meeting of NAS-AEC, Nov. 18, 1949," and A. N. Richards to Members of the Council of the Academy, November 19, 1949, Fellowships 1949: AEC-NRC Fellowship Boards: Security Clearance: Resultant Modification of Program, NAS Archives; A. N. Richards to Carroll L. Wilson, November 30, 1949, Carroll L. Wilson to A. N. Richards, December 9, 1949, and "AEC Announces Provisions for Operation of Fellowship Program during 1950–1951 Academic Year," December 16, 1949, RG 128, Records of the JCAE, General Correspondence, box 28, folder "AEC Fellowship General."

71. Press release, "AEC to Sponsor New Postdoctoral Fellowship Program," February 12, 1950, box 173, folder "Fellowship Program/AEC/Press Releases," Oppenheimer Papers; Federation of American Scientists, "Resolution addressed to the Presidents of Universities participating in the National Laboratory Program of the Atomic Energy Commission," February 7, 1950, Fellowships 1950: AEC-NRC Fellowship Boards: Predoctoral, NAS Archives; and W. A. Higinbotham to Trustees, Associated Universities, Inc.; Board of Governors, Argonne National Laboratory; Council, Oak Ridge Institute of Nuclear Studies, Presidents of universities participating in the AEC National Laboratory Program, and chairmen of university departments in physics, chemistry, and biology, September 5, 1950, Institutions Associations Individuals 1950: Federation of American Scientists, NAS Archives.

72. Schweber, "Mutual Embrace of Science and the Military," 33. On military research in the universities, see Leslie, *Cold War and American Science.*

73. On the impact of military funding on university research programs, consult Leslie, *Cold War and American Science.*

74. For accounts of the incident, see Hewlett and Duncan, *Atomic Shield*, 8, and Lilienthal, *Journals*, 139.

CHAPTER EIGHT

1. The Cold War political consensus could be described as hegemonic. On the meaning of cultural hegemony, see Lears, "The Concept of Cultural Hegemony." Yergin's description of anticommunism and national security as the "commanding ideas" behind the national security state seems implicitly to invoke a notion of cultural hegemony. Yergin, *Shattered Peace*, esp. prologue, and chs. 8–9. Carmichael's interpretation of the Cold War as the rise of a "totalizing myth" expressed through a national narrative also seems to lie within the limits of cultural hegemony. Carmichael, *Framing History*, esp. ch. 1.

2. As legal scholar Ralph S. Brown observed, "the shifts in wording . . . represented a hardening of the standard: from reasonable grounds for belief of disloyalty, to reasonable doubts as to loyalty; from discharge if advisable in the interest of national security, to retention only if clearly consistent with national security." Brown, *Loyalty and Security*, 31.

3. Howard A. Meyerhoff to Douglas E. Scates, July 19, 1949, box 2, folder "National Science Foundation (II) 1943–1950," Borras Files, American Association for the Advancement of Science, 1st Accession.

4. House Subcommittee of the Committee on Interstate and Foreign Commerce, *National Science Foundation*, 81st Cong., 1st sess., April 5, 1949, 146–54. The quotations are from p. 151.

5. Ibid., April 26, 1949, 170, 171.

6. Ibid., 174–77. The quotation is from p. 174. See also the discussion in England, *Patron for Pure Science*, 96.

7. House Subcommittee of the Committee on Interstate and Foreign Commerce, *National Science Foundation*, 81st Cong., 1st sess., April 5, 1949, 177, 178.

8. "National Science Foundation," *FAS Newsletter*, May 21, 1949, 7, and Dael Wolfle to Members of the Inter-Society Committee for a National Science Foundation, April 15, 1949, box 2, folder "National Science Foundation (II) 1943–1950," Borras Files, 1st Accession.

9. See, for example, Detlev Bronk's comments in *Atomic Energy Commission Fellowship Program*, May 17, 1949, 63–64, and telegram to FAS chapters at Chicago, Berkeley, Madison, Cornell, Oak Ridge, Los Alamos, Schenectady, Rochester, in William A. Higinbotham to FAS chapters, May 24, 1949, box 11E, [folder 5], Shapley Papers, Harvard University.

10. Dael Wolfle to Members of the Inter-Society Committee for a National Science Foundation, June 29, 1949, box 2, folder "National Science Foundation (II) 1943–1950," Borras Files, 1st Accession, and Howard A. Meyerhoff to Douglas E. Scates, July 19, 1949. See also Dael Wolfle to Members of the Executive Committee of the Inter-Society Committee for a National Science Foundation, January 23, 1950, box 2, folder "National Science Foundation (I) 1950–1981," Borras Files, 1st Accession.

11. "Special FAS Newsletter," June 17, 1949, box 2, folder "National Science Foundation (II) 1943–1950," Borras Files, 1st Accession.

12. For the results of the questionnaire, see Scientists' Committee on Loyalty Problems, "Questionnaire," box 28, folder "Scientists' Committee on Loyalty," Spitzer Papers, Princeton University. For the SCLP's letters to members of Congress, see, for example, Lyman Spitzer Jr. to Representative Robert Crosser, February 2, 1950; Lyman Spitzer Jr. to Representive J. Percy Priest, February 2, 1950; and Lyman Spitzer Jr. to Senator H. Alexander Smith, February 2, 1950, in box 28, folder "Scientists' Committee on Loyalty," Spitzer Papers. Crosser (D.-Ohio) was chairman of the House Interstate and Foreign Commerce Committee, and Priest (D.-Tenn.) chaired the subcommittee of the Interstate and Foreign Commerce Committee responsible for science legislation. In 1946, Smith (R.-N.J.) was the only Republican member of the Senate Military Affairs Committee not to sign the minority report on the Kilgore-Magnuson compromise bill, and in the following years, he remained a key supporter of the NSF legislation.

13. House, Representative Howard W. Smith (D.-Va.) speaking, 81st Cong., 2d sess., *Congressional Record* (February 28, 1950), 96, pt. 2:2529.

14. "National Science Foundation — How Far Can We Compromise?" *FAS Newsletter*, March 1, 1950, 2.

15. See, for example, Lyman Spitzer Jr. to Representative J. Percy Priest, March 2,

1950, box 28, folder "Scientists' Committee on Loyalty," Spitzer Papers. The SCLP sent letters to the conference committee members, as well as other members of Congress with influence over the NSF legislation.

16. Clifford Grobstein, "Time Bomb in the Science Foundation," *Washington Post*, March 11, 1950, copy in House, 81st Cong., 2d sess., *Congressional Record* (March 13, 1950), 96, pt. 14:A1863.

17. "Statement of the Council of the National Academy of Sciences Relating to Certain Provisions of the National Science Foundation Bill H.R. 4846," March 7, 1950, RG 128, Records of the JCAE, General Correspondence, box 29, folder "AEC Fellowships National Science Foundation," National Archives.

18. A. N. Richards to Harry S. Truman, March 9, 1950, NAS: Congress 1950: General, NAS Archives.

19. Peyton Ford to Robert Crosser, March 6, 1950, in House, 81st Cong., 2nd sess., *Congressional Record* (April 27, 1950), 96, pt. 5:5904.

20. England, *Patron for Pure Science*, 118–19, 152–60.

21. "National Science Foundation—Realization!" *FAS Newsletter*, May 3, 1950, 1, 4.

22. England, *Patron for Pure Science*, chs. 8, 10–12.

23. Daniel J. Kevles, "K_1S_2: Korea, Science, and the State," in Galison and Hevly, *Big Science*, 329.

24. Forman, "Behind Quantum Electronics," 194.

25. Melba Phillips, "Dangers Confronting American Science," *Science*, October 24, 1952, 442.

26. On the Rosenbergs and Julius Rosenberg's espionage ring, consult Radosh and Milton, *Rosenberg File*, esp. chs. 3, 5, 6. On Barr's and Sarant's careers in the Soviet Union, see also Lamphere, *FBI-KGB War*, 309–11. On Barr's and Sarant's involvement in the Rosenberg spy ring, see New York to Moscow, No. 1600, November 14, 1944, Venona Project releases, 1st installment (July 1995), and Moscow to New York, Nos. 1749, 1750, December 13, 1944, Venona Project releases, 3d installment (March 1996), vol. 3, National Security Archives.

27. Press release, SCLS, "Summary of 'The Fort Monmouth Security Investigations, August 1953–April 1954,'" and SCLS, "The Fort Monmouth Security Investigations, August 1953–April 1954," April 25, 1954, esp. sec. 4, pp. 15–32, both in box 242, folder "UA [Un-American] Inves. Correspondence and Articles, 1952–1959, F–I," Records of the Office of the President, 1930–1959 (Compton-Killian), Massachusetts Institute of Technology. The quotation about the FAS is from sec. 4, p. 24, of "The Fort Monmouth Security Investigations." On the employment of Jews at Fort Monmouth, see Oshinsky, *Conspiracy So Immense*, 331.

28. The Venona Project intercepts released thus far show no signs of Soviet penetration of Fort Monmouth. Not even Barr and Sarant committed espionage at Fort Monmouth—neither was recruited for spying until after they had left the Signal Corps Engineering Laboratories. On Sarant's recruitment, see New York to Moscow, No. 628, May 5, 1944, Venona Project releases, 1st installment (July 1995). Sarant did provide the Soviet Union with information about an airborne radar system, but it came from Bell Laboratories, not Fort Monmouth. New York to Moscow, Nos. 1749, 1750, Venona Project releases, 3d installment (March 1996). Granted, the Venona Project intercepts constitute only a small fraction of Soviet cables sent between the

United States and Moscow during the war and immediate postwar years. The full story of Soviet acquisition of American high technology through espionage and intelligence will not be known until scholars have had the opportunity to comb through Soviet intelligence records.

29. On wartime atomic espionage, see Rhodes, *Dark Sun*. The Venona Project intercepts also identify two more individuals who were personnel at Los Alamos as sources of information about the Manhattan Project. Both had probably been members of the Young Communist League. See New York to Moscow, No. 1585, November 12, 1944, Venona Project releases, 1st installment (July 1995), and Moscow to New York, No. 298, March 31, 1945; New York to Moscow, No. 799, May 26, 1945, and Moscow to New York, No. 709, July 5, 1945, all in Venona Project releases, 3d installment (March 1996), vol. 3.

30. Klehr, Haynes, and Firsov, *Secret World of American Communism*, 323. The Venona Project intercepts identify approximately 200 Americans who participated in Soviet-sponsored espionage. Most, but not all, had connections to the Communist Party. Isserman, *Which Side Were You On?*, notes that Communist Party membership in 1944 stood at close to 80,000. Turnover in party membership was rapid; it is not unreasonable to suppose that perhaps 100,000–150,000 persons passed through the party between 1942 and 1946, the years covered by the Venona Project. If the Venona intercepts provide a relatively complete picture of espionage by Americans, Communist Party membership indicated participation in espionage at a rate of, at most, 1 in 500 persons.

31. Scientists' Committee on Loyalty and Security, "Some Individual Cases," *BAS* 11 (April 1955): 151–55, 158; Clifford Grobstein to John Phelps, February 15, 1954, box 40, folder 6, FAS Papers, University of Chicago; and Robert L. Kenngott to Alan H. Shapley, April 14, 1951, box 23, folder 6, FAS Papers.

32. John W. French to Dael Wolfle, May 6, 1954, box 4, folder "Security (AAAS Membership), 1954," Borras Files, 2d Accession. General correspondence relating to the use of AAAS membership to deny security clearances is available throughout this folder.

33. John Phelps to Dorothy Higinbotham, August 16, 1954, box 40, folder 6, FAS Papers.

34. Leonard Hersher to Robert D. Bur, Manager, Veterans Administration Regional Office and Ray R. Adams, Chairman, VA Loyalty Board, January 10, 1951, and Edwin Rothschild to Edward Meyerding, ACLU, April 8, 1952, both in box 46, folder 7, American Civil Liberties Union, Illinois Division Papers, University of Chicago.

35. Ernest Pollard to Samuel A. Goudsmit, January 7, 1954, box 40, ser. 9, folder 4, Goudsmit Papers, American Institute of Physics.

36. Stern, *Oppenheimer Case*, 491–92.

37. John P. Blewett, interview by the author.

38. W. E. Kelley, Manager, New York Operations Office, AEC, to John P. Blewett, January 2, 1952, personal possession of John P. Blewett.

39. "Condon Abandons Clearance Fight," *New York Times*, December 14, 1954, 20.

40. Other scholars have explored the Oppenheimer case in detail, so I will not do so here. For extensive accounts, see Bernstein, "Oppenheimer Loyalty-Security Case"; Bernstein, " 'J. Robert Oppenheimer' "; Rhodes, *Dark Sun*, ch. 26; and Stern, *Oppenheimer Case*.

41. Bernstein, "Oppenheimer Loyalty-Security Case," 1385–88. The quotation is from p. 1388.

42. Stern, *Oppenheimer Case,* 15–36, 44–46, 53–70; Bernstein, "Oppenheimer Loyalty-Security Case," 1393–94, 1396–1401, 1404; and Rhodes, *Dark Sun,* 544–49.

43. Rhodes, *Dark Sun,* ch. 26, esp. 531, sets out this argument most clearly.

44. For scientists' reactions, see, for example, "Statement by FAS Executive Committee," April 17, 1954, NAS: Institutions Associations Individuals 1954: Federation of American Scientists, NAS Archives; John T. Edsall to the Editor of the *New York Times,* June 8, 1954, box 5, folder "Edsall, John T.—Personal 1952–53," Edsall Papers, Harvard University, 1st Accession; FAS, Information Bulletin No. 51 [Statements of Los Alamos Scientific Laboratory personnel, professors at the University of Washington, American Physical Society Council, and University of Illinois FAS chapter], June 14, 1954, box 70, folder 9, FAS Papers; "U.S. Atomic Energy Commission Holds That Moral Scruples Are 'Derogatory Information,'" *SSRS [Society for Social Responsibility in Science] Newsletter* 5 (June 1954): 1–2, box 4, folder "Security Regulations—Oppenheimer," Borras Files, 1st accesesion; and press release, FAS, "Scientists' Group Criticizes Oppenheimer Decision," July 5, 1954, box 69, folder 7, FAS Papers.

45. Schatz, *Electrical Workers,* 177–79; Rosswurm, *CIO's Left-Led Unions,* 1–2.

46. Schrecker, *No Ivory Tower,* esp. pp. 127–30; Lewis, *Cold War on Campus,* esp. 49; and Holmes, *Stalking the Academic Communist.*

47. Schrecker, *No Ivory Tower,* esp. chs. 9, 10; Diamond, *Compromised Campus*; and Hershberg, *James B. Conant,* chs. 21–23, 31.

48. Schrecker, *No Ivory Tower,* 10, 44.

49. More than fifty academic scientists can be identified by name in cases discussed in Schrecker, *No Ivory Tower,* and Lewis, *Cold War on Campus,* plus other cases of which I am aware. Well over half lost their jobs or were suspended for a significant period of time. This compilation is by no means complete.

50. On Melba Phillips, see "5 Teachers Called by Senate Inquiry," *New York Times,* October 9, 1952, 46; "Ex-Red Describes City Teacher Blocs," *New York Times,* October 14, 1952, 1, 19; "3 Teachers Ousted in 2 City Colleges," *New York Times,* October 29, 1952, 10; "Holman, 2 Others Suspended in College Basketball Inquiry," *New York Times,* November 18, 1952, 1, 38; "Court Upholds Ousters," *New York Times,* April 23, 1954, 17; "13 Ex-Teachers Here Lose Red Case Plea in High Court," *New York Times,* February 8, 1955, 1, 14; and "Ex-Teachers Lost in Supreme Court," *New York Times,* December 18, 1956, 26. On Philip Morrison, consult Schrecker, *No Ivory Tower,* 149–60, and Lewis, *Cold War on Campus,* 211, 219–20. Regarding Urey, see George Eckel, "Red Inquiry Hears Attack on Dr. Urey," *New York Times,* April 24, 1949, 30.

51. "Passport Refusals for Political Reasons: Constitutional Issues and Judicial Review," *Yale Law Journal* 61 (February 1952): 171–203, esp. 171–74; Leonard B. Boudin, "The Constitutional Right to Travel," *Columbia Law Review* 56 (January 1956): 47–75; and Kutler, *American Inquisition,* ch. 4.

52. "The McCarran Act," *BAS* 8 (October 1952): 257.

53. Quoted in Boudin, "Constitutional Right to Travel," 62.

54. On the impact of the Pauling case, see Geoffrey Chew, chairman, FAS Passport Committee, "Memorandum on new passport appeal procedure and the Kamen

case," [undated draft, August 1954], box 5, folder 7, Kamen Papers, University of California at San Diego. For Pauling's political activities, consult Hagen, *Force of Nature*, chs. 13, 15.

55. Linus Pauling, "My Efforts to Obtain a Passport," *BAS* 8 (October 1952): 253.

56. Some scholars speculate that had he been able to make the trip, consult with colleagues, and review important data, Pauling might have discovered the structure of DNA before James D. Watson and Francis Crick. See, for example, Horace Freeland Judson, *The Eighth Day of Creation: The Makers of the Revolution in Biology* (New York: Simon and Schuster, 1979), 132.

57. Linus Pauling to Farrington Daniels, June 19, 1952, box 18, folder "Pauling, Linus, 1952," Daniels Papers, University of Wisconsin.

58. Hagen, *Force of Nature*, ch. 16. For an example of Pauling's newly temperate political style, see Pauling, "My Efforts to Obtain a Passport."

59. Hagen, *Force of Nature*, chs. 17–22.

60. Kutler, *American Inquisition*, 95; Schrecker, *No Ivory Tower*, 296.

61. Harlow Shapley to J. B. Koepfli, May 5, 1953, box 11C, folder "United States Gov.," Shapley Papers.

62. Ruth B. Shipley to Oswald Veblen, April 22, 1953, box 22, folder "Communist Charges against Veblen; J. Volpe corres. 1953–54," Veblen Papers, Library of Congress. For details on Veblen's passport case, consult the relevant materials in boxes 21 and 22, Veblen Papers.

63. Ruth B. Shipley to Harold C. Urey, August 12, 1952; Harold C. Urey to Ruth B. Shipley, August 18, 1952; handwritten note by Urey, undated, notes "Passport received Aug. 20, 1952"; and Harold C. Urey to Senator Alexander Wiley, March 31, 1955, all in box 94, folder 13, Urey Papers, University of California at San Diego. The quotation is from Urey to Shipley, August 18, 1952.

64. Schrecker, *No Ivory Tower*, 151.

65. Russell B. Olwell, "Princeton, David Bohm, and the Cold War: A Case Study in McCarthyism and Physics," unpublished ms., personal possession of the author, 39–43, 46, 50. For instances of scientists and other academics who emigrated (some voluntarily, others after they became stranded because of passport difficulties), see Schrecker, *No Ivory Tower*, 291–97.

66. Norman Ramsey to Robert F. Bacher, February 11, 1949, RG 326, Records of the AEC, Records of the Commissioners, Office Files of Robert F. Bacher, Subject File, 1947–49, box 4, folder "Security," National Archives; Caute, *Great Fear*, 251–60.

67. Victor F. Weisskopf, "Report on the Visa Situation," 221–22; "Some British Experiences," 223–32, esp. 229–30; "Some French Experiences," 236–46, esp. 236; "Some European Experiences," 247–49; and "The Treatment of Good Neighbors," 250–52, all in *BAS* 8 (October 1952).

68. Sidney Painter, H. A. Meyerhoff, and Alan T. Waterman, "The Visa Problem," *Scientific Monthly* 76 (January 1953): 14, and John T. Edsall, writing for the Committee on International Relations, American Academy of Arts and Sciences, to Representative Francis E. Walter, July 15, 1955, box 1, folder "AAAS [American Academy of Arts and Scientists] Comm. on Int'l. Relations," Edsall Papers, 1st Accession.

69. The story of Tsien's deportation is told in Iris Chang, *Thread of the Silkworm* (New York: Basic Books, 1995), chs. 16–21.

70. Scientists' Committee on Loyalty and Security, "Loyalty and U.S. Public Health Service Grants," *BAS* 11 (May 1955): 196–97.

71. [Statement by Harry Grundfest (2)], undated, probably April 1954, box 12, folder "Civil Liberties," Visscher Papers, University of Minnesota; Bernard D. Davis to John T. Edsall, January 18, 1955, box 7, folder "Hobby, Oveta Culp," Edsall Papers, 2d accession; and Warren M. Sperry to John T. Edsall, May 2, 1955, box 6, folder " 'Government and Freedom of Science,' " Edsall Papers, 2d accession.

72. Victor F. Weisskopf to Samuel A. Goudsmit, July 2, 1953, and July 22, 1953, box 24, ser. 3, folder 257, Goudsmit Papers. See also Theodore Lidz to Detlev Bronk, February 22, 1955, Organization 1955: NAS: Committee on Loyalty in Relation to Government Support of Unclassified Research, General, NAS Archives.

73. Peter Kihss, "Columbia Professor Gets McCarthy Contempt Threat," *New York Times*, November 26, 1953, 1, 37; H. Houston Merritt to Harry Grundfest, December 5, 1953, and December 11, 1953, box 12, folder "Civil Liberties," Visscher Papers; and [Statement by Harry Grundfest (1)], undated, probably April 1954, box 12, folder "Civil Liberties," Visscher Papers. For biographical background on Grundfest, see John P. Reuben, "Harry Grundfest," in *Biographical Memoirs of the National Academy of Sciences* (Washington, D.C.: National Academy of Sciences, 1995), 66:151–63. Grundfest became a member of the National Academy of Sciences in 1976.

74. William J. Hoff, NSF General Counsel, to Alan T. Waterman, May 18, 1954, Agencies and Departments 1954: National Science Foundation: Board, General, NAS Archives.

75. On the NSF's grant policy as compared with the Public Health Service, see England, *Patron for Pure Science*, 329–37. On Martin D. Kamen's turn to the NSF for funding after he was denied a grant renewal from the National Institutes of Health in 1954, see Kamen, *Radiant Science, Dark Politics*, 269, 274–75.

76. Kamen, *Radiant Science, Dark Politics*, 164; Schrecker, *No Ivory Tower*, 134.

77. Martin Kamen to A. N. Richards, November 24, 1947, box 5, folder 1, Kamen Papers.

78. Martin Kamen to Jesse P. Greenstein, August 14, 1947, and Greenstein to Kamen, October 24, 1947, box 2, folder 5, Kamen Papers.

79. For details regarding the Kamen case, consult Kamen, *Radiant Science, Dark Politics*, chs. 11–14, and Schrecker, *No Ivory Tower*, 144–45.

80. Maurice B. Visscher, "Scientists in a Mad World," *Nation*, January 24, 1953, 69–71.

81. John T. Edsall, "Government and the Freedom of Science," *Science*, April 29, 1955, 615–19. For reactions to Edsall's declaration, see box 6, folder " 'Government and Freedom of Science,' " Edsall Papers, 2d Accession.

82. Maurice B. Visscher to Howard A. Meyerhoff, January 28, 1953, box 2, folder "Civil Liberties—3/73; 1952–," Borras Files, 2d Accession; Jules Halpern, Chairman, FAS, to Howard A. Meyerhoff, February 11, 1953; William A. Higinbotham to Robert Marshak and Robert R. Wilson, March 8, 1953; and Alan H. Shapley to Geoffrey Chew, June 2, 1953, all in box 46, folder 3, FAS Papers.

83. On the American Psychological Association Committee on Freedom of Enquiry, consult the materials in box 469, Records of the American Psychological Association, Library of Congress.

84. "Proposal for a Study of Security in Science," November 19, 1954, box 40, ser. 9, folder 4, Goudsmit Papers; John B. Phelps to Dorothy Higinbotham, Novem-

ber 26, 1955, box 40, folder 6, FAS Papers; and Minutes of Meeting, Scientists' Committee on Security, Inc., May 28, 1956, box 41, ser. 9, folder 6, Goudsmit Papers. I have not come across any evidence of the existence of the Scientists' Committee on Security, Inc., after May 1956.

85. On the AEC's decision, see Sterling Cole to Robert S. Rochlin, June 22, 1954, box 655, folder "Scientists—General," and George Norris to Corbin Allardice, June 16, 1954, box 496, folder "Oppenheimer, Dr. J. Robert; Misc. Articles, Memos, Press Releases, Etc.," both in RG 128, Records of the JCAE, General Correspondence; and Anthony Lewis, "AEC Eases 'Risk' Rules; Will Question Informants," *New York Times*, May 10, 1956, 1, 14. See also the materials in box 72, folder "AEC Security 1955," and box 73, folder "AEC Security," RG 128, Records of the JCAE, General Correspondence. For the Eisenhower administration's directive on unclassified research, consult "Security Guards Eased on Science," *New York Times*, August 15, 1956, 8.

86. Broadwater, *Eisenhower and the Anti-Communist Crusade*, 190, 195, 197, 205–7; Heale, *American Anticommunism*, 195–96.

87. "Security Guards Eased on Science." Details on the NAS Committee on Loyalty in Relation to Government Support of Unclassified Research are available in the files of the NAS Archives under the headings Organization 1955: NAS: CLRGSUR and Organization 1956: NAS: CLRGSUR.

88. Broadwater, *Eisenhower and the Anti-Communist Crusade*, 195, and *Peters v. Hobby*, 349 U.S. 331 (1955).

89. Kamen, *Radiant Science, Dark Politics*, ch. 14, esp. 275–77, 290–95; Geoffrey Chew to "Fellow Scientist," March 22, 1955, box 69, folder 7, FAS Papers; and Boudin, "Constitutional Right to Travel," 56–58.

90. Kutler, *American Inquisition*, 112–13, and press release, "FAS Passport Committee Gratified over Supreme Court Passport Decision," June 17, 1958, box 69, folder 7, FAS Papers. See also *Kent v. Dulles*, 357 U.S. 116 (1958).

91. Donner, *Age of Surveillance*, 5.

92. Bryce Nelson, "Scientists Increasingly Protest HEW Investigation of Advisers," *Science*, June 27, 1969, 1499–1504, and Bryce Nelson, "HEW Security Checks Said to Bar Qualified Applicants to PHS," *Science*, July 18, 1969, 269–71.

93. Schrecker, *No Ivory Tower*, 338. Melba Phillips was among the scientists and other academics who returned to the campuses that had expelled them as guests of honor. In 1987, Brooklyn College faculty members passed a resolution acknowledging the wrong committed against Phillips and the other faculty members dismissed in the 1950s. Phillips and the other faculty members who had lost their jobs visited Brooklyn College that same year to be honored at a special symposium. Francis T. Bonner, "Melba Newell Phillips," in Louise S. Grinstein, Rose K. Rose, and Miriam H. Rafailovich, eds., *Women in Chemistry and Physics: A Biobibliographic Sourcebook* (Westport, Conn.: Greenwood Press, 1993), 490.

CONCLUSION

1. As historian Richard Pells observes, Cold War liberals "praised the tactics of bargaining and compromise, the virtues of pluralism and interest-group competition, the superiority of Keynesian economics and a limited welfare state. They agreed that the problems of modern America were no longer ideological but technical and administrative, and that these could be solved by knowledgeable experts rather than by mass movements." Pells, *Liberal Mind in a Conservative Age*, 130.

2. Gillon, *Politics and Vision.*

3. Sudoplatov and Sudoplatov, *Special Tasks*, ch. 7. For critiques of *Special Tasks*, see Thomas Powers, "Were the Atomic Scientists Spies?" *New York Review of Books*, June 9, 1994, 10–17; Roald Sagdeev, "The Long Hand of the KGB," *Washington Post Book World*, May 1, 1994; Priscilla Johnson McMillan, "'Atom Spies,'" *Washington Post*, May 3, 1994, A23; Hans A. Bethe, "Atomic Slurs," *Washington Post*, May 27, 1994, A25; George Kennan, "In Defense of Oppenheimer," *New York Review of Books*, June 23, 1994, 8; and the entire issue of *FAS Public Interest Report* 47 (May/June 1994): 1–16. The publishers of *Special Tasks* soon backed away from full support of the book's claims: David Streitfeld, "The Book at Ground Zero," *Washington Post*, May 27, 1994, D1, D8.

4. Allen Weinstein, *Perjury: The Hiss-Chambers Case* (New York: Alfred A. Knopf, 1978), and Radosh and Milton, *Rosenberg File*, constituted the major exceptions. Together, they argued that the tumultuous politics surrounding the Hiss and Rosenberg cases should not obscure the virtually certain guilt of Hiss and the Rosenbergs of the crimes of which they were convicted.

5. Klehr and Radosh, *Amerasia Spy Case*, and Klehr, Haynes, and Firsov, *Secret World of American Communism*. See also Haynes, *Red Scare or Red Menace?*

6. Haynes, *Red Scare or Red Menace?*, and Powers, *Not without Honor.*

7. Thomas Powers, "Were the Atomic Scientists Spies?," 17.

8. Holloway, *Stalin and the Bomb*, esp. chs. 4, 5, 7; Rhodes, *Dark Sun*, esp. pts. 1, 2.

9. Quoted in Radosh and Milton, *Rosenberg File*, 284.

10. Holloway, *Stalin and the Bomb*, 138–41, 222–23; Rhodes, *Dark Sun*, 181–82, 196–97.

11. Klehr and Radosh note that with respect to the *Amerasia* case, both the anticommunist Right and the New Deal Left were more interested in scoring political points than getting to the truth of the matter. Klehr and Radosh, *Amerasia Spy Case*, ch. 8, esp. 218.

12. For an estimate of probabilities, see chapter 8, n. 30. A comparison with more recent cases in which foreign governments recruited American spies by appealing to their ethnic heritage is useful. In September 1996, the FBI arrested Robert C. Kim, a Korean American employee of the navy, who apparently passed classified information to South Korea out of a sense of loyalty to the land of his birth. The case was reminiscent of the convictions of Jonathan Jay Pollard in 1985 and Larry Wu-Tai in 1986, who spied for Israel and China, respectively. The *Washington Post* reported in February 1996 that a Defense Department study discussed the reality of foreign nations' efforts "to exploit ethnic or religious ties" but stressed that "'ethnic targeting' does not equate to 'ethnic susceptibility' to cultivation by a foreign economic power." Substitute the word "ideological" for "ethnic," and the relevance to the Cold War becomes clear. Rather than attempt to exclude persons of a particular ethnic heritage from classified work, an effort that would violate antidiscrimination laws as well as deprive U.S. intelligence agencies of the intimate knowledge of language and culture that persons of particular ethnic backgrounds can often provide, the CIA and other agencies concentrate on cautioning employees of the tactics foreign powers use to try to recruit agents. See R. Jeffrey Smith and Peter Pae, "Navy Worker's Case Raises Issue of Ethnic Sympathy," *Washington Post*, September 26, 1996, A15.

13. For an example of the fortuitous application of FBI files, see Lamphere, *FBI-KGB War*, regarding the identification of "Raymond" as Harry Gold.

14. Transcript, National Public Radio, "The Ames Case Explored," *All Things Considered*, February 23, 1994, 2. Commentator Daniel Schorr notes that of the five Americans known to have been recruited for espionage by the Soviet Union since 1985, all became spies for money, not out of ideological commitment. Transcript, National Public Radio, "Schorr Says Spy Case Presents Headaches for Clinton," *All Things Considered*, February 23, 1994, 5. John Earl Haynes also notes the 1950s and 1960s cases of Jack Dunlap, William Martin, and Bernon Mitchell: Haynes, *Red Scare or Red Menace?*, 174.

SELECT BIBLIOGRAPHY

MANUSCRIPT COLLECTIONS

American Association for the Advancement of Science, Washington, D.C.
Board/Council Files
Catherine Borras Files
American Institute of Physics, College Park, Maryland
American Institute of Physics Public Information Office Files
Official Files, Director's Office, Brookhaven National Laboratory (on microfilm)
Samuel A. Goudsmit Papers
American Philosophical Society, Philadelphia, Pennsylvania
Edward U. Condon Papers
Leslie C. Dunn Papers
Curt Stern Papers
Federal Bureau of Investigation, Washington, D.C.
File on Edward U. Condon, FBI HQ 121-2673
File on the Federation of American Scientists, FBI HQ 100-344452
Washington Field Office File on the Federation of American Scientists, FBI WFO 65-4736
Harvard University, Cambridge, Massachusetts
John T. Edsall Papers
Harlow Shapley Papers
The Johns Hopkins University, Baltimore, Maryland
Isaiah Bowman Papers
Library of Congress, Washington, D.C.
Vannevar Bush Papers
J. Robert Oppenheimer Papers
Records of the American Psychological Association
Records of the National Committee on Atomic Information
Oswald Veblen Papers
Massachusetts Institute of Technology, Cambridge, Massachusetts
Philip Morrison Papers
Philip Morse Papers
Records of the Office of the President, 1930–1959 (Compton-Killian)
National Academy of Sciences, Washington, D.C.
[General Records]
Jewett File
National Archives, Washington, D.C.
Records of the Atomic Energy Commission
AEC Minutes, 1947 and 1950
Office Files of Robert F. Bacher

Office Files of David E. Lilienthal
Secretariat Files
Records of the Department of Commerce
General Records
Records of the Joint Committee on Atomic Energy, 1946–1977
Executive Session Transcripts
General Correspondence
National Security Archives, Washington, D.C.
Venona Project Releases
Princeton University, Princeton, New Jersey
Lyman Spitzer Papers
Philip Wylie Papers
State University of New York at Albany, Albany, New York
Eugene Rabinowitch Papers
Harry S. Truman Library, Independence, Missouri
Harry S. Truman Papers
President's Secretary's Files (also consulted on microfilm)
University of California at Berkeley, Berkeley, California
Ernest O. Lawrence Papers
University of California at San Diego, San Diego, California
Martin D. Kamen Papers
Harold C. Urey Papers
University of Chicago, Chicago, Illinois
American Civil Liberties Union, Illinois Division Papers
Association of Oak Ridge Engineers and Scientists Papers
Atomic Scientists of Chicago Papers
Atomic Scientists' Printed and Near-Print Material
Charles D. Coryell Papers
Emergency Committee of Atomic Scientists Papers
Federation of American Scientists Papers
Joseph H. Rush Papers
Washington Association of Scientists Papers
University of Minnesota, Minneapolis, Minnesota
Maurice B. Visscher Papers
University of Wisconsin, Madison, Wisconsin
Farrington Daniels Papers

INTERVIEWS

Blewett, John P. Interview by the author. August 14, 1995. Tape and transcribed notes in possession of the author.

——. Interviews by Lillian Hoddeson. March 22 and May 11, 1979. Tape. Oral History Collection, Niels Bohr Library, American Institute of Physics, College Park, Md.

Condon, Edward U. Interviews by Charles Weiner. October 17, 1967, April 27, 1968, and September 11–12, 1973. Transcript. Oral History Collection, Niels Bohr Library, American Institute of Physics, College Park, Md.

Rabinowitch, Eugene. Interview by Govindjee. January 5, 1964. Tape. Oral History Collection, Niels Bohr Library, American Institute of Physics, College Park, Md.

Vought, Robert H. Interview by the author. September 9, 1993. Tape and transcribed notes in possession of the author.

DISSERTATIONS

Carlson, Lewis H. "J. Parnell Thomas and the House Committee on Un-American Activities, 1938–1948." Ph.D. diss., Michigan State University, 1967.

Hall, Harry S. "Congressional Attitudes toward Science and Scientists: A Study of Legislative Reactions to Atomic Energy and the Political Participation of Scientists." Ph.D. diss., University of Chicago, 1961; published New York: Arno Press, 1979.

Hodes, Elizabeth. "Precedents for Social Responsibility among Scientists: The American Association of Scientific Workers and the Federation of American Scientists, 1938–1948." Ph.D. diss., University of California at Santa Barbara, 1982.

Jones, Kenneth MacDonald. "Science, Scientists, and Americans: Images of Science and the Formation of Federal Science Policy, 1945–1950." Ph.D. diss., Cornell University, 1975.

McGrath, Patrick. "American Scientists at War: The Scientific Elite and the State, 1930–1960." Ph.D. diss., New York University, 1996.

Rowan, Carl Milton. "Politics and Pure Research: The Origins of the National Science Foundation, 1942–1954." Ph.D. diss., Miami University (Ohio), 1985.

ARTICLES AND ESSAYS

Balogh, Brian. "Reorganizing the Organizational Synthesis: Federal-Professional Relations in Modern America." *Studies in American Political Development* 5 (Spring 1991): 119–72.

Bernstein, Barton J. "Four Physicists and the Bomb: The Early Years, 1945–1950." *Historical Studies in the Physical and Biological Sciences* 18, no. 2 (1988): 231–63.

———. " 'In the Matter of J. Robert Oppenheimer.' " *Historical Studies in the Physical Sciences* 12, no. 2 (1982): 195–252.

———. "The Oppenheimer Loyalty-Security Case Reconsidered." *Stanford Law Review* 42 (July 1990): 1383–1484.

Brickman, Jane Pacht. " 'Medical McCarthyism': The Physicians Forum and the Cold War." *Journal of the History of Medicine* 49 (July 1994): 380–415.

Doel, Ronald E. "Evaluating Soviet Lunar Science in Cold War America." *Osiris* 7 (1992): 44–70.

———. "Redefining a Mission: The Smithsonian Astrophysical Observatory on the Move." *Journal for the History of Astronomy* 21 (1990): 137–53.

Edsforth, Ronald. "Affluence, Anti-Communism, and the Transformation of Industrial Unionism among Automobile Workers, 1933–1973." In *Popular Culture and Political Change in Modern America*, edited by Ronald Edsforth and Larry Bennett, 101–25. Albany: State University of New York Press, 1991.

Feffer, Stuart M. "Atoms, Cancer, and Politics: Supporting Atomic Science at the University of Chicago, 1944–1950." *Historical Studies in the Physical and Biological Sciences* 22, no. 2 (1992): 233–61.

Forman, Paul. "Behind Quantum Electronics: National Security as Basis for Physical Research in the United States, 1940–1960." *Historical Studies in the Physical and Biological Sciences* 18, no. 1 (1987): 149–229.

Galison, Peter, and Barton J. Bernstein. "In Any Light: Scientists and the Decision to Build the Superbomb, 1952–1954." *Historical Studies in the Physical and Biological Sciences* 19, no. 2 (1989): 267–347.
———. "Physics between War and Peace." In *Science, Technology, and the Military*, edited by Everett Mendelsohn, Merritt Roe Smith, and Peter Weingart, 1:47–86. Dordrecht, the Netherlands: Kluwer Academic Publishers, 1988.
Haynes, John Earl. "The New History of the Communist Party in State Politics: The Implications for Mainstream Political History." *Labor History* 27 (Fall 1986): 549–63.
Hoch, Paul K. "The Crystallization of a Strategic Alliance: The American Physics Elite and the Military in the 1940s." In *Science, Technology, and the Military*, edited by Everett Mendelsohn, Merritt Roe Smith, and Peter Weingart, 1:87–111. Dordrecht, the Netherlands: Kluwer Academic Publishers, 1988.
Kevles, Dan [Daniel J.]. "Cold War and Hot Physics: Science, Security, and the American State, 1945–56." *Historical Studies in the Physical and Biological Sciences* 20, no. 2 (1989): 239–64.
———. "The National Science Foundation and the Debate over Postwar Research Policy, 1942–1945." *Isis* 68 (March 1977): 5–26.
———. "Scientists, the Military, and the Control of Postwar Defense Research: The Case of the Research Board for National Security, 1944–46." *Technology and Culture* 16 (January 1975): 20–47.
Kimball, Warren F. "The Cold War Warmed Over." *American Historical Review* 79 (October 1974): 1119–36.
Kleinman, Daniel Lee. "Layers of Interests, Layers of Influence: Business and the Genesis of the National Science Foundation." *Science, Technology, and Human Values* 19 (Summer 1994): 259–82.
Kozhievnikov, Aleksei. "Piotr Kapitza and Stalin's Government: A Study in Moral Choice." *Historical Studies in the Physical and Biological Sciences* 22, no. 1 (1991): 131–64.
Lears, T. J. Jackson. "The Concept of Cultural Hegemony: Problems and Possibilities." *American Historical Review* 90 (June 1985): 567–93.
Leffler, Melvyn P. "The American Conception of National Security and the Beginnings of the Cold War, 1945–48." *American Historical Review* 89 (April 1984): 346–81.
Lieberman, Robbie. " 'Does That Make Peace a Bad Word?': American Responses to the Communist Peace Offensive, 1949–1950." *Peace and Change* 17 (April 1992): 198–228.
Maddox, Robert Franklin. "The Politics of World War II Science: Senator Harley M. Kilgore and the Legislative Origins of the National Science Foundation." *West Virginia History* 40 (Fall 1979): 20–39.
Needell, Allan A. "Nuclear Reactors and the Founding of Brookhaven National Laboratory." *Historical Studies in the Physical Sciences* 14, no. 1 (1983): 93–122.
O'Reilly, Kenneth. "A New Deal for the FBI: The Roosevelt Administration, Crime Control, and National Security." *Journal of American History* 69 (December 1982): 638–58.
Pleasants, Julian M. "A Question of Loyalty: Frank Porter Graham and the Atomic Energy Commission." *North Carolina Historical Review* 69 (October 1992): 414–37.

Reingold, Nathan. "Vannevar Bush's New Deal for Research: Or the Triumph of the Old Order." *Historical Studies in the Physical and Biological Sciences* 17, no. 2 (1987): 299–344.

Rosswurm, Steven, and Toni Gilpin. "The FBI and the Farm Equipment Workers: FBI Surveillance Records as a Source for CIO Union History." *Labor History* 27 (Fall 1986): 485–505.

Schweber, Silvan S. "The Mutual Embrace of Science and the Military: ONR and the Growth of Physics in the United States after World War II." In *Science, Technology, and the Military*, edited by Everett Mendelsohn, Merritt Roe Smith, and Peter Weingart, 1:3–46. Dordrecht, the Netherlands: Kluwer Academic Publishers, 1988.

Seidel, Robert W. "A Home for Big Science: The Atomic Energy Commission's Laboratory System." *Historical Studies in the Physical and Biological Sciences* 16, no. 1 (1986): 135–75.

Wang, Jessica. "Liberals, the Progressive Left, and the Political Economy of Postwar American Science: The National Science Foundation Debate Revisited." *Historical Studies in the Physical and Biological Sciences* 26, no. 1 (1995): 139–66.

———. "Science, Security, and the Cold War: The Case of E. U. Condon." *Isis* 83 (June 1992): 238–69.

BOOKS

Ambrose, Stephen E. *Rise to Globalism: American Foreign Policy since 1938*. 4th rev. ed. New York: Penguin Books, 1985.

Balogh, Brian. *Chain Reaction: Expert Debate and Public Participation in American Commercial Nuclear Power, 1945–1975*. Cambridge: Cambridge University Press, 1991.

Bell, Daniel, ed. *The Radical Right*. New York: Doubleday & Co., 1963; expanded and updated edition of *The New American Right*, 1955.

Bernstein, Barton J., ed. *Politics and Policies of the Truman Administration*. Chicago: Quadrangle Books, 1970.

Bess, Michael. *Realism, Utopia, and the Mushroom Cloud: Four Activist Intellectuals and Their Strategies for Peace, 1945–1989*. Chicago: University of Chicago Press, 1993.

Bontecou, Eleanor. *The Federal Loyalty-Security Program*. Ithaca, N.Y.: Cornell University Press, 1953.

Boyer, Paul. *By the Bomb's Early Light: American Thought and Culture at the Dawn of the Atomic Age*. New York: Pantheon Books, 1985.

Brands, H. W. *The Devil We Knew: Americans and the Cold War*. New York: Oxford University Press, 1993.

Brinkley, Alan. *The End of Reform: New Deal Liberalism in Recession and War*. New York: Alfred A. Knopf, 1995.

Britten, Wesley E., and Halis Odabasi, *Topics in Modern Physics: A Tribute to Edward U. Condon*. Boulder: Colorado Associated University Press, 1971.

Broadwater, Jeff. *Eisenhower and the Anti-Communist Crusade*. Chapel Hill: University of North Carolina Press, 1992.

Brown, Ralph S., Jr. *Loyalty and Security: Employment Tests in the United States*. New Haven: Yale University Press, 1958.

Carmichael, Virginia. *Framing History: The Rosenberg Story and the Cold War*. Minneapolis: University of Minnesota Press, 1993.

Carr, Robert K. *The House Committee on Un-American Activities, 1945–1950*. Ithaca, N.Y.: Cornell University Press, 1952.
Caute, David. *The Great Fear: The Anti-Communist Purge under Truman and Eisenhower*. New York: Simon and Schuster, 1978.
Cochrane, Rexmond C. *The National Academy of Sciences: The First Hundred Years, 1863–1963*. Washington, D.C.: National Academy of Sciences, 1978.
Cohen, Warren I. *The Cambridge History of American Foreign Relations*. Vol. 4, *America in the Age of Soviet Power, 1945–1991*. Cambridge: Cambridge University Press, 1993.
Diamond, Sigmund. *Compromised Campus: The Collaboration of Universities with the Intelligence Community, 1945–1955*. New York: Oxford University Press, 1992.
Donner, Frank. *The Age of Surveillance: The Aims and Methods of America's Political Intelligence System*. New York: Alfred A. Knopf, 1980.
Donovan, Robert J. *Conflict and Crisis: The Presidency of Harry S Truman, 1945–1948*. New York: W. W. Norton and Co., 1977.
Dyson, Lowell K. *Red Harvest: The Communist Party and American Farmers*. Lincoln: University of Nebraska Press, 1982.
Engelhardt, Tom. *The End of Victory Culture: Cold War America and the Disillusioning of a Generation*. New York: Basic Books, 1995.
England, J. Merton. *A Patron for Pure Science: The National Science Foundation's Formative Years, 1945–57*. Washington, D.C.: National Science Foundation, 1982.
Fraser, Steve, and Gary Gerstle, eds. *The Rise and Fall of the New Deal Order, 1930–1980*. Princeton: Princeton University Press, 1989.
Freeland, Richard M. *The Truman Doctrine and the Origins of McCarthyism: Foreign Policy, Domestic Politics, and Internal Security, 1946–1948*. New York: Alfred A. Knopf, 1972.
Fried, Richard M. *Nightmare in Red: McCarthyism in Perspective*. New York: Oxford University Press, 1990.
Gaddis, John Lewis. *The United States and the Origins of the Cold War, 1941–1947*. New York: Columbia University Press, 1972.
Galison, Peter, and Bruce Hevly. *Big Science: The Growth of Large-Scale Research*. Stanford: Stanford University Press, 1992.
Gardner, David P. *The California Oath Controversy*. Berkeley and Los Angeles: University of California Press, 1967.
Gellhorn, Walter. *Security, Loyalty, and Science*. Ithaca, N.Y.: Cornell University Press, 1950.
Gillon, Steven M. *Politics and Vision: The ADA and American Liberalism, 1947–1985*. New York: Oxford University Press, 1985.
Ginger, Ann Fagan, and Eugene M. Tobin, eds. *The National Lawyers Guild: From Roosevelt through Reagan*. Philadelphia: Temple University Press, 1988.
Goldstein, Robert Justin. *Political Repression in Modern America, from 1870 to the Present*. Cambridge, Mass.: Schenkman Publishing Co., 1978.
Goodman, Walter. *The Committee: The Extraordinary Career of the House Committee on Un-American Activities*. New York: Farrar, Straus and Giroux, 1968.
Green, Harold P., and Alan Rosenthal. *Government of the Atom: The Integration of Powers*. New York: Atherton Press, 1963.

Griffith, Robert, and Athan Theoharis, eds. *The Specter: Original Essays on the Cold War and the Origins of McCarthyism.* New York: New Viewpoints, 1974.
Hagen, Thomas. *Force of Nature: The Life of Linus Pauling.* New York: Simon and Schuster, 1995.
Halpern, Martin. *UAW Politics in the Cold War Era.* Albany: State University of New York Press, 1988.
Hamby, Alonzo. *Beyond the New Deal: Harry S. Truman and American Liberalism.* New York: Columbia University Press, 1973.
Hawley, Ellis W. *The New Deal and the Problem of Monopoly.* Princeton: Princeton University Press, 1966.
Haynes, John E. *Red Scare or Red Menace? American Communism and Anticommunism in the Cold War Era.* Chicago: Ivan R. Dee, 1996.
Heale, M. J. *American Anticommunism: Combating the Enemy Within, 1830–1970.* Baltimore: Johns Hopkins University Press, 1990.
Healey, Dorothy Ray, and Maurice Isserman. *California Red: A Life in the American Communist Party.* Urbana: University of Illinois Press, 1993; reprint of *Dorothy Healey Remembers: A Life in the American Communist Party,* New York: Oxford University Press, 1990.
Herken, Gregg. *The Winning Weapon: The Atomic Bomb in the Cold War, 1945–1950.* Princeton: Princeton University Press, 1981.
Hershberg, James G. *James B. Conant: Harvard to Hiroshima and the Making of the Nuclear Age.* New York: Alfred A. Knopf, 1993.
Hewlett, Richard G., and Oscar E. Anderson Jr. *The New World, 1939/1946.* Vol. 1 of *A History of the United States Atomic Energy Commission.* 1962. Reprint, Washington, D.C.: U.S. Atomic Energy Commission, 1972.
Hewlett, Richard G., and Francis Duncan. *Atomic Shield, 1947/1952.* Vol. 2 of *A History of the United States Atomic Energy Commission.* 1969. Reprint, Washington, D.C.: U.S. Atomic Energy Commission, 1972.
Higham, John. *Strangers in the Land: Patterns of American Nativism, 1860–1925.* 2d ed. New Brunswick, N.J.: Rutgers University Press, 1963. Reprint, New York: Atheneum, 1971.
Hofstadter, Richard. *The Paranoid Style in American Politics and Other Essays.* New York: Alfred A. Knopf, 1965. Reprint, New York: Vintage Books, 1967.
Holloway, David. *Stalin and the Bomb: The Soviet Union and Atomic Energy, 1939–1956.* Stanford: Stanford University Press, 1994.
Holmes, David R. *Stalking the Academic Communist: Intellectual Freedom and the Firing of Alex Novikoff.* Hanover, N.H.: University Press of New England, 1989.
Isserman, Maurice. *Which Side Were You On? The American Communist Party during the Second World War.* Middletown, Conn.: Wesleyan University Press, 1982.
Kamen, Martin D. *Radiant Science, Dark Politics: A Memoir of the Nuclear Age.* Berkeley and Los Angeles: University of California Press, 1985.
Keller, William W. *The Liberals and J. Edgar Hoover: Rise and Fall of a Domestic Intelligence State.* Princeton: Princeton University Press, 1989.
Kevles, Daniel J. *The Physicists: The History of a Scientific Community in Modern America.* New York: Alfred A. Knopf, 1977. Reprint, Cambridge: Harvard University Press, 1987.
Klehr, Harvey, and Ronald Radosh. *The Amerasia Spy Case: Prelude to McCarthyism.* Chapel Hill: University of North Carolina Press, 1996.

Klehr, Harvey, John Earl Haynes, and Fridrikh Igorevich Firsov. *The Secret World of American Communism*. New Haven: Yale University Press, 1995.

Kleinman, Daniel Lee. *Politics on the Endless Frontier: Postwar Research Policy in the United States*. Durham, N.C.: Duke University Press, 1995.

Kofsky, Frank. *Harry S. Truman and the War Scare of 1948: A Successful Campaign to Deceive the Nation*. New York: St. Martin's Press, 1993.

Kutler, Stanley I. *The American Inquisition: Justice and Injustice in the Cold War*. New York: Hill and Wang, 1982.

Kuznick, Peter J. *Beyond the Laboratory: Scientists as Political Activists in 1930s America*. Chicago: University of Chicago Press, 1987.

LaFeber, Walter. *America, Russia, and the Cold War, 1945–1975*. 3d ed. New York: John Wiley and Sons, 1976.

——. *The American Age: United States Foreign Policy at Home and Abroad since 1750*. New York: W. W. Norton and Co., 1989.

Lamphere, Robert J., and Tom Schachtman. *The FBI-KGB War: A Special Agent's Story*. New York: Random House, 1986. Reprint with new afterword, Macon, Ga.: Mercer University Press, 1995.

Larson, Deborah Welch. *Origins of Containment: A Psychological Explanation*. Princeton: Princeton University Press, 1985.

Layton, Edwin T., Jr. *Revolt of the Engineers: Social Responsibility and the American Engineering Profession*. Cleveland: Press of Case Western Reserve University, 1971.

Lazarsfeld, Paul F., and Wagner Thielens Jr. *The Academic Mind: Social Scientists in a Time of Crisis*. Glencoe, Ill.: Free Press, 1958.

Leslie, Stuart W. *The Cold War and American Science: The Military-Industrial-Academic Complex at MIT and Stanford*. New York: Columbia University Press, 1993.

Levenstein, Harvey A. *Communism, Anticommunism, and the CIO*. Westport, Conn.: Greenwood Press, 1981.

Lewis, Lionel S. *Cold War on Campus: A Study of the Politics of Organizational Control*. New Brunswick, N.J.: Transaction Books, 1988.

Lilienthal, David E. *The Journals of David E. Lilienthal*. Vol. 2, *The Atomic Energy Years, 1945–1950*. New York: Harper and Row, 1964.

Lipsitz, George. *Class and Culture in Cold War America: "A Rainbow at Midnight."* New York: Praeger Publishers, 1981.

Lowi, Theodore J. *The End of Liberalism: The Second Republic of the United States*. 2d ed. New York: W. W. Norton and Co., 1979.

May, Elaine Tyler. *Homeward Bound: American Families in the Cold War Era*. New York: Basic Books, 1988.

May, Lary, ed. *Recasting America: Culture and Politics in the Age of Cold War*. Chicago: University of Chicago Press, 1989.

McAuliffe, Mary Sperling. *Crisis on the Left: Cold War Politics and American Liberals, 1947–1954*. Amherst: University of Massachusetts Press, 1978.

McCormick, Thomas J. *America's Half-Century: United States Foreign Policy in the Cold War*. Baltimore: Johns Hopkins University Press, 1989.

McCullough, David. *Truman*. New York: Simon and Schuster, 1992.

Mills, C. Wright. *The Causes of World War III*. New York: Ballantine Books, 1960.

——. *The Power Elite*. New York: Oxford University Press, 1956.

Morse, Philip M. *In at the Beginnings: A Physicist's Life*. Cambridge: MIT Press, 1977.

Murray, Robert K. *Red Scare: A Study of National Hysteria, 1919–1920.* 1955. Reprint, New York: McGraw-Hill Book Co., 1964.
Naison, Mark. *Communists in Harlem during the Depression.* Urbana: University of Illinois Press, 1983.
Nathanson, Jerome, ed. *Science for Democracy.* New York: King's Crown Press, 1946.
Navasky, Victor S. *Naming Names.* New York: Penguin Books, 1981.
Newman, James R., and Byron S. Miller. *The Control of Atomic Energy: A Study of Its Social, Economic, and Political Implications.* New York: McGraw-Hill Book Co., 1948.
O'Reilly, Kenneth. *FBI File on the House Committee on Un-American Activities.* Wilmington, Del.: Scholarly Resources, 1985 (microfilm edition).
———. *Hoover and the Un-Americans: The FBI, HUAC, and the Red Menace.* Philadelphia: Temple University Press, 1983.
Oshinsky, David M. *A Conspiracy So Immense: The World of Joe McCarthy.* New York: Free Press, 1983.
Patterson, James T. *Congressional Conservatism and the New Deal: The Growth of the Conservative Coalition in Congress, 1933–1939.* Lexington: University of Kentucky Press, 1967.
Pells, Richard H. *The Liberal Mind in a Conservative Age: American Intellectuals in the 1940s and 1950s.* New York: Harper and Row, 1985.
Powers, Richard Gid. *Not without Honor: The History of American Anticommunism.* New York: Free Press, 1995.
———. *Secrecy and Power: The Life of J. Edgar Hoover.* New York: Free Press, 1987.
Radosh, Ronald, and Joyce Milton. *The Rosenberg File: A Search for Truth.* 1983. Reprint, New York: Vintage Books, 1984.
Reeves, Thomas C. *The Life and Times of Joe McCarthy: A Biography.* New York: Stein and Day, 1982.
Reuben, William A. *Atom Spy Hoax.* New York: Action Books, 1955.
Rhodes, Richard. *Dark Sun: The Making of the Hydrogen Bomb.* New York: Simon and Schuster, 1995.
Rosswurm, Steve. *The CIO's Left-Led Unions.* New Brunswick, N.J.: Rutgers University Press, 1992.
Schatz, Ronald W. *The Electrical Workers: A History of Labor at General Electric and Westinghouse, 1923–60.* Urbana: University of Illinois Press, 1983.
Schlesinger, Arthur M., Jr. *The Vital Center: The Politics of Freedom.* Boston: Houghton Mifflin Co., 1949.
Schrecker, Ellen W. *No Ivory Tower: McCarthyism and the Universities.* New York: Oxford University Press, 1986.
Shapley, Harlow. *Through Rugged Ways to the Stars.* New York: Charles Scribner's Sons, 1969.
Sherwin, Martin J. *A World Destroyed: The Atomic Bomb and the Grand Alliance.* New York: Alfred A. Knopf, 1975. Reprint, New York: Vintage Books, 1977.
Smith, Alice Kimball. *A Peril and a Hope: The Scientists' Movement in America, 1945–47.* Chicago: University of Chicago Press, 1965.
Stern, Philip M. *The Oppenheimer Case: Security on Trial.* New York: Harper and Row, 1969.
Stouffer, Samuel A. *Communism, Conformity, and Civil Liberties.* 1955. New Brunswick, N.J.: Transaction Publishers, 1992.

Strickland, Donald A. *Scientists in Politics: The Atomic Scientists Movement, 1945–46.* [Indianapolis]: Purdue University Studies, 1968.

Sudoplatov, Pavel, and Anatoli Sudoplatov, with Jerrold L. and Leona P. Schecter. *Special Tasks: The Memoirs of an Unwanted Witness—A Soviet Spymaster.* Boston: Little, Brown and Co., 1994.

Taubman, William. *Stalin's American Policy: From Entente to Détente to Cold War.* New York: W. W. Norton and Co., 1982.

Theoharis, Athan G. *Seeds of Repression: Harry S. Truman and the Origins of McCarthyism.* New York: Quadrangle Books, 1971.

———. *Spying on Americans: Political Surveillance from Hoover to the Huston Plan.* Philadelphia: Temple University Press, 1978.

———, ed. *Beyond the Hiss Case: The FBI, Congress, and the Cold War.* Philadelphia: Temple University Press, 1982.

Theoharis, Athan G., and John Stuart Cox. *The Boss: J. Edgar Hoover and the Great American Inquisition.* Philadelphia: Temple University Press, 1988.

Thomas, Morgan, with Robert M. Northrop. *Atomic Energy and Congress.* Ann Arbor: University of Michigan Press, 1956.

Tomlins, Christopher L. *The State and the Unions: Labor Relations, Law, and the Organized Labor Movement in America, 1880–1960.* Cambridge: Cambridge University Press, 1985.

Whitfield, Stephen J. *The Culture of the Cold War.* Baltimore: Johns Hopkins University Press, 1991.

Williams, Robert Chadwell. *Klaus Fuchs, Atom Spy.* Cambridge: Harvard University Press, 1987.

Williams, Selma R. *Red-Listed: Haunted by the Washington Witch Hunt.* Reading, Mass.: Addison-Wesley Publishing Co., 1993.

Williams, William Appleman. *The Tragedy of American Diplomacy.* Rev. ed. New York: Dell Publishing Co., 1962.

Wilson, Joan Hoff. *Herbert Hoover: Forgotten Progressive.* Boston: Little, Brown and Co., 1975.

Wittner, Lawrence S. *The Struggle against the Bomb.* Vol. 1, *One World or None: A History of the World Nuclear Disarmament Movement through 1953.* Stanford: Stanford University Press, 1993.

Wolfle, Dael. *Renewing a Scientific Society: The American Association for the Advancement of Science from World War II to 1970.* Washington, D.C.: American Association for the Advancement of Science, 1989.

Yergin, Daniel. *Shattered Peace: The Origins of the Cold War.* Rev. ed. New York: Penguin Books, 1990.

INDEX